T0212971

Lecture Notes in Bioinformatics 9838

Subseries of Lecture Notes in Computer Science

More information about this series at http://www.springer.com/series/5381

Martin Frith
Christian Nørgaard Storm Pedersen (Eds.)

Algorithms in Bioinformatics

16th International Workshop, WABI 2016
Aarhus, Denmark, August 22–24, 2016
Proceedings

 Springer

Editors
Martin Frith
AIST and University of Tokyo
Tokyo
Japan

Christian Nørgaard Storm Pedersen
Aarhus University
Aarhus
Denmark

ISSN 0302-9743
Lecture Notes in Bioinformatics
ISBN 978-3-319-43680-7
DOI 10.1007/978-3-319-43681-4

ISSN 1611-3349 (electronic)

ISBN 978-3-319-43681-4   (eBook)

Library of Congress Control Number: 2016945963

LNCS Sublibrary: SL8 – Bioinformatics

Printed on acid-free paper

This Springer imprint is published by Springer Nature
The registered company is Springer International Publishing AG Switzerland

Preface

This proceedings volume contains papers presented at the 16th Workshop on Algorithms in Bioinformatics (WABI 2016) that was held at Aarhus University, Aarhus, Denmark, August 22–24, 2016. WABI 2016 was one of eight conferences that were organized as part of ALGO 2016. The Workshop on Algorithms in Bioinformatics was established in 2001, and is an annual conference on all aspects of algorithmic work in bioinformatics, computational biology, and systems biology. The emphasis is on discrete algorithms and machine-learning methods that address important problems in molecular biology, that are founded on sound models, that are computationally efficient, and that have been implemented and tested in simulations and on real datasets. The goal is to present recent research results, including significant work-in-progress, and to identify and explore directions of future research. WABI 2016 was sponsored by the European Association for Theoretical Computer Science (EATCS).

In 2016, a total of 56 manuscripts were submitted to WABI from which 27 were selected for presentation at the conference. Among them, 25 are included in this proceedings volume as full papers presenting novel results not previously published in journals, and two are included as short abstracts of papers that are in the process of being published simultaneously in journals. The 27 papers were selected based on thorough reviewing, usually involving three independent reviewers per submitted paper, followed by discussions in the WABI Program Committee. The selected papers cover a wide range of topics from networks, to phylogenetic studies, sequence and genome analysis, comparative genomics, and mass spectrometry data analysis. Extended versions of selected papers will be published in a thematic series in the journal *Algorithms for Molecular Biology* (AMB), published by BioMed Central.

We thank all the authors of submitted papers and the members of the WABI Program Committee and their reviewers for their efforts that made this conference possible, and the WABI Steering Committee for their help and advice. We also thank all the conference participants and speakers. In particular, we are indebted to the keynote speaker of the conference, Kiyoshi Asai, for his presentation. Finally, we thank Gerth Stølting Brodal and the local ALGO Organizing Committee for their hard work.

June 2016

Martin Frith
Christian N. S. Pedersen

Organization

Program Chairs

Martin Frith   AIST and University of Tokyo, Japan
Christian N.S. Pedersen   Aarhus University, Denmark

Program Committee

Tatsuya Akutsu   Kyoto University, Japan
Timothy L. Bailey   University of Queensland, Australia
Jan Baumbach   University of Southern Denmark, Denmark
Anne Bergeron   Université du Québec à Montréal, Canada
Paola Bonizzoni   Università di Milano-Bicocca, Italy
Alessandra Carbone   Université Pierre et Marie Curie, France
Rita Casadio   UNIBO, Italy
Nadia El-Mabrouk   University of Montreal, Canada
Anna Gambin   Warsaw University, Poland
Raffaele Giancarlo   Università di Palermo, Italy
Michiaki Hamada   Waseda University, Japan
Thomas Hamelryck   University of Copenhagen, Denmark
Fereydoun Hormozdiari   University of Washington, USA
Katharina Huber   University of East Anglia, UK
Carl Kingsford   Carnegie Mellon University, USA
Hisanori Kiryu   University of Tokyo, Japan
Gregory Kucherov   CNRS/LIGM, France
Timo Lassmann   Telethon Kids, Australia
Ming Li   University of Waterloo, Canada
Zsuzsanna Liptak   University of Verona, Italy
Stefano Lonardi   University of California at Riverside, USA
Gerton Lunter   University of Oxford, UK
Thomas Mailund   Aarhus University, Denmark
Paul Medvedev   Pennsylvania State University, USA
Daniel Merkle   University of Southern Denmark, Denmark
István Miklós   Rényi Institute, Hungary
Bernard Moret   EPFL, Switzerland
Burkhard Morgenstern   University of Göttingen, Germany
Vincent Moulton   University of East Anglia, UK
Veli Mäkinen   University of Helsinki, Finland
Luay Nakhleh   Rice University, USA
William Noble   University of Washington, USA

Nadia Pisanti	Università di Pisa, Italy, and Inria, France
Mihai Pop	University of Maryland, USA
Teresa Przytycka	NIH, USA
Sven Rahmann	University of Duisburg-Essen, Germany
Marie-France Sagot	Inria, France
Kengo Sato	Keio University, Japan
Michael Schatz	Cold Spring Harbor Laboratory, USA
Russell Schwartz	Carnegie Mellon University, USA
Kana Shimizu	Waseda University, Japan
Anish Man Singh Shrestha	University of Tokyo, Japan
Peter F. Stadler	University of Leipzig, Germany
Jens Stoye	Bielefeld University, Germany
Krister Swenson	CNRS, Université de Montpellier, France
Hélène Touzet	CNRS, University of Lille and Inria, France
Lusheng Wang	City University of Hong Kong, China
Siu Ming Yiu	University of Hong Kong, SAR China
Louxin Zhang	National University of Singapore, Singapore
Michal Ziv-Ukelson	Ben-Gurion University of the Negev, Israel

WABI Steering Committee

Bernard Moret	EPFL, Switzerland
Vincent Moulton	University of East Anglia, UK
Jens Stoye	Bielefeld University, Germany
Tandy Warnow	University of Illinois at Urbana-Champaign, USA

ALGO Organizing Committee

Gerth Stølting Brodal (Chair)	Trine Ji Holmgaard
Marianne Dammand Iversen	Katrine Østerlund Rasmussen

Additional Reviewers

Nicolas Alcaraz	Pietro Di Lena
Hind Alhakami	Daniel Doerr
Eloi Araujo	Norbert Dojer
Matthias Bernt	Mikhail Dubov
Karel Brinda	Oliver Eulenstein
Laurent Bulteau	Pedro Feijao
Victoria Cepeda	Jay Ghurye
Daniel Cooke	Krzysztof Gogolewski
Phuong Dao	Roberto Grossi
Gianluca Della Vedova	Laurent Gueguen
Alex Di Genova	Marc Hellmuth

Donna Henderson
Farhad Hormozdiari
Jeff Howbert
Alex Hu
Laurent Jacob
Mateusz Krzysztof Łącki
Manuel Lafond
Cong Ma
Guillaume Marcais
Damon May
Blazej Miasojedow
Ibrahim Numanagic
Rachid Ounit
Solon Pissis

Raffaella Rizzi
Abbas Roayaei Ardakany
Giovanna Rosone
Yutaka Saito
Guillaume Scholz
Marcel Schulz
Celine Scornavacca
Mingfu Shao
Grzegorz Skoraczyński
Bianca Stöcker
Peng Sun
Hao Wang
Lusheng Wang
Martin Weigt

Abstracts

Mass Graphs and Their Applications in Top-Down Proteomics

Qiang Kou[1], Si Wu[2], Nikola Tolić[3], Yunlong Liu[4,5],
Ljiljana Paša-Tolić[3] and Xiaowen Liu[1,5(✉)]

[1] Department of BioHealth Informatics,
Indiana University-Purdue University Indianapolis, Indianapolis, Indiana
[2] Department of Chemistry and Biochemistry,
University of Oklahoma, Norman, Oklahoma
[3] Biological Science Division, Pacific Northwest National Laboratory,
Richland, USA
[4] Department of Medical and Molecular Genetics,
Indiana University School of Medicine, Indianapolis, Indiana
[5] Center for Computational Biology and Bioinformatics,
Indiana University School of Medicine, Indianapolis, Indiana
xwliu@iupui.edu

Abstract. Although proteomics has rapidly developed in the past decade, researchers are still in the early stage of exploring the world of complex proteoforms, which are protein products with various primary structure alterations resulting from gene mutations, alternative splicing, post-translational modifications, and other biological processes. Proteoform identification is essential to mapping proteoforms to their biological functions as well as discovering novel proteoforms and new protein functions. Top-down mass spectrometry is the method of choice for identifying complex proteoforms because it provides a "bird's eye view" of intact proteoforms. Fragment ion series in top-down tandem mass spectra provide essential information for identifying primary sequence alterations in proteoforms. Extended proteoform databases and spectral alignment are the two main approaches for proteoform identification. However, due to the combinatorial explosion of various alterations on a protein and the limitations of available spectral alignment algorithms, the proteoform identification problem has still not been fully solved.

We propose a new data structure, called the mass graph, for efficient representation of proteoforms of a protein with variable post-translational modifications and/or terminal truncations. The proteoform identification problem is transformed to the mass graph alignment problem, and dynamic programming algorithms are proposed for a restricted version of the problem. The proposed algorithms were tested on two top-down tandem mass spectrometry data sets. Experimental results showed that the proposed algorithms were efficient in identifying proteoforms with variable post-translational modifications and outperformed MS-Align-E in running time and sensitivity for identifying complex proteoforms, especially those with terminal truncations.

Acknowledgement. The research was supported by the National Institute of General Medical Sciences, National Institutes of Health (NIH) through Grant R01GM118470.

Safely Filling Gaps with Partial Solutions Common to all Solutions

Leena Salmela and Alexandru I. Tomescu

Department of Computer Science, Helsinki Institute for Information
Technology HIIT, University of Helsinki, Helsinki, 00014, Finland
{leena.salmela,tomescu}@cs.helsinki.fi

Abstract. Gap filling has emerged as a natural sub-problem of many *de novo* genome assembly projects (e.g., filling gaps in scaffolds). Several methods have addressed it, but only few have focused on strategies for dealing with multiple gap filling solutions and for guaranteeing reliable results. Such strategies include reporting only unique solutions, or exhaustively enumerating all filling solutions and heuristically creating their consensus.

The gap filling problem is usually formulated as finding an *s-t* path in the assembly graph, whose length matches the gap length estimate. In this paper we address it with the "safe and complete" framework proposed in [Tomescu and Medvedev, RECOMB 2016] for the contig assembly problem. In terms of gap filling, a *safe solution* is a path of the assembly graph that is a sub-path of all possible *s-t* paths whose length matches the gap length estimate.

We give an efficient safe algorithm for the gap filling problem, running in time $O(dm)$, where d is the gap length estimate and m is the number of edges of the assembly graph. To transform the safe paths into a single filling sequence usable in downstream analysis, we fill the gap with an arbitrary filling path, in which we mark the safe subsequences. Experiments on the GAGE bacterial datasets show that our method retrieves over 90 % more safe and correct bases as compared to previous methods differentiating between ambiguous and unambiguous positions, with a precision similar to the one of previous methods.

We implemented this method as version 2.0 of our gap filler of scaffolds, Gap2Seq, available at www.cs.helsinki.fi/u/lmsalmel/Gap2Seq/.

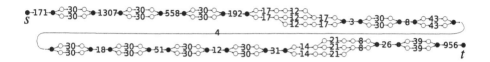

Fig. 1. A de Bruijn graph ($k = 31$) built on *S.aureus* data. We represent unary paths by numbers indicating their length. The estimated gap length is $d = 3774$, and there are 9216 different *s-t* paths of length d. The safe sub-paths (in black) have length 3337 and the precision of our method on these sub-paths is 99.9 %. Notice that most of the bubbles of this graph are caused by SNPs.

Contents

Optimal Computation of Avoided Words

Yannis Almirantis[1], Panagiotis Charalampopoulos[2], Jia Gao[2],
Costas S. Iliopoulos[2], Manal Mohamed[2], Solon P. Pissis[2(✉)],
and Dimitris Polychronopoulos[3]

[1] National Center for Scientific Research Demokritos, Athens, Greece
yalmir@bio.demokritos.gr
[2] Department of Informatics, King's College London, London, UK
{panagiotis.charalampopoulos,jia.gao,costas.iliopoulos,
manal.mohamed,solon.pissis}@kcl.ac.uk
[3] Computational Regulatory Genomics, MRC Clinical Sciences Centre,
Imperial College London, London W12 0NN, UK
d.polychronopoulos@csc.mrc.ac.uk

Abstract. The deviation of the observed frequency of a word w from its expected frequency in a given sequence x is used to determine whether or not the word is *avoided*. This concept is particularly useful in DNA linguistic analysis. The value of the standard deviation of w, denoted by $std(w)$, effectively characterises the extent of a word by its edge contrast in the context in which it occurs. A word w of length $k > 2$ is a ρ-avoided word in x if $std(w) \leq \rho$, for a given threshold $\rho < 0$. Notice that such a word may be completely *absent* from x. Hence computing all such words naïvely can be a very time-consuming procedure, in particular for large k. In this article, we propose an $\mathcal{O}(n)$-time and $\mathcal{O}(n)$-space algorithm to compute all ρ-avoided words of length k in a given sequence x of length n over a fixed-sized alphabet. We also present a time-optimal $\mathcal{O}(\sigma n)$-time algorithm to compute all ρ-avoided words (of any length) in a sequence of length n over an integer alphabet of size σ. We provide a tight asymptotic upper bound for the number of ρ-avoided words over an integer alphabet and the expected length of the longest one. We make available an implementation of our algorithm. Experimental results, using both real and synthetic data, show the efficiency of our implementation.

1   Introduction

The one-to-one mapping of a DNA molecule to a sequence of letters suggests that DNA analysis can be modelled within the framework of formal language theory [13]. For example, a region within a DNA sequence can be considered as a "word" on a fixed-sized alphabet in which some of its natural aspects can be described by means of certain types of automata or grammars. However, a linguistic analysis of the DNA needs to take into account many distinctive physical and biological characteristics of such sequences: DNA contains coding regions

---

This research was partially supported by the Leverhulme Trust.

M. Frith and C.N.S. Pedersen (Eds.): WABI 2016, LNBI 9838, pp. 1–13, 2016.
DOI: 10.1007/978-3-319-43681-4_1

that encode for polypeptide chains associated with biological functions; and non-coding regions, most of which are not linked to any particular function. Both appear to have many statistical features in common with natural languages [10].

A computational tool oriented towards the systematic search for avoided words is particularly useful for *in silico* genomic research analyses. The search for *absent words* is already undertaken in the recent past and several results exist [1]. However, words which may be present in a genome or in genomic sequences of a specific role (e.g., protein coding segments, regulatory elements, conserved non-coding elements etc.) but they are strongly underrepresented—as we can estimate on the basis of the frequency of occurrence of their longest proper factors—may be of particular importance. They can be words of nucleotides which are hardly tolerated because they negatively influence the stability of the chromatin or, more generally, the functional genomic conformation; they can represent targets of restriction endonucleases which may be found in bacterial and viral genomes; or, more generally, they may be short genomic regions whose presence in wide parts of the genome are not tolerated for less known reasons. The understanding of such avoidances is becoming an interesting line of research (for recent studies, see [5, 12]).

On the other hand, short words of nucleotides may be systematically avoided in large genomic regions or whole genomes for entirely different reasons: just because they play important signaling roles which restrict their appearance only in specific positions: consensus sequences for the initiation of gene transcription and of DNA replication are well-known such oligonucleotides. Other such cases may be insulators, sequences anchoring the chromatin on the nuclear envelope like lamina-associated domains, short sequences like dinucleotide repeat motifs with enhancer activity, and several other cases. Again, we cannot exclude that this area of research could lead to the identification of short sequences of regulatory activities still unknown.

Brendel et al. in [6] initiated research into the linguistics of nucleotide sequences that focuses on the concept of words in continuous languages—languages devoid of blanks—and introduced an operational definition of words. The authors suggested a method to measure, for each possible word w of length k, the deviation of its observed frequency from the expected frequency in a given sequence. The values of the standard deviation, denoted by $std(w)$, were then used to identify words that are avoided among all possible words of length k. The typical length of avoided (or of overabundant) words of the nucleotide language was found to range from 3 to 5 (tri- to pentamers). The statistical significance of the avoided words was shown to reflect their biological importance. This work, however, was based on the very limited sequence data available at the time: only DNA sequences from two viral and one bacterial genomes were considered. Also note that k might change when considering eukaryotic genomes, the complex dynamics and function of which might impose a more demanding analysis.

Our Contribution. The computational problem can be described as follows. Given a sequence x of length n, an integer k, and a real number $\rho < 0$, compute the set of ρ-avoided words of length k, i.e. all words w of length k for which

$std(w) \leq \rho$. We call this set the ρ-*avoided words* of length k in x. Brendel et al. did not provide an efficient solution for this computation [6]. Notice that such a word may be completely absent from x. Hence the set of ρ-avoided words can be naïvely computed by considering all possible σ^k words, where σ is the size of the alphabet. Here we present an $\mathcal{O}(n)$-time and $\mathcal{O}(n)$-space algorithm for computing all ρ-avoided words of length k in a sequence x of length n over a fixed-sized alphabet. We also present a time-optimal $\mathcal{O}(\sigma n)$-time algorithm to compute all ρ-avoided words (of any length) in a sequence of length n over an integer alphabet of size σ. We provide a tight asymptotic upper bound for the number of ρ-avoided words over an integer alphabet and the expected length of the longest one. We make available an open-source implementation of our algorithm. Experimental results, using both real and synthetic data, show its efficiency and applicability. Specifically, using our method we confirm that restriction endonucleases which target self-complementary sites are not found in eukaryotic sequences [12].

2   Terminology and Technical Background

Definitions and Notation. We begin with basic definitions and notation generally following [7]. Let $x = x[0]x[1] \ldots x[n-1]$ be a *word* of *length* $n = |x|$ over a finite ordered *alphabet* Σ of fixed size, i.e. $\sigma = |\Sigma| = \mathcal{O}(1)$. We also consider the case of an *integer alphabet*; in this case each letter is replaced by its rank such that the resulting string consists of integers in the range $\{1, \ldots, n\}$. For two positions i and j on x, we denote by $x[i \ldots j] = x[i] \ldots x[j]$ the *factor* (sometimes called *subword*) of x that starts at position i and ends at position j (it is empty if $j < i$), and by ε the *empty word*, word of length 0. We recall that a prefix of x is a factor that starts at position 0 ($x[0 \ldots j]$) and a suffix is a factor that ends at position $n-1$ ($x[i \ldots n-1]$), and that a factor of x is a *proper* factor if it is not x itself. A factor of x that is neither a prefix nor a suffix of x is called an *infix* of x.

Let $w = w[0]w[1] \ldots w[m-1]$ be a word, $0 < m \leq n$. We say that there exists an *occurrence* of w in x, or, more simply, that w *occurs in* x, when w is a factor of x. Every occurrence of w can be characterised by a starting position in x. Thus we say that w occurs at the *starting position* i in x when $w = x[i \ldots i + m - 1]$. Further let $f(w)$ denote the *observed frequency*, that is, the number of occurrences of a non-empty word w in word x. If $f(w) = 0$ for some word w, then w is called *absent*, otherwise, w is called *occurring*.

By $f(w_p)$, $f(w_s)$, and $f(w_i)$ we denote the observed frequency of the longest proper prefix w_p, suffix w_s, and infix w_i of w in x, respectively. We can now define the *expected frequency* of word w, $|w| > 2$, in x as in Brendel et al. [6]:

$$E(w) = \frac{f(w_p) \times f(w_s)}{f(w_i)}, \text{ if } f(w_i) > 0; \text{ else } E(w) = 0. \qquad (1)$$

The above definition can be explained intuitively as follows. Suppose we are given $f(w_p)$, $f(w_s)$, and $f(w_i)$. Given an occurrence of w_i in x, the probability of it being preceded by $w[0]$ is $\frac{f(w_p)}{f(w_i)}$ as $w[0]$ precedes exactly $f(w_p)$ of the $f(w_i)$

occurrences of w_i. Similarly, this occurrence of w_i is also an occurrence of w_s with probability $\frac{f(w_s)}{f(w_i)}$. Although these two events are not always independent, the product $\frac{f(w_p)}{f(w_i)} \times \frac{f(w_s)}{f(w_i)}$ gives a good approximation of the probability that an occurrence of w_i at position j implies an occurrence of w at position $j - 1$. It can be seen then that by multiplying this product by the number of occurrences of w_i we get the above formula for the expected frequency of w.

Moreover, to measure the deviation of the observed frequency of a word w from its expected frequency in x, we define the *standard deviation* (χ^2 test) of w as:

$$std(w) = \frac{f(w) - E(w)}{\max\{\sqrt{E(w)}, 1\}}. \tag{2}$$

For more details on the *biological* justification of these definitions see [6].

Using the above definitions and a given threshold, we are in a position to classify a word w as either *avoided* or *common* in x. In particular, for a given threshold $\rho < 0$, a word w is called ρ-*avoided* if $std(w) \leq \rho$. In this article, we consider the following computational problem.

AVOIDEDWORDSCOMPUTATION
Input: A word x of length n, an integer $k > 2$, and a real number $\rho < 0$
Output: All ρ-avoided words of length k in x

Suffix Trees. In our algorithms, suffix trees are used extensively as computational tools. For a general introduction to suffix trees, see [7].

The *suffix tree* $\mathcal{T}(x)$ of a non-empty word x of length n is a compact trie representing all suffixes of x, the nodes of the trie which become nodes of the suffix tree are called *explicit* nodes, while the other nodes are called *implicit*. Each edge of the suffix tree can be viewed as an upward maximal path of implicit nodes starting with an explicit node. Moreover, each node belongs to a unique path of that kind. Then, each node of the trie can be represented in the suffix tree by the edge it belongs to and an index within the corresponding path.

We use $\mathcal{L}(v)$ to denote the *path-label* of a node v, i.e., the concatenation of the edge labels along the path from the root to v. We say that v is path-labelled $\mathcal{L}(v)$. Additionally, $\mathcal{D}(v) = |\mathcal{L}(v)|$ is used to denote the *word-depth* of node v. Node v is a *terminal* node, if and only if, $\mathcal{L}(v) = x[i..n-1]$, $0 \leq i < n$; here v is also labelled with index i. It should be clear that each occurring word w in x is uniquely represented by either an explicit or an implicit node of $\mathcal{T}(x)$. The *suffix-link* of a node v with path-label $\mathcal{L}(v) = \alpha y$ is a pointer to the node path-labelled y, where $\alpha \in \Sigma$ is a single letter and y is a word. The suffix-link of v exists if v is a non-root internal node of $\mathcal{T}(x)$.

In any standard implementation of the suffix tree, we assume that each node of the suffix tree is able to access its parent. Note that once $\mathcal{T}(x)$ is constructed, it can be traversed in a depth-first manner to compute the word-depth $\mathcal{D}(v)$ for each node v. Let u be the parent of v. Then the word-depth $\mathcal{D}(v)$ is computed by adding $\mathcal{D}(u)$ to the length of the label of edge (u, v). If v is the root then

$\mathcal{D}(v) = 0$. Additionally, a depth-first traversal of $\mathcal{T}(x)$ allows us to count, for each node v, the number of terminal nodes in the subtree rooted at v, denoted by $\mathcal{C}(v)$, as follows. When internal node v is visited, $\mathcal{C}(v)$ is computed by adding up $\mathcal{C}(u)$ of all the nodes u, such that u is a child of v, and then $\mathcal{C}(v)$ is incremented by 1 if v itself is a terminal node. If a node v is a leaf then $\mathcal{C}(v) = 1$.

3   Useful Properties

In this section, we provide some useful insights of combinatorial nature which were not considered by Brendel et al. [6]. By the definition of ρ-avoided words it follows that a word w may be ρ-avoided even if it is absent from x. In other words, $std(w) \leq \rho$ may hold for either $f(w) > 0$ (occurring) or $f(w) = 0$ (absent).

This means that a naïve computation should consider all possible σ^k words. Then for each possible word w, the value of $std(w)$ can be computed via pattern matching on the suffix tree of x. In particular we can search for the occurrences of w, w_p, w_s, and w_i in x in time $\mathcal{O}(k)$ [7]. In order to avoid this inefficient computation, we exploit the following crucial lemmas.

Definition 1 [3]. *An absent word w of x is minimal if and only if all its proper factors occur in x.*

Lemma 1. *Any absent ρ-avoided word w in x is a minimal absent word of x.*

Proof. For w to be a ρ-avoided word it must hold that

$$std(w) = \frac{f(w) - E(w)}{\max\{\sqrt{E(w)}, 1\}} \leq \rho < 0.$$

This implies that $f(w) - E(w) < 0$, which in turn implies that $E(w) > 0$ since $f(w) = 0$. From $E(w) = \frac{f(w_p) \times f(w_s)}{f(w_i)} > 0$, we conclude that $f(w_p) > 0$ and $f(w_s) > 0$ must hold. Since $f(w) = 0$, $f(w_p) > 0$, and $f(w_s) > 0$, w is a minimal absent word of x: all proper factors of w occur in x.   □

Lemma 2. *Let w be a word occurring in x and $\mathcal{T}(x)$ be the suffix tree of x. Then, if w_p is a path-label of an implicit node of $\mathcal{T}(x)$, $std(w) \geq 0$.*

Proof. For any w that occurs in x it holds that $f(w_i) \geq f(w_s)$, which implies that $f(w_p) \geq \frac{f(w_p) \times f(w_s)}{f(w_i)} = E(w)$. Furthermore, by the definition of the suffix tree, if w occurs in x and w_p is a path-label of an implicit node then $f(w_p) = f(w)$. It thus follows that $f(w) - E(w) = f(w_p) - E(w) \geq 0$, and since $\max\{1, \sqrt{E(w)}\} > 0$, the claim holds.   □

Lemma 3. *The number of ρ-avoided words of length $k > 2$ in a word of length n over an alphabet of size σ is $\mathcal{O}(\sigma n)$; in particular, this number is no more than $(\sigma + 1)n - k + 1$.*

Proof. By Lemma 1, every ρ-avoided word is either occurring or a minimal absent word. It is known that the number of minimal absent words in a word of length n is smaller than or equal to σn [11]. Clearly, the occurring ρ-avoided words in a word of length n are at most $n - k + 1$. Therefore the lemma holds.   □

4   Avoided Words Algorithm

In this section, we present Algorithm AVOIDEDWORDS for computing all ρ-avoided words of length k in a given word x. The algorithm builds the suffix tree $\mathcal{T}(x)$ for word x, and then prepares $\mathcal{T}(x)$ to allow constant-time observed frequency queries. This is mainly achieved by counting the terminal nodes in the subtree rooted at node v for every node v of $\mathcal{T}(x)$. Additionally during this preprocessing, the algorithm computes the word-depth of v for every node v of $\mathcal{T}(x)$. By Lemma 1, ρ-avoided words are classified as either occurring or (minimal) absent, therefore Algorithm AVOIDEDWORDS calls Routines ABSENTAVOIDEDWORDS and OCCURRINGAVOIDEDWORDS to compute both classes of ρ-avoided words in x. The outline of Algorithm AVOIDEDWORDS is as follows.

AVOIDEDWORDS(x, k, ρ)

   1   $\mathcal{T}(x) \leftarrow$ SUFFIXTREE(x)
   2   **for** each node $v \in \mathcal{T}(x)$ **do**
   3        $\mathcal{D}(v) \leftarrow$ word-depth of v
   4        $\mathcal{C}(v) \leftarrow$ number of terminal nodes in the subtree rooted at v
   5   ABSENTAVOIDEDWORDS(x,k,ρ)
   6   OCCURRINGAVOIDEDWORDS(x,k,ρ)

4.1   Computing Absent Avoided Words

In Lemma 1, we showed that each absent ρ-avoided word is a minimal absent word. Thus, Routine ABSENTAVOIDEDWORDS starts by computing all minimal absent words in x; this can be done in time and space $\mathcal{O}(n)$ for a fixed-sized alphabet or in time $\mathcal{O}(\sigma n)$ for integer alphabets [3,4]. Let $<(i,j),\alpha>$ be a tuple representing a minimal absent word in x, where for some minimal absent word w of length $|w| > 2$, $w = x[i \mathinner{.\,.} j]\alpha$, $\alpha \in \Sigma$; this representation is clearly unique.

ABSENTAVOIDEDWORDS(x, k, ρ)

   1   $\mathcal{A} \leftarrow$ MINIMALABSENTWORDS(x)
   2   **for** each tuple $< (i,j),\alpha >\in \mathcal{A}$ such that $k = j - i + 2$ **do**
   3        $u_p \leftarrow$ NODE(i,j)
   4   **if** ISIMPLICIT(u_p) **then**
   5            $(u,v) \leftarrow$ EDGE(u_p)
   6            $f_p \leftarrow \mathcal{C}(v)$
   7   **else** $f_p \leftarrow \mathcal{C}(u_p)$
   8        $u_i \leftarrow$ NODE($i + 1,j$)
   9   **if** ISIMPLICIT(u_i) **then**
  10            $(u,v) \leftarrow$ EDGE(u_i)
  11            $f_i \leftarrow f_s \leftarrow \mathcal{C}(v)$
  12   **else** $f_i \leftarrow \mathcal{C}(u_i)$
  13        $u_s \leftarrow$ CHILD(u_i,α)
  14        $f_s \leftarrow \mathcal{C}(u_s)$
  15        $E \leftarrow f_p \times f_s/f_i$
  16   **if** $(0 - E)/(\max\{1, \sqrt{E}\}) \leq \rho$ **then**
  17   REPORT($x[i \mathinner{.\,.} j]\alpha$)

Intuitively, the idea is to check the length of every minimal absent word. If a tuple $<(i,j), \alpha>$ represents a minimal absent word w of length $k = j-i+2$, then the value of $std(w)$ is computed to determine whether w is an absent ρ-avoided word. Note that, if $w = x[i\mathbin{..}j]\alpha$ is a minimal absent word, then $w_p = x[i\mathbin{..}j]$, $w_i = x[i+1\mathbin{..}j]$, and $w_s = x[i+1\mathbin{..}j]\alpha$ occur in x by Definition 1. Thus, there are three (implicit or explicit) nodes in $\mathcal{T}(x)$ path-labelled w_p, w_i, and w_s, respectively. The observed frequencies of w_p, w_i, and w_s are already computed during the preprocessing of $\mathcal{C}$, which stores the number of terminal nodes in the subtree rooted at v, for each node v.

Notice that for an explicit node v path-labelled $w' = x[i'\mathbin{..}j']$, the value $\mathcal{C}(v)$ represents the number of occurrences (observed frequency) of w' in x; whereas for an implicit node along the edge (u, v) path-labelled w'', the number of occurrences of w'' is equal to $\mathcal{C}(v)$ (and not $\mathcal{C}(u)$). The implementation of this procedure is given in Routine ABSENTAVOIDEDWORDS.

4.2   Computing Occurring Avoided Words

Lemma 2 suggests that for each occurring ρ-avoided word w, w_p is a path-label of an explicit node v of $\mathcal{T}(x)$. Thus, for each internal node v such that $\mathcal{D}(v) = k-1$ and $\mathcal{L}(v) = w_p$, Routine OCCURRINGAVOIDEDWORDS computes $std(w)$, where $w = w_p\alpha$, $\alpha \in \Sigma$, is a path-label of a child (explicit or implicit) node of v. Note that if w_p is a path-label of an explicit node v then w_i is a path-label of an explicit node u of $\mathcal{T}(x)$; node u is well-defined and it is the node pointed at by the suffix-link of v. The implementation of this procedure is given in Routine OCCURRINGAVOIDEDWORDS.

OCCURRINGAVOIDEDWORDS(x, k, ρ)

```
 1   N ← an empty stack
 2   PUSH(N, root(T(x)))
 3   while N is not empty do
 4       u ← POP(N)
 5       for each edge (u, v) of T(x) do
 6           if D(v) < k − 1 then
 7               PUSH(N, v)
 8           elseif D(v) = k − 1 then
 9               fₚ ← C(v)
10               fᵢ ← C(suffix-link[v])
11               for each child v′ of v do
12                   f_w ← C(v′)
13                   α ← L(v′)[k − 1]
14                   f_s ← C(CHILD(suffix-link[v], α))
15                   E ← fₚ × f_s/fᵢ
16                   if (f_w − E)/(max{1, √E}) ≤ ρ then
17                       REPORT(L(v′)[0 .. k − 1])
```

4.3   Analysis of the Algorithm

Lemma 4. *Given a word x, an integer $k > 2$, and a real number $\rho < 0$, Algorithm* AVOIDEDWORDS *computes all ρ-avoided words of length k in x.*

Proof. By definition, a ρ-avoided word w is either an absent ρ-avoided word or an occurring one. Hence, the proof of correctness relies on Lemmas 1 and 2. First, Lemma 1 indicates that an absent ρ-avoided word in x is necessarily a minimal absent word. Routine ABSENTAVOIDEDWORDS considers each minimal absent word w and verifies if w is a ρ-avoided word of length k.

Second, Lemma 2 indicates that for each occurring ρ-avoided word w, w_p is a path-label of an explicit node v of $\mathcal{T}(x)$. Routine OCCURRINGAVOIDEDWORDS considers every child of each such node of word-depth k, and verifies if its path-label is a ρ-avoided word. □

Lemma 5. *Given a word x of length n over a fixed-sized alphabet, an integer $k > 2$, and a real number $\rho < 0$, Algorithm* AVOIDEDWORDS *requires time and space $\mathcal{O}(n)$; for integer alphabets, it requires time $\mathcal{O}(\sigma n)$.*

Proof. Constructing the suffix tree $\mathcal{T}(x)$ of the input word x takes time and space $\mathcal{O}(n)$ for a word over a fixed-sized alphabet [7]. Once the suffix tree is constructed, computing arrays $\mathcal{D}$ and $\mathcal{C}$ by traversing $\mathcal{T}(x)$ requires time and space $\mathcal{O}(n)$. Note that the path-labels of the nodes of $\mathcal{T}(x)$ can by implemented in time and space $\mathcal{O}(n)$ as follows: traverse the suffix tree to compute for each node v the smallest index i of the terminal nodes of the subtree rooted at v. Then $\mathcal{L}(v) = x[i \mathinner{.\,.} i + \mathcal{D}(v) - 1]$.

Next, Routine ABSENTAVOIDEDWORDS requires time $\mathcal{O}(n)$. It starts by computing all minimal absent words of x, which can be achieved in time and space $\mathcal{O}(n)$ over a fixed-sized alphabet [3,4]. The rest of the procedure deals with checking each of the $\mathcal{O}(n)$ minimal absent words of length k. Checking each minimal absent word w to determine whether it is a ρ-avoided word or not requires time $\mathcal{O}(1)$. In particular, an $\mathcal{O}(n)$-time preprocessing of $\mathcal{T}(x)$ allows the retrieval of the (implicit or explicit) node in $\mathcal{T}(x)$ corresponding to the longest proper prefix of w in time $\mathcal{O}(1)$ [9]. Finally, Routine OCCURRINGAVOIDEDWORDS requires time $\mathcal{O}(n)$. It traverses the suffix tree $\mathcal{T}(x)$ to consider all explicit nodes of word-depth $k - 1$. Then for each such node, the procedure checks every (explicit or implicit) child of word-depth k. The total number of these children is at most $n - k + 1$. For every child node, the procedure checks whether its path-label is a ρ-avoided word in time $\mathcal{O}(1)$ via the use of suffix-links.

For integer alphabets, the suffix tree can be constructed in time $\mathcal{O}(n)$ [8] and all minimal absent words can be computed in time $\mathcal{O}(\sigma n)$ [3,4]. The efficiency of Algorithm AVOIDEDWORDS is then limited by the total number of words to be considered, which, by Lemma 3, is $\mathcal{O}(\sigma n)$. □

Lemmas 4 and 5 imply the first result of this article.

Theorem 1. *Algorithm* AVOIDEDWORDS *solves Problem* AVOIDEDWORDSCOMPUTATION *in time and space $\mathcal{O}(n)$. For integer alphabets, the algorithm solves the problem in time $\mathcal{O}(\sigma n)$.*

4.4   Optimal Computation of all ρ-Avoided Words

Although the biological motivation is yet to be shown for this, we present here how we can modify Algorithm AVOIDEDWORDS so that it computes *all* ρ-avoided words (of all lengths) in a given word x of length n over an integer alphabet of size σ in time $\mathcal{O}(\sigma n)$. We further show that this algorithm is in fact time-optimal. All omitted proofs will be presented in the full version of this article.

Lemma 6. *The upper bound $\mathcal{O}(\sigma n)$ on the number of minimal absent words of a word of length n over an alphabet of size σ is tight if $2 \leq \sigma \leq n$.*

Lemma 7. *The number of ρ-avoided words in a word of length n over an alphabet of size $2 \leq \sigma \leq n$ is $\mathcal{O}(\sigma n)$ and this bound is tight.*

It is clear that if we just remove the condition on the length of each minimal absent word in Line 2 of ABSENTAVOIDEDWORDS we then compute all absent ρ-avoided words in time $\mathcal{O}(\sigma n)$. In order to compute all occurring ρ-avoided words in x it suffices by Lemma 2 to investigate the children of explicit nodes. We can thus traverse the suffix tree $\mathcal{T}(x)$ and for each explicit internal node, check for all of its children (explicit or implicit) whether their path-label is a ρ-avoided word. We can do this in $\mathcal{O}(1)$ time as described. The total number of these children is at most $2n - 1$, as this is the bound on the number of edges of $\mathcal{T}(x)$ [7]. This modified algorithm is clearly time-optimal for fixed-sized alphabets as it then runs in time $\mathcal{O}(n)$. The time optimality for integer alphabets follows directly from Lemmas 6 and 7. Hence we obtain the second result of this article.

Theorem 2. *Given a word x of length n over an integer alphabet of size σ and a real number $\rho < 0$, all ρ-avoided words in x can be computed in time $\mathcal{O}(\sigma n)$. This is time-optimal if $2 \leq \sigma \leq n$.*

Lemma 8. *The expected length of the longest ρ-avoided word in a word of length n over an alphabet Σ of size $\sigma > 1$ is $\mathcal{O}(\log_\sigma n)$ when the letters are independent and identically distributed random variables uniformly distributed over Σ.*

5   Implementation and Experimental Results

Algorithm AVOIDEDWORDS was implemented as a program to compute the ρ-avoided words of length k in one or more input sequences. The program was implemented in the C++ programming language and developed under GNU/Linux operating system. The input parameters are a (Multi)FASTA file with the input sequences(s), an integer $k > 2$, and a real number $\rho < 0$. The output is a file with the set of ρ-avoided words of length k per input sequence. The implementation is distributed under the GNU General Public License, and it is available at http://github.com/solonas13/aw. The experiments were conducted on a Desktop PC using one core of Intel Core i5-4690 CPU at 3.50 GHz under GNU/Linux. The program was compiled with g++ version 4.8.4 at optimisation level 3 (-O3). We also implemented a brute-force approach for the computation

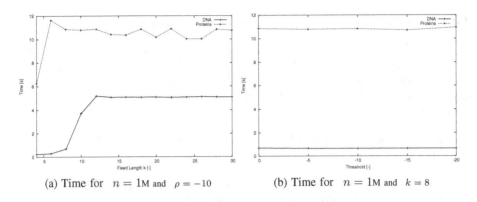

(a) Time for $n = 1\text{M}$ and $\rho = -10$ (b) Time for $n = 1\text{M}$ and $k = 8$

Fig. 1. Elapsed time of Algorithm AVOIDEDWORDS using synthetic DNA ($\sigma = 4$) and proteins ($\sigma = 20$) data of length 1M for variable k and variable ρ.

of ρ-avoided words. We mainly used it to confirm the correctness of our implementation. Here we do not plot the results of the brute-force approach as it is easily understood that it is orders of magnitude slower than our approach.

To evaluate the time performance of our implementation, synthetic DNA ($\sigma = 4$) and proteins ($\sigma = 20$) data were used. The input sequences were generated using a randomised script. In the first experiment, our task was to establish that the performance of the program does not essentially depend on k and ρ; i.e., the elapsed time of the program remains unchanged up to some constant with increasing values of k and decreasing values of ρ. As input datasets, for this experiment, we used a DNA and a proteins sequence both of length 1M (1 Million letters). For each sequence we used different values of k and ρ. The results, for elapsed time are plotted in Fig. 1. It becomes evident from the results that the time performance of the program remains unchanged up to some constant. The longer time required for the proteins sequences for some value of k is explained by the increased number of branching nodes in this depth in the corresponding suffix tree due to the size of the alphabet ($\sigma = 20$). To confirm this we counted the number of nodes considered by the algorithm to compute the ρ-avoided words for $k = 4$ and $\rho = -10$ for both sequences. The number of considered nodes for the DNA sequence was 260 whereas for the proteins sequence it was $1,585,510$.

In the second experiment, our task was to establish the fact that the elapsed time and memory usage of the program grow linearly with n, the length of the input sequence. As input datasets, for this experiment, we used synthetic DNA and proteins sequences ranging from 1 to 128 M. For each sequence we used constant values for k and ρ: $k = 8$ and $\rho = -10$. The results, for elapsed time and peak memory usage, are plotted in Fig. 2. It becomes evident from the results that the elapsed time and memory usage of the program grow linearly with n. The longer time required for the proteins sequences compared to the DNA sequences for increasing n is explained by the increased number of branching nodes in this depth ($k = 8$) in the corresponding suffix tree due to the size of the alphabet ($\sigma = 20$). To confirm this we counted the number of nodes considered by the algorithm to compute the ρ-avoided words for $n = 64M$ for both the DNA and the proteins sequence. The number of nodes for the DNA sequence was $69,392$ whereas for the proteins sequence it was $43,423,082$.

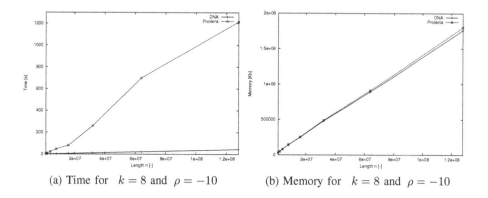

(a) Time for $k = 8$ and $\rho = -10$ (b) Memory for $k = 8$ and $\rho = -10$

Fig. 2. Elapsed time and peak memory usage of Algorithm AVOIDEDWORDS using synthetic DNA ($\sigma = 4$) and proteins ($\sigma = 20$) data of length 1M to 128M.

In the next experiment, our task was to evaluate the time and memory performance of our implementation with real data. As input datasets, for this experiment, we used all chromosomes of the human genome. Their lengths range from around 46M (chromosome 21) to around 249M (chromosome 1). For each sequence we used $k = 8$ and $\rho = -10$. The results, for elapsed time and peak memory usage, are plotted in Fig. 3. The results with real data confirm that the elapsed time and memory usage of the program grow linearly with n.

As last experiment, we computed the set of avoided words for $k = 6$ (hexamers) and $\rho = -10$ in the complete genome of *E. coli* and sorted the output in increasing order of their standard deviation. The most avoided words were extremely enriched in self-complementary (palindromic) hexamers. In particular, within the output of 28 avoided words, 23 were self-complementary; and the 17 most avoided ones were *all* self-complementary. For comparison, we computed the set of avoided words for $k = 6$ and $\rho = -10$ from an eukaryotic sequence: a segment of the human chromosome 21 (its leftmost segment devoid of N's) equal to the length of the *E. coli* genome. In the output of 10 avoided words, no self-complementary hexamer was found. Our results confirm that the restriction endonucleases which target self-complementary sites are not found in eukaryotic sequences [12].

Our immediate target is to investigate the avoidance of words in the context of Genomic Regulatory Blocks (GRBs), chromosomal regions spanned by highly conserved non-coding elements (HCNEs), most of which serve as regulatory inputs of one target gene in the region [2].

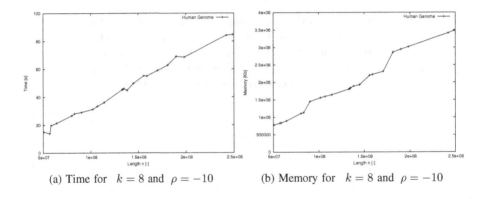

(a) Time for $k = 8$ and $\rho = -10$ (b) Memory for $k = 8$ and $\rho = -10$

Fig. 3. Elapsed time and peak memory usage of Algorithm AVOIDEDWORDS using all chromosomes of the human genome.

References

1. Acquisti, C., Poste, G., Curtiss, D., Kumar, S.: Nullomers: really a matter of natural selection? PLoS ONE **2**(10), e1022 (2007)
2. Akalin, A., Fredman, D., Arner, E., Dong, X., Bryne, J., Suzuki, H., Daub, C., Hayashizaki, Y., Lenhard, B.: Transcriptional features of genomic regulatory blocks. Genome Biol. **10**(4), 1 (2009)
3. Barton, C., Heliou, A., Mouchard, L., Pissis, S.P.: Linear-time computation of minimal absent words using suffix array. BMC Bioinform. **15**(1), 1–10 (2014)
4. Barton, C., Heliou, A., Mouchard, L., Pissis, S.P.: Parallelising the computation of minimal absent words. In: Wyrzykowski, R., Deelman, E., Dongarra, J., Karczewski, K., Kitowski, J., Wiatr, K. (eds.) PPAM 2015, Part II. LNCS, vol. 9574, pp. 243–253. Springer, Heidelberg (2016)
5. Belazzougui, D., Cunial, F.: Space-efficient detection of unusual words. In: Iliopoulos, C., Puglisi, S., Yilmaz, E. (eds.) SPIRE 2015. LNCS, vol. 9309, pp. 222–233. Springer, Heidelberg (2015)
6. Brendel, V., Beckmann, J.S., Trifonov, E.N.: Linguistics of nucleotide sequences: morphology and comparison of vocabularies. J. Biomol. Struct. Dyn. **4**(1), 11–21 (1986)
7. Crochemore, M., Hancart, C., Lecroq, T.: Algorithms on Strings. Cambridge University Press, New York (2007)
8. Farach, M.: Optimal suffix tree construction with large alphabets. In: FOCS, pp. 137–143. IEEE (1997)
9. Gawrychowski, P., Lewenstein, M., Nicholson, P.K.: Weighted ancestors in suffix trees. In: Schulz, A.S., Wagner, D. (eds.) ESA 2014. LNCS, vol. 8737, pp. 455–466. Springer, Heidelberg (2014)
10. Mantegna, R.N., Buldyrev, S.V., Goldberger, A.L., Havlin, S., Peng, C.K., Simons, M., Stanley, H.E.: Linguistic features of noncoding DNA sequences. Phys. Rev. Lett. **73**, 3169–3172 (1994)

11. Mignosi, F., Restivo, A., Sciortino, M.: Words and forbidden factors. Theoret. Comput. Sci. **273**(1–2), 99–117 (2002)
12. Rusinov, I., Ershova, A., Karyagina, A., Spirin, S., Alexeevski, A.: Lifespan of restriction-modification systems critically affects avoidance of their recognition sites in host genomes. BMC Genom. **16**(1), 1–15 (2015)
13. Searls, D.B.: The linguistics of DNA. Am. Sci. **80**(6), 579–591 (1992)

A Biclique Approach to Reference Anchored Gene Blocks and Its Applications to Pathogenicity Islands

Arnon Benshahar[1], Vered Chalifa-Caspi[3], Danny Hermelin[2($\boxtimes$)], and Michal Ziv-Ukelson[1($\boxtimes$)]

[1] Department of Computer Science,
Ben-Gurion University of the Negev, Beersheba, Israel
{benarnon,michaluz}@bgu.ac.il
[2] Department of Industrial Engineering and Management,
National Institute for Biotechnology in the Negev, Beersheba, Israel
hermelin@bgu.ac.il
[3] National Institute for Biotechnology in the Negev,
Ben-Gurion University of the Negev, Beersheba, Israel
veredcc@bgu.ac.il

Abstract. We formalize a new problem variant in gene-block discovery, denoted *Reference-Anchored Gene Blocks* (RAGB). Given a query sequence Q of length n, representing the gene-array of a DNA element, a window size bound d on the length of a substring of interest in Q, and a set of target gene sequences $T = \{T_1 \ldots T_c\}$. Our objective is to identify gene-blocks in T that are centered in a subset q of co-localized genes from Q, and contain genomes from T in which the corresponding orthologs of the genes from q are also co-localized. We cast RAGB as a variant of a (colored) biclique problem in bipartite graphs, and analyze its parameterized complexity, as well as the parameterized complexity of other related problems. We give an $O(nm + 2^d nm / \lg m)$ time algorithm for the uncolored variant of our biclique problem, where m is the number of areas of interest that are parsed from the target sequences, and n and d are as defined above. Our algorithm can be adapted to compute all maximal bicliques in the graph within the same time complexity, and to handle edge-weights with a slight $O(\lg d)$ increase to its time complexity. For the colored version of the problem, our algorithm has a time complexity of $O(2^d nm)$. We implement the algorithm and exemplify its application to LEE, a well-known pathogenicity island from the e. coli genome harboring virulence genes. *Our code and supplementary materials, including omitted proofs and figures, are available at* https://www.cs.bgu.ac.il/~negevcb/RAGB/.

1 Introduction

Genomes of bacterial species can evolve through a variety of processes including mutations, rearrangements and horizontal gene transfer. Information gathered over the past few years from a rapidly increasing number of sequenced genomes

© Springer International Publishing Switzerland 2016
M. Frith and C.N.S. Pedersen (Eds.): WABI 2016, LNBI 9838, pp. 14–26, 2016.
DOI: 10.1007/978-3-319-43681-4_2

has shown that besides the core genes which encode essential metabolic functions, bacterial genomes also harbour a variable number of accessory genes acquired by horizontal gene transfer, encoding adaptative traits that might be beneficial for bacteria under certain growth or environmental conditions [15]. Many of the accessory genes acquired by horizontal gene transfer form syntenic blocks recognized as genomic islands - discrete DNA segments that are transferred between closely related strains. During island evolution, several genetic elements have been acquired independently at different time points and from different hosts. Thus, genomic islands often represent mosaic-like structures rather than homogeneous segments of horizontally acquired DNA.

Genomic islands are known to carry genes offering a selective advantage for host bacteria. They play a key role in the emergence of highly virulent and highly resistant pathogenic strains of bacteria, which are of major concern worldwide [15]. This concern motivates new studies, such as the one proposed in this paper, aimed to develop new tools that will help explore the evolution and spreading patterns of pathogenicity, virulence and resistance components harbored within genomic islands, across the bacterial kingdom. Furthermore, applying gene-block discovery approaches to these studies may shed light on the function of unknown proteins which are consistently co-transferred with functional gene cascades.

1.1   The Reference Anchored Gene Block Problem

We propose a new bioinformatic approach that is based on interrogating a given reference DNA element from a specific genome for subsets of genes that are conserved as proximity blocks across other microbial genomes. The subsequent computational problem is called the *Reference Anchored Gene Blocks* problem (RAGB). The input to this problem consists of the gene sequence of a reference DNA element, and a set of target genome sequences. The target sequences are then parsed, either via a simple sliding window approach or according to some a priori biological data, into areas of interest or segments of small proximity. The output of our problem are gene blocks that are clustered together in small vicinity in the reference element, and that have orthologous genes clustered together in segments of sufficiently many target genomes. Note that our model allows paralogous occurrences of genes from the reference elements, and moreover, we do not require that all input genomes be represented in an output block.

Our framework is based on a bipartite graph modulation. Following the phase where the input element and genomes are parsed into (gene cluster encompassing) segments, a bipartite graph is constructed according to these segments. In this graph, vertices of one side represent subsets of reference genes, nodes in the other side represent segments from the target genomes, and edges connect the subsets of reference genes to segments from the target genomes that contain corresponding orthologs. Based on this, we cast the problem of enumerating reference-anchored gene blocks as a special type of *biclustering problem*: Compute appropriately constrained bicliques in an input bipartite graph. The constraint is a bound that ensures the co-localization of the reference genes participating in a block. When it is necessary to distinguish between segments of different genomes, we color-code the vertices corresponding to segments in our bipartite graph, and require colorful bicliques as solutions.

1.2   Results

Since our problem translates to a computationally hard problem, we use the theory of parameterized complexity [13] which provides a convenient theoretical framework for analyzing exact algorithms for hard problems. In particular, given a bipartite graph with n vertices corresponding to references genes, and m vertices corresponding to target segments, we show how to compute all maximal constrained bicliques in $O(nm + 2^d nm/\lg m)$ time, where d is the bound on the genomic distance between two genes in a cluster. We then show how to extend this algorithm to a weighted variant of the problem with an $O(\lg d)$ increase to the time complexity. Finally, our algorithm can also be extended to the more challenging vertex-colored variant corresponding to the case where segments are overlapping sliding windows in the target genomes, yet we allow only one segment per each genome in the output. The time complexity increases in this case to $O(2^d nm)$. We also use the theory of parameterized complexity to analyze closely related biclique problems, and show that these are unlikely to admit efficient algorithms with respect to their natural parameterizations.

We implement our algorithm in a program called *RAGB Monitor* (Reference-Anchored Gene Block Monitor). This program enumerates conserved blocks that are centered in small components of a given input DNA element and ranks them according to a probabilistic p-value. The program is exemplified by applying it to the analysis of LEE, a well-known pathogenicity island from the *E. coli* genome, where it identifies components of type III secretion system from LEE that are conserved across several proteobacterial genomes.

1.3   Related Work

On the biological front, previous related works studied the evolution of operons across different species by using either experimentally validated operons, or sets of genes from a pathway of interest, as anchors. A computational method was recently proposed for generalizing such studies in [12]. This method uses alignment-based approaches to measure the distance between the gene maps of orthologous gene clusters in various species and then interprets this information against the phylogeny of the target genomes. The method assumes a model where a chromosome is considered as a permutation of distinct genes. Similarly to these works, we also base our search on the gene map of an anchor DNA element. However, in contrast to these previous works, we consider several orthologs in each target genome, per each gene in the reference element. More importantly, our biological objective is quite different: We apply an exhaustive approach, aimed to discover *all* (possibly overlapping) co-localized *subsets* of genes from the anchor reference element that are conserved as orthologous gene clusters in (possibly overlapping) *subsets* of genomes from the input set.

In this respect, our problem is related to the well-studied *gene team* discovery problem (thoroughly reviewed in [2]) that seeks conserved gene clusters in an input set of genomes. Several models were considered for this problem. In the most basic one, a chromosome is considered as a permutation of distinct genes, and a gene team is defined to be a set of genes that appear in *all* the prespecified species, possibly in a different order, yet with the distance between adjacent genes in the team for each chromosome bounded by a certain threshold. This model is generalized to consider paralogous copies of the same gene in [16]. Polynomial time exact algorithms exist for the problem variants mentioned above. The next step to further generalize the gene team problem is to find teamsthat only occur in a subset of a given set of genomes. This step makes the problem NP-complete, and several heuristic approaches were proposed for this variant [10,11]. Chateau et al. [3] modeled approximate gene clusters as cliques in a graph, where nodes represent intervals (sets of genes co-located in a genomic region) and an edge connecting two nodes indicates that their set-distance is bounded by some predefined constant. The problem introduced in our paper could be viewed as a special variant of gene team discovery, where the sought gene teams are clustered around a predefined team of "centroid" genes. The model we follow in our solution to the problem is the most general one: It allows paralogous occurrences of genes in input strings, does not require gene order conservation, and does not require that all input genomes participate in a candidate solution.

2 Problem Definition and Formulations

Let Σ denote a finite set of characters representing genes. A genome is represented by a *sequence* $S = \sigma_1 \cdots \sigma_n$ of concatenated characters $\sigma_1, \ldots, \sigma_n \in \Sigma$. For a sequence $S = \sigma_1 \cdots \sigma_n$, we use $|S| = n$ to denote the *length* of S, and $S[i] = \sigma_i$ to denote the i'th character of S. A *subsequence* of S is any non-empty sequence S' that can be obtained by deleting zero or more characters from S. An *interval* of S is a subsequence of S with consecutive characters. For $1 \leq i \leq j \leq |S|$, we let $S[i,j] = \sigma_i \cdots \sigma_j$ denote the interval of S beginning at position i and ending at position j. We call a sequence where all characters are different a *permutation*. Two sequences S_1 and S_2 are said to be *equivalent*, denoted $S_1 \equiv S_2$, if $|\{S_1[i] = \sigma : 1 \leq i \leq |S_1|\}| = |\{S_2[i] = \sigma : 1 \leq i \leq |S_2|\}|$ for all $\sigma \in \Sigma$. In other words, $S_1 \equiv S_2$ if both sequences have the same number of occurrences of each character $\sigma \in \Sigma$. Clearly, for two equivalent sequences S_1 and S_2 we have $|S_1| = |S_2|$.

Let Q denote a sequence representing our designated reference element, and let $\mathcal{T} = \{T_1, \ldots, T_C\}$ denote a set of sequences representing the target genomes. An instance of our problem is defined by a triplet $(Q, \mathcal{I}, d)$, where d is a positive integer, and $\mathcal{I} = \{\mathcal{I}_1, \ldots, \mathcal{I}_C\}$ is a family of interval sets where each $\mathcal{I}_i = \{T_i^1, \ldots, T_i^{t_i}\}$ contains intervals of T_i. Each interval T_i^j represents an area of interest in the target genome T_i, and d represents the length of intervals that are of interest in Q. Our goal is to find subsequences in intervals of length d

in Q, representing operons in the reference element modeled by Q, that have equivalent occurrences in areas of interest of the target genomes. We formalized this in the following definition:

Definition 1 (Block). A *block* in $(Q, \mathcal{I}, d)$ is a set of sequences $\{q, t_{i_1}, \ldots, t_{i_k}\}$ satisfying:

1. q is a subsequence of some interval of length d in Q,
2. t_{i_j} is a subsequence of some interval in $\mathcal{I}_{i_j}$ for each $1 \leq j \leq k$,
3. $i_1 \neq i_2 \neq \cdots \neq i_k$, and
4. $q \equiv t_{i_j}$ for each $1 \leq j \leq k$.

We say that a block $\{q, t_{i_1}, \ldots, t_{i_k}\}$ is *maximal* in $(Q, \mathcal{I}, d)$ if there is no other block $\{q', t'_{j_1}, \ldots, t'_{j_\ell}\}$ in this instance where q is a subsequence of q' and $\{i_1, \ldots, i_k\} \subseteq \{j_1, \ldots, j_\ell\}$.

Definition 2 (Reference Anchored Gene Blocks Problem (RAGB)). The REFERENCE ANCHORED GENE BLOCKS problem is the problem of computing all maximal blocks in a given problem instance $(Q, \mathcal{I}, d)$.

We consider two distinct approaches to parse the intervals of our target genomes. The first approach, which we call the *sliding window* approach, is an exhaustive approach where each target genome in T_i is parsed into all its substrings of length d, i.e. $T_i[1, d], T_i[2, d+1], \ldots, T_i[n-d, n]$, and each such substring yields an interval in $\mathcal{I}_i$. The second approach takes into account biological signals to parse the genome into non-overlapping intervals. Another modeling option to be considered is whether we allow one or more orthologous genes in each of our input genomes; that is, whether or not our input sequences are permutations. This leads to the following two RAGB problem variants:

RAGB1. Compute reference anchored gene clusters in the following model: Intervals in $\mathcal{I}$ are parsed biologically into non-overlapping intervals. All input sequences are permutations.

RAGB2. Compute reference anchored gene clusters in the following model: Intervals in $\mathcal{I}$ are parsed via the sliding window approach. The input sequences are not necessarily permutations.

We cast both RAGB problem variants as biclique enumeration problems in bipartite graphs. The input to our framework consists of a sequence Q representing our designated genome, and $T_1, \ldots, T_C$ sequences representing the target genomes. Each genome T_i is parsed into intervals, and the ensemble of intervals from all the genomes yields the interval set $\mathcal{I} = \{\mathcal{I}_1, \ldots, \mathcal{I}_C\}$.

Based on Q and the intervals in $\mathcal{I}$ a bipartite graph $G = (A \uplus B, E)$ is constructed: Each node in $A = \{a_1, \ldots, a_n\}$ represents a single character in Q, such that node $a_i \in A$ corresponds to character $Q[j]$. Each node in $B = \{b_1, \ldots, b_m\}$ represents a distinct interval in $\mathcal{I}$. We then connect vertex $a_i \in A$ to vertex $b_j \in B$ iff the character $Q[i]$ appears in the interval corresponding to b_j.

We can also add a *weight measure* to this edge to indicate the level of similarity between the gene $Q[i]$ and its occurrence in the interval corresponding to b_j.

Next, for the second variant of RAGB, RAGB2, we further need to distinguish between intervals of different genomes. For this, we introduce a *coloring function* $c : B \to \{1, \ldots, C\}$ for the vertices of B, where $c(b_j) = i$ iff b_j corresponds to an interval of T_i. The reason we do not need this function for RAGB1 is that each sequence T_i is a permutation, so any character of Q can appear at most once in any of these sequences.

A *biclique* in G is a pair of non-empty vertex subsets $A' \subseteq A$ and $B' \subseteq B$ where $\{a, b\} \in E$ for each pair of vertices $a \in A'$ and $b \in B'$. We say that a biclique (A', B') is *maximal* if for any biclique (A'', B'') in G with $A' \subseteq A''$ and $B' \subseteq B''$ we have $A' = A''$ and $B' = B''$. In case G is equipped with a coloring function for the vertices in B, we say that a biclique (A', B') is *colorful* if no two distinct vertices in B' have the same color. For $1 \le i \le n - d$, let $A[i, i + d]$ denote the subset of vertices $\{a_i, a_{i+1}, \ldots, a_{i+d}\} \subseteq A$.

Observation 1. *There is one-to-one bijection between maximal (colorful) bicliques (A', B') in G with $A' \subseteq A[i, i+d]$ for some $1 \le i \le n - d$ and maximal blocks in $(Q, \mathcal{I}, d)$.*

3   Block Bicliques

In this section we present algorithms for our model for the REFERENCE ANCHORED GENE BLOCKS problem, as well as analyze related possible models. We are interested in computing bicliques of certain properties in a bipartite graph. Since computing a biclique with a certain number of edges or vertices in a bipartite graph is NP-complete [7], any meaningful model for our problem will be NP-hard as well. Thus, we use the theory of parameterized complexity [13] to cope with this hardness.

Recall that $G = (A \uplus B, E)$ denotes a bipartite graph with $A = \{a_1, \ldots, a_n\}$ and $B = \{b_1, \ldots, b_m\}$. For a vertex $v \in A \uplus B$, let $N(v)$ denote the set of neighbors of v, i.e. $N(v) = \{u : \{u, v\} \in E\}$. For a subset of vertices $A' \subseteq A$, denote the set of *common neighbors* of A' by $B_{A'} = \bigcap_{a \in A'} N(a)$. Similarly, let $A_{B'} = \bigcap_{b \in B'} N(b)$ denote the set of common neighbors of any $B' \subseteq B$. In this way, a pair of non-empty subsets $A' \subseteq A$ and $B' \subseteq B$ is a biclique in G iff $B_{A'} = B'$ and $A_{B'} = A'$. Clearly, the number of edges in a biclique (A', B') is $|A'||B'|$.

3.1   Three Biclique Problems

We next consider three possible candidates for biclique computation problems. For the sake of simplicity, we consider only decision problems for now.

BIPARTITE BICLIQUE :
Input: A bipartite graph $G = (A \uplus B, E)$ and an integer k.
Question: Is there a biclique (A', B') in G with $|A'||B'| \geq k$?

BIPARTITE BALANCED BICLIQUE :
Input: A bipartite graph $G = (A \uplus B, E)$ and an integer k.
Question: Is there is a biclique (A', B') in G with $|A'| = |B'| \geq k$?

For a fixed positive integer d, a biclique (A', B') is called a *d-block biclique* if $A' \subseteq A[i, i+d]$ for some $1 \leq i \leq n - d$.

BLOCK BIPARTITE BICLIQUE :
Input: A bipartite graph $G = (A \uplus B, E)$ and two positive integers d and k.
Question: Is there a d-block biclique (A', B') in G with with $|A'||B'| \geq k$?

Clearly, the latter of these problems is tailor suited for the RAGB problem, but the other two might *a priori* be of use in this context as well. In BIPARTITE BICLIQUE we wish to find a biclique with at least k edges, and in BIPARTITE BALANCED BICLIQUE we wish to find a biclique where each side has at least k vertices. Solutions to both of these problems are clearly meaningful in our context. Note that we could have also considered a third variant where the goal is to find a biclique with k vertices altogether (*i.e.* on both sides), but in the setting of parameterized complexity of which we analyze all our problems, this problem is quite similar to BIPARTITE BICLIQUE.

Lemma 1. BIPARTITE BICLIQUE can be solved in $O(2^k n)$ time.

Lemma 2. BIPARTITE BALANCED BICLIQUE *is* W[1]-hard with respect to parameter k.

Thus, the BIPARTITE BICLIQUE problem is FPT with respect to parameter k, while BALANCED BIPARTITE BICLIQUE is not (under the widely believed assumption that FPT $\neq$ W[1]). Note however that the main issue with the BIPARTITE BICLIQUE problem is that we assume that the number of edges in a solution biclique will be rather small, and can thus be taken as a parameter. This is not the case for the BLOCK BIPARTITE BICLIQUE problem. As we will see in the next section, this latter problem is fixed-parameter tractable with respect to d, which for our purposes is much smaller than the number of edges in a solution biclique. The biological motivation for RAGB1 and RAGB2 naturally yields small bounds on d.

3.2 Solving RAGB1: An Algorithm for Computing d-block Bicliques

The BLOCK BIPARTITE BICLIQUE problem trivially has a fixed parameter algorithm with respect k, the number of edges in the solution biclique, by using the same arguments used in proving Lemma 1. We next show that this problem also has a fixed parameter algorithm with respect to parameter d, the size of the block, which is expected to be much smaller in practice than k. In fact, we will show a much stronger result in that we can compute in FPT time the set of all maximal d-block bicliques of our input graph.

Lemma 3. *Given a bipartite graph $G = (A \uplus B, E)$ with $|A| = n$ and $|B| = m$, and an integer d, one can compute the set of all maximal d-block bicliques of G in $O(nm + 2^d nm / \lg m)$ time.*

Algorithm for computing d-block bicliques:

- For each $i \in \{1, \ldots, n - d\}$ and $A' \subseteq A[i, i + d]$ do
  a. Compute the set $B_{A'}$ of common neighbors of A'.
  b. Return (A', B').

Note that the set of all bicliques produced by this algorithm contains all maximal bicliques of G. These can be easily weeded out at a post-processing stage, or during the computation of the algorithm.

To bound the running time of the algorithm above, first observe that we need $O(nm)$ time just to read the entire input. Next, notice that the algorithm has $O(n)$ iterations, where in each iteration it computes 2^d bicliques (A', B'). Starting in each iteration with bicliques (A', B') where $|A'| = 1$, and increasing the size of A' by one each time, each set of common neighbors $B_{A'}$ can be computed with a single *set intersection* operation between $N(a) \subseteq B$ and $B_{A' \setminus \{a\}} \subseteq B$ for some $a \in A'$. This set intersection operation can be naively performed in $O(m)$ time, giving a total running time of $O(2^d nm)$ to our algorithm. However, using standard bit-tricks of the RAM model, we can improve the running time of each such operation to $O(m / \lg m)$, reducing the total running time of our algorithm to the one stated in Lemma 3.

In the full version of the paper, we show how to use the "four russians technique" in order to adapt the algorithm above the case where the edges of G are weighted. This allows us to compute all maximal d-block bicliques, along with their weight, with only a factor of $O(\lg d)$ increase to the time complexity of the algorithm. Details are omitted due to space constraints.

Lemma 4. *Given a bipartite graph $G = (A \uplus B, E)$ with $|A| = n$ and $|B| = m$, a function $w : E \to \{1, \ldots, x\}$ assigning weights to the edges of G, and an integer d, one can compute the set of all weighted maximal d-block bicliques of G in $O(nm + 2^d \lg d \cdot nm / \log m)$ time.*

3.3   Solving RAGB2: The Colorful Variant

For the purposes of solving RAGB2, we consider the *colorful variant* of the BLOCK BIPARTITE BICLIQUE problem where the vertices in B have colors, and we wish to find a biclique that contains at most one vertex $b \in B$ of each color. For this purpose, let $c : B \to \{1, \ldots, C\}$ be a coloring function of the vertices in B. Recall that a biclique (A', B') is said to be *colorful* if $c(b_1) = c(b_2)$ implies $b_1 = b_2$, for every $b_1, b_2 \in B'$. The COLORFUL BLOCK BIPARTITE BICLIQUE problem is the variant of BLOCK BIPARTITE BICLIQUE where we wish to find a colorful block biclique with a certain number of edges.

Unfortunately, we can no longer apply dynamic programming and the four russians trick in this case. This is because the colors make the problem harder to handle. In fact, it turns out that the colorful variants of the two other bipartite biclique problems discussed above are W[1]-hard. This is not surprising for COLORFUL BALANCED BIPARTITE BICLIQUE, as BALANCED BIPARTITE BICLIQUE is W[1]-hard by Lemma 2, and there is a generic parameterized reduction from any problem to its colorful variant using the color-coding technique [1]. For COLORFUL BIPARTITE BICLIQUE we need a slightly more elaborate argument:

Lemma 5. COLORFUL BIPARTITE BICLIQUE *is* W[1]-*hard when parameterized by* k.

Regardless of the above, we can still compute in $O(m)$ time a maximum sized subset $B' \subseteq B_{A'}$ in a set of common neighbors of some $A' \subseteq A$. This means that we can adapt the algorithm above to compute maximum size colorful biclique (A', B') in $O(m)$ time per each biclique. Moreover, note that in the same amount of time we can actually count the number of different colorful bicliques (A', B') corresponding to some $A' \subseteq A[i, i + d]$. This is done by a simple combinatorial computation that considers all possibilities of picking a single vertex out of each color in $B_{A'}$.

Lemma 6. COLORFUL BLOCK BIPARTITE BICLIQUE *can be solved in* $O(2^d nm)$ *time.*

4 Methods

We implemented the algorithm for the RAGB2 problem in a program called *RAGB Monitor* (Reference-Anchored Gene Blocks Monitor). Given the gene map of a reference element and a set of target genomes, both in GenBank file format, our program first BLASTs each gene from the reference element against each gene from the target genomes, and considers the two genes to be orthologous if their BLAST score is below 10^{-8}. Upon a successful BLAST result, genes from the target genome are re-labeled with the gene id of the corresponding gene from the reference element. If a gene in a target genome is found to be orthologous with more than one gene from the reference element, we map it to the one with the highest BLAST score. Genes from the reference element, on the other hand, are allowed to be mapped to more than one gene in each target genome. For each target genome, a sequence of gene ids is then created, consisting only of the genes that were labeled by an ortholog from the reference element genes, and preserving the gene order in the original target genome.

Our program also takes as input several parameters, including an upper bound d on the length (measured as number of genes) of an interval in the reference element, an upper bound d' (measured as number of genes) on the length of an interval in the target genomes, and quorums q_1 and q_2 on the minimal number of anchor genes and target genomes, respectively, required in a bicluster. For segmenting the target genomes into intervals, biological segmentation is applied: The distance between two consecutive genes in an interval is bounded from above by 2000bp, and in addition, an interval length is bounded from above by parameter d'. The tool was implemented in Python 2.8.3 and the experiments performed on an Intel Xeon X5680 machine with 192 GB RAM. For a query reference element consisting of 42 genes versus 33 target proteobacterial genomes (see Sect. 5), the running time of our program ranged from 0.19 s for $d = 2$, up to 379.8 s for $d = 20$.

Finally, we define a p-value that determines the probabilistic likelihood of each gene block found. Let m denote the number of target genomes, and let n denote the length of each target genome. We define our p-value as the probability that k genes appear together in d-blocks of c out of the m genomes. We denote the probability of this event by $\Pr[k, d, c]$. Here, we assume that each genome is a permutation on $\{1, \ldots, n\}$ drawn uniformly and independently at random.

Theorem 1. *The following bound holds:*

$$\Pr[k, d, c] \leq \binom{m}{c} \left(\frac{\binom{n-k}{d-k}}{\binom{n}{d}} (n - d) \right)^c .$$

5   Preliminary Bioinformatics Results

Enteropathogenic Escherichia coli (EPEC) is a major cause of food poisoning, leading to significant morbidity and mortality. EPEC virulence is dependent on a type III secretion system (T3SS), a molecular syringe employed by EPEC to inject effector proteins into host cells [8]. The hallmark of T3SS is the needle apparatus it forms, also called "injectisome". Bacterial effector proteins that need to be secreted pass from the bacterial cytoplasm through the needle directly into the host cytoplasm. Three membranes separate the two cytoplasms: the double membrane (inner and outer membranes) of the Gram-negative bacterium and the eukaryotic membrane. The needle provides a smooth passage through those highly selective and almost impermeable membranes. The injected effector proteins subvert host cellular functions to the benefit of the infecting bacteria. A single bacterium can have several hundred needle complexes spread across its membrane. It has been proposed that the needle complex is a universal feature of all T3SSs of pathogenic bacteria. More than 15 proteins are needed to build the T3SS, some of which are highly conserved in all known T3SSs. In EPEC, the T3SS and related genes reside in several operons clustered in the Locus of Enterocyte Effacement (LEE), which is a stable pathogenicity island [5]. We exemplify our tool based on LEE (EPEC) as the reference element and on representative proteobacteria species as the target genomes.

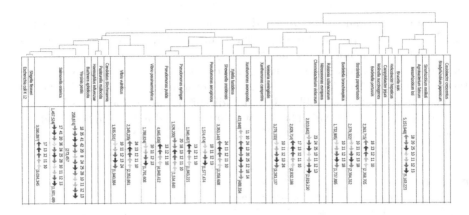

Fig. 1. The main result bicluster, anchored by genes 10–13 (escR, escS, escT, and escU) from the LEE pathogenicity Island. For each co-localized instance of the anchor genes in a target proteobacterial genome, the figure also shows additional homologs of genes from the query island that were identified within the same genomic interval.

Sequences of LEE proteins from EPEC strain O127:H6 (str. E2348/69) [6], NC 011601, were downloaded from the IslandViewer database [9], based on prediction by SIGI-HMM [17], spanning genomic coordinates 4103271-4138312 of the main chromosome. Islandviewer predicted 42 genes for this island. 33 target genomes of proteobacteria were downloaded from NCBI, as described in Ream et al. [12], who used this set of genomes in a study of the evolution of E. coli operons.

Our objective was to identify biclusters: sets of genes, that are located in LEE within a small proximity (i.e. contained within a window of size d in the gene map of LEE), and that have orthologous genes conserved within small proximity in some of the genomes of the proteobacteria species in our target set. For this, we ran sliding windows of increasing sizes on the gene map of LEE, starting with $d = 2$ (two genes) and converging at $d = 17$, after three runs in a row gave consistent clusters. For each value of d, we considered all sliding windows of length d (i.e. d consecutive genes) in the gene-map of LEE. For each sliding window, we computed all biclusters, and selected the one with the lowest p-value, computed according to the method described in Sect. 4.

The results shown in Fig. 1 were obtained for $d = 17$. The top-scoring cluster, consistently across all window sizes, was the cluster containing the four genes: 10,11,12, and 13. The p-value for this bicluster (e^{-78} for $d = 17$) was significantly lower than the p-values for all the other biclusters, based on its orthologous occurrences in 14 out of the 33 genomes in the target set. Furthermore, in some of these genomes: Chromobacterium violaceum, Pseudomonas aeruginosa, Vibrio parahaemolyticus, and Salmonella enterica serovar Typhi CT18, our tool identified two or three copies of this bicluster in distinct locations on the chromosome. To interpret this result, we checked the literature and the Virulence Factor Data Base [4]. The genes participating in the main bicluster predicted by our program were identified as four consecutive genes within the first operon of LEE: escR, escS, escT, and escU. These genes are annotated as conserved T3SS proteins: assembly of an inner membrane complex containing these proteins might represent a critical early step in the biogenesis of the "syringe" apparatus mentioned above [14]. The other (less significant) biclusters yielded by our program were also combinations of T3SS genes.

Acknowledgements. The research of D.H. and A.B. was partially supported by the People Programme (Marie Curie Actions) of the European Union's Seventh Framework Programme (FP7/2007-2013) under REA grant agreement number 631163.11, and by the Israel Science Foundation (grant No. 551145/). The research of M.Z-U. and A.B. was partially supported by the Israel Science Foundation (grant No. 179/14.) and by the Frankel Center for Computer Science at Ben Gurion University.

References

1. Alon, N., Yuster, R., Zwick, U.: Color-coding. J. ACM **42**(4), 844–856 (1995)
2. Bergeron, A., Chauve, C., Gingras, Y.: Formal models of gene clusters. Bioinform. Algorithms Tech. Appl. **8**, 177–202 (2008)
3. Chateau, A., Riou, P., Rivals, E.: Approximate common intervals in multiple genome comparison. In: IEEE International Conference on Bioinformatics and Biomedicine (BIBM), pp. 131–134. IEEE (2011)
4. Chen, L., et al.: VFDB: a reference database for bacterial virulence factors. Nucleic Acids Res. **33**(suppl 1), D325–D328 (2005)
5. Deng, W., et al.: Dissecting virulence: systematic and functional analyses of a pathogenicity island. Proc. Natl. Acad. Sci. U.S.A. **101**(10), 3597–3602 (2004)
6. Elliott, S.J., et al.: The complete sequence of the locus of enterocyte effacement (LEE) from enteropathogenic escherichia coli e2348/69. Mol. Microbiol. **28**(1), 1–4 (1998)
7. Garey, M.R., Johnson, D.S.: Computers and Intractability: A Guide to the Theory of NP-Completeness. W. H Freeman, New York (1979)
8. Hazen, T.H., et al.: Refining the pathovar paradigm via phylogenomics of the attaching and effacing escherichia coli. Proc. Natl. Acad. Sci. U.S.A. **110**(31), 12810–12815 (2013)
9. Dhillon, B.K.: Islandviewer 3: more flexible, interactive genomic island discovery, visualization and analysis. Nucleic Acids Res. **43**, W104–W108 (2015). gkv401
10. Kim, S., et al.: A hybrid gene team model and its application to genome analysis. J. Bioinform. Comput. Biol. **4**(02), 171–196 (2006)
11. Ling, X., He, X., Xin, D.: Detecting gene clusters under evolutionary constraint in a large number of genomes. Bioinform. **25**(5), 571–577 (2009)
12. Ream, D.C., et al.: An event-driven approach for studying gene block evolution in bacteria. Bioinform. **31**(13), 2075–2083 (2015)
13. Downey, R.G., Fellows, M.R.: Parameterized Complexity. Springer, New York (1999)
14. Samuel, W., et al.: Organization and coordinated assembly of the type III secretion export apparatus. Proc. Nat. Acad. Sci. **107**(41), 17745–17750 (2010)

15. Schmidt, H., Hensel, M.: Pathogenicity islands in bacterial pathogenesis. Clin. Microbiol. Rev. **17**(1), 14–56 (2004)
16. Schmidt, T., Stoye, J.: Quadratic time algorithms for finding common intervals in two and more sequences. In: Sahinalp, S.C., Muthukrishnan, S.M., Dogrusoz, U. (eds.) CPM 2004. LNCS, vol. 3109, pp. 347–358. Springer, Heidelberg (2004)
17. Waack, S., et al.: Score-based prediction of genomic islands in prokaryotic genomes using hidden Markov models. BMC Bioinform. **7**(1), 142 (2006)

An Efficient Branch and Cut Algorithm to Find Frequently Mutated Subnetworks in Cancer

Anna Bomersbach[1], Marco Chiarandini[1], and Fabio Vandin[1,2,3(✉)]

[1] Department of Mathematics and Computer Science,
University of Southern Denmark, Odense, Denmark
anna.bomersbach@gmail.com, marco@imada.sdu.dk
[2] Department of Information Engineering, University of Padova, Padova, Italy
vandinfa@dei.unipd.it
[3] Department of Computer Science, Brown University, Providence, USA

Abstract. Cancer is a disease driven mostly by somatic mutations appearing in an individual's genome. One of the main challenges in large cancer studies is to identify the handful of driver mutations responsible for cancer among the hundreds or thousands mutations present in a tumour genome. Recent approaches have shown that analyzing mutations in the context of interaction networks increases the power to identify driver mutations.

In this work we propose an ILP formulation for the exact solution of the combinatorial problem of finding subnetworks mutated in a large fraction of cancer patients, a problem previously proposed to identify important mutations in cancer. We show that a branch and cut algorithm provides exact solutions and is faster than previously proposed greedy and approximation algorithms. We test our algorithm on real cancer data and show that our approach is viable and allows for the identification of subnetworks containing known cancer genes.

Keywords: Cancer mutations · Branch and cut · Combinatorial optimization · Network analysis

1 Introduction

Recent advances in DNA sequencing technologies have allowed the study of cancer genomes at an unprecedented level of detail. In particular, it is now possible to measure all *somatic mutations*, changes in the DNA arising during the lifetime of an individual and causing the disease, in a large number of cancer patients [12,28]. These large cancer studies have shown that each individual tumour harbours hundreds or thousands somatic mutations, with two tumours showing a large diversity in the complement of somatic mutations they exhibit [9,27]. This phenomenon is commonly referred to as (intertumor) cancer heterogeneity.

© Springer International Publishing Switzerland 2016
M. Frith and C.N.S. Pedersen (Eds.): WABI 2016, LNBI 9838, pp. 27–39, 2016.
DOI: 10.1007/978-3-319-43681-4_3

Cancer heterogeneity is explained by the fact that only a handful of all the somatic mutations in a cancer genome are *driver* mutations related to the diseases, while the majority of mutations are *passenger* mutations not related to cancer progression and development. Moreover, driver mutations target regulatory and signaling *pathways*, groups of interacting genes that perform specific functions in the cell [10,26] and that may be altered by mutating any of the genes in the group. Therefore, to identify all driver mutations and the genes they affect one cannot focus on genes in isolation, but has to study mutations in the context of interaction networks [5].

In recent years, several methods have been proposed to identify significantly mutated pathways in cancer [21]. Some of these methods work on known pathways [4], thus limiting our ability to identify novel pathways as well as subnetworks connecting two pathways that are important for cancer. Other methods identify significantly mutated pathways by combining mutation data with a large protein-protein interaction network [14,15,18,23,25]. A common formulation is to look for connected subnetworks that are mutated in a large number of patients, that is equivalent to identifying connected subnetworks whose vertices *cover* a large number of elements (i.e., patients) from a universe. In particular [25] defined the connected maximum coverage problem (CMCP) as finding a connected subnetwork of cardinality k that covers the maximum number of patients, proven to be NP-hard in [25] where an approximation algorithm was also presented.

In this paper, we propose an integer linear programming formulation for CMCP. Our formulation draws from an analogous formulation for Steiner tree problems that have been the object of a recent DIMACS challenge [13]. In particular, the connectivity constraint leads to an exponential number of constraints that we handle within a branch and cut framework. We show that our algorithm allows for the identification of the optimal solution of CMCP on real cancer datasets, and that the identified solutions cover more patients compared to previously proposed heuristic approaches or approximation algorithms. We also show that the subnetworks identified by our approach have higher statistical significance, estimated through permutation testing, compared to solutions found by the approximation algorithm. We generalize our formulation to the weighted version of the problem, and show that our branch and cut strategy can be used to solve this formulation as well, and also show that our approach identifies subnetworks of genes known to be associated with cancer.

Related Work. The computational problem of identifying connected subnetworks with vertices covering a large number of elements have been studied in bioinformatics [14,15,24,25] as well as in wireless network design [16]. As mentioned above, [25] studied the CMCP and provided an approximation algorithm for its solution (see Sect. 2.1); [14,15,24] studied related, but different, problems. In gene expression studies, [24] studied the problem of finding the smallest connected subnetwork such that at least k genes in the subnetwork are differentially expressed in all patients but at most ℓ. [14,15] study the problem of finding the minimum cost collection of modules (i.e., subgraphs) covering each patient at least k times, where a patient is covered by a module if at least one gene in the module is altered in the patient. The cost of a collection of modules is a function of the size of the modules and other pairwise properties of the genes in the module (e.g., their distance in a network; the degree of exclusivity of alterations).

The CMCP has some similarity with the cardinality constrained Maximum Weight Connected Subgraph Problem [1,6], that asks to find a connected subgraph with maximum total weight in a node-weighted graph, and with the prize-collecting Steiner Tree Problem [8,13,19], that asks to find a subtree of minimum costs that spans all vertices from a set of terminals. Different ILP formulations for these problems have recently been studied both theoretically and in practice [1,6,29]. The main issue from an ILP perspective is modeling the connectivity requirement. In the formulations of [1,6], the connectivity constraints are formulated by means of a root node and a generalized form of node separators. With this approach, an additional set of node variables is needed to locate the root. Later [8] proposes a thinned formulation that does not need a root and variables to locate it and that uses node separators in non-generalized form. Both formulations exhibit an exponential number of connectivity constraints; therefore, branch and cut (B&C) algorithms have been used to solve these models. In this approach connectivity constraints are not explicitly declared, rather they are introduced during the search when they are needed. The B&C algorithm by [1,6] finds violated connectivity cuts by identifying a minimum cut in a support digraph. The B&C algorithm by [8] uses a lazy approach in which violated constraints are only searched when an integer solution is found and employs a linear time algorithm to discover minimal node separators that are facet-defining. [3] introduced a flow based, polynomial size formulation for connectivity constraints for the problem of finding colorful connected subgraphs.

We note that while the connectivity requirement of CMCP is the same as the connectivity requirements in the Maximum Weight Connected Subgraph Problem and the prize-collecting Steiner Tree Problem, the latter two problems have objective functions that are additive in the vertices chosen in the solution (and also in the edges for the prize-collecting Steiner Tree Problem), while the objective function for the CMCP is the more complicated coverage function, that is a submodular set function [16].

2   Model and Algorithms

We are given a graph $G = (V, E)$, with vertices $V = \{1, \ldots, n\}$ representing genes and edges E representing interactions among genes (or the associated proteins). Let P denotes the set of patients for which mutations have been assayed. Let $P_i \subseteq P$ be the set of patients in which gene $i \in V$ is mutated. We say that a patient $j \in P$ is *covered* by a subset of vertices $S \subseteq V$, if there exists at least one vertex v in S such that $j \in P_v$.

Our goal is to identify connected subgraphs of G that are mutated in a large number of patients, where a subgraph is mutated in a patient if at least one of the vertices in the subgraph covers the patient. More formally, we consider the following problem, defined in [25].

 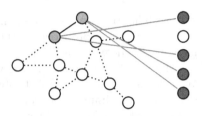

Fig. 1. Left: an instance of CMCP, with blue vertices and black edges representing G, red vertices representing patients P and gray lines linking patients to their mutated genes. Right: an optimal solution (in blue) to the instance on the left for $k = 2$. Patients covered by the optimal solution are in red. (Color figure online)

Connected Maximum Coverage Problem (CMCP). Given a graph G defined on a set of n vertices V, an integer $k > 0$, a set P, a family of subsets $\mathcal{P} = \{P_1, \ldots, P_n\}$ where for each i, $P_i \subseteq P$ is associated with $i \in V$, find the connected subgraph $S^* \subseteq G$ with k vertices that maximizes the coverage $|\cup_{i \in S^*} P_i|$.

Figure 1 shows an instance of CMCP. If G is a complete graph, the connected maximum coverage problem is the maximum coverage problem [11], where, given a set U of elements, a family of subsets $\mathcal{F} \subset 2^U$, and a value k, one needs to find a collection of k sets in $\mathcal{F}$ that covers the maximum number of elements in U. The maximum coverage problem is NP-hard [11], thus, the connected maximum coverage problem is NP-hard for a general graph and even for star graphs [25].

Preprocessing. Only connected components of G of size $\geq k$ need to be considered. The problem can be solved for each of those components in turn, returning the best solution found. Nodes $v \in V$ that cannot be in an optimal solution are removed using the following rules: (i) v has degree 1, $P_v = \emptyset$, and after removal the number of nodes in the connect component is $\geq k$; (ii) there is a node $v' \neq v$ whose set of neighbors is a superset of the set of neighbors of v and the set $P_{v'}$ is a superset of the set P_v.

The rest of the section is organized as follows: Sect. 2.1 reviews previously proposed algorithms for the CMCP; Sect. 2.2 presents our ILP formulation for the CMCP and the corresponding branch and cut algorithm; Sect. 2.3 extends the ILP to a weighted version of the CMCP; and Sect. 2.4 describes the permutation test used to assess the statistical significance of the solutions.

2.1 Previous Methods: Approximation Algorithm and Greedy Algorithms

In this section we present previously proposed methods as well as simple greedy algorithms for the CMCP. [7] proposed a polynomial time $1/(cr)$-approximation algorithm, where $c = (2e-1)/(e-1)$ and r is the radius of optimal solution in the graph. The algorithm starts by computing and storing all pairs shortest paths. Then the algorithm finds a solution starting from each node v of the graph, and at the end reports the best solution found. Given the current solution S obtained starting from vertex v, the algorithm augments it by adding the (shortest) path to a vertex u that maximizes the ratio between the number of newly covered patients and the number of new vertices added to the solution, while keeping the number of vertices in the solution $\leq k$. We also implemented two variants of the approximation algorithm that do not compute all pairs shortest paths and differ in the way they define *candidate paths* to extend the current solution. The first variant (bfs) performs a BFS (of depth $\leq k$) starting from the node v from which the solution is grown. The second variant (ratio) considers *all* paths (not only shortest ones) among pairs of vertices, and it does not guarantee to run in polynomial time but has been shown to run efficiently on real datasets [7].

We also consider a simple greedy algorithm that builds a solution starting from each vertex v and reports the best solution found. The algorithm builds a solution by always adding to the current solution S the neighboring node providing the maximum increase in the coverage.

2.2   ILP Formulation and Branch and Cut Algorithm

Our ILP formulation for CMCP is analogous to a recent formulation [8] for the prize-collecting Steiner Tree problem. It involves only node variables and is based on node separator inequalities, some of which can be proved to be facet defining for the connected subgraph polytope [29].

Given the graph $G = (V, E)$ and two distinct nodes h and ℓ from V, a subset of nodes $N \subseteq V \setminus \{h, \ell\}$ is an (h, ℓ) *(node) separator* if and only if after removing N from V there is no path between h and ℓ in G. Let $\mathcal{N}(h, \ell)$ denote the family of all (h, ℓ) separators. A separator $N \in \mathcal{N}(h, \ell)$ is minimal if $N \setminus \{i\}$ is not an (h, ℓ) separator, for any $i \in N$. We use binary coefficients a_{ij} for $i \in V$ and $j \in P$ to indicate whether i covers j (i.e., gene i is mutated in patient j), that is, $a_{ij} = 1$ if $j \in P_i$, and $a_{ij} = 0$ otherwise. Let x_i for $i \in V$ be binary variables such that $x_i = 1$ if the vertex i is in the solution $S \subseteq V$ and $x_i = 0$ otherwise, and let y_j for $i \in V$ be binary variables such that $y_j = 1$ if patient j is covered by S and $y_j = 0$ otherwise. We formulate the CMCP as an ILP as follows:

$$\max \sum_{j \in P} y_j \tag{1}$$

$$\sum_{i \in V} x_i = k \tag{2}$$

$$\sum_{i \in V} a_{ij} x_i \geq y_j \qquad \forall j \in P \tag{3}$$

$$\sum_{i \in N} x_i \geq x_h + x_\ell - 1 \qquad \forall h, \ell \in V, h \neq \ell, \forall N \in \mathcal{N}(h, \ell) \tag{4}$$

$$x_i \in \{0, 1\} \qquad \forall i \in V \tag{5}$$

$$y_j \in \{0, 1\} \qquad \forall j \in P \tag{6}$$

Constraints (2) impose that exactly k nodes of V are in the solution S. Constraints (3) ensure that the variables y_j are set to one only when the corresponding $j \in P$ is covered. Constraints (4) are the node separator constraints for the connectivity requirement. They ensure that for any pair of nodes h, ℓ in S there is a path in the graph induced by S, i.e., for any node separator $N \in \mathcal{N}(h, \ell)$ at least one node in N must also be selected. [29] shows that the constraints (4) are facets defining for the connected subgraph polytope if and only if N is a minimal node separator separating h and ℓ.

The Branch and Cut Algorithm. Our B&C algorithm (B&C) is analogous to the one for the Steiner tree problem in [8]. The connectivity cuts (4) are treated as lazy constraints, that is, they are not explicitly represented in the initial ILP model but they are introduced only when an integer solution that violates any of these constraints is found. B&C starts by including only a special case of inequalities (4), i.e., $x_i \leq \sum_{\ell \in V:(i,\ell) \in E} x_\ell$ for all $i \in V$. They state that a node, if selected, must have a neighbor also selected. At any node of the branch and bound tree an integral solution $\tilde{x}$ to the model (1)–(6) with a subset of the constraints (4) gives a set of nodes $\tilde{S} = \{i \in V \mid \tilde{x}_i = 1\}$. If the solution is not feasible with respect to the full model, then in the subgraph of G induced by $\tilde{S}$ there are disjoint connected components. Let C_h and C_ℓ be two such components, containing the nodes h and ℓ, respectively. Then, the linear time algorithm from [8] reported in Fig. 2 and exemplified in Fig. 3 finds the minimal node separator that must be added to the model. In our implementation, for an integer solution with m disjoint connected components we find the minimal node separator for all $m \times (m-1)$ combinations and add all corresponding constraints. B&C terminates when an integer solution that does not violate any lazy constraint and whose value is proven optimal is found.

---

1 **Function** FINDMINNODESEPARATOR$(G, \tilde{S}, \{h, \ell\} \in \tilde{S}, C_h)$
2 $\quad$ $A(C_h) \leftarrow$ neighbors of nodes of C_h in G
3 $\quad$ $G' \leftarrow G$ with all edges between vertices in $C_h \cup A(C_h)$ removed
4 $\quad$ $R_\ell \leftarrow$ nodes that can be reached from ℓ in G'
5 $\quad$ **return** $N = A(C_h) \cap R_\ell$

Fig. 2. Linear time algorithm for finding a minimal node separator N. The input is G, two nodes h and ℓ of an infeasible solution $\tilde{S}$, and the connected component C_h of the subgraph of G induced by $\tilde{S}$ containing h and not containing ℓ.

2.3 Weighted Model

We extend the ILP formulation to the case where for each (gene) vertex $i \in V$ and each patient $j \in P$ we have a weight w_{ij} that we gain when i is used to cover j; the objective is to maximize the weight of the covered patients with at most one gene that can be picked to cover a patient. Without loss of generality

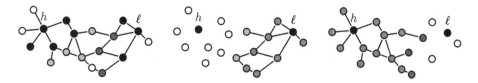

Fig. 3. Left: example of disconnected solution $\tilde{S}$ (black nodes). Blue nodes are nodes $A(C_h)$, neighbors of C_h, and constitute a non minimal node separator. Middle: grey and blue nodes represent the nodes in R_ℓ that are reachable from ℓ in G' and the blue nodes constitute the minimal node separator. Right: the red nodes represent the minimal node separator determined repeating the same procedure for component C_ℓ. (Color figure online)

we assume that for all $i \in V$ and all $j \in P : w_{ij} \le 1$ (weights can be normalized by dividing them by the highest value). We introduce binary variables z_{ij} for all pairs $i \in V$ and $j \in P_i$. The interpretation of z_{ij} is that $z_{ij} = 1$ if gene i is chosen to cover patient j, and $z_{ij} = 0$ otherwise. For $j \in P$, let M_j be the set of genes mutated in j. We define the following model: we keep the objective function (1) and the constraints (2), (4), (5) and add the following constraints: (7) $x_i \ge z_{ij}$, $\forall i \in V, j \in P_i$; (8) $y_j \le w_{ij}z_{ij} + (1 - z_{ij})$, $\forall i \in V, j \in P_i$; (9) $y_j \le \sum_{i \in M_j} z_{ij}$, $\forall j \in P$; (10) $0 \le y_j \le 1$, $\forall j \in P$. Note that y_j are now continuous variables and that for a feasible solution we may have that for some $j \in P : \sum_{i \in M_j} z_{ij} > 1$, however at the optimal solution $\sum_{i \in M_j} z_{ij} \le 1$ (assuming all weights are different).

2.4 Permutation Test

We use a permutation test to assess the statistical significance of the subnetworks identified by the methods above. In particular, we generate datasets under the null hypothesis by permuting the identity of the genes in the network. The test statistic used to compute the (empirical) p-value is the value of the objective function of the solution. Given a permuted dataset and the value X of the test statistic obtained from the solution found using the real dataset, we are only interested in knowing if there is solution with test statistic $\ge X$ in a permuted dataset. Hence, we can stop an algorithm as soon as we are sure that the statistic on the current permuted dataset will be either certainly larger or certainly lower than the observed value X. This can be easily implemented in the branch and bound framework of B&C by adding a constraint on the value of a feasible solution, so that we can halt the search when either the lower bound of B&C becomes $\ge X$ or when the upper bound becomes $< X$.

We note that when a significant correlation between the degree of a node and its coverage is present, permuting the identity of the genes in the network may overestimate the significance of the subnetworks identified. We therefore checked if in our instances there was a high correlation between degree and coverage of the nodes: in all instances the absolute value of such correlation is low (≤ 0.08).

3   Results

In this section we describe the cancer data and the computing environment used in our experiments and the results obtained on the cancer datasets.

Data. We use the HIPPIE network[1] [22]. The corresponding interaction graph G consists of 15094 nodes and 188891 edges. Mutation data is obtained from the TCGA Pan-Cancer analysis[2] [18,28], with mutations of all genes measured in 3425 patients from 11 cancer types. We considered datasets of individual cancer types as well as all samples together, the latter referred to as pancan dataset (Table 1).

In all our experiments we performed the preprocessing described in Sect. 2, and we report running times for methods after the preprocessing.

Table 1. Cancer datasets. For each dataset, we report the number *genes* of gene nodes in graph G after preprocessing and the number $|P|$ of patients in the instance.

Dataset	Genes	$|P|$	Dataset	Genes	$|P|$	Dataset	Genes	$|P|$	Dataset	Genes	$|P|$
pancan	12310	3412	coadread	12088	495	kirc	11611	424	lusc	11752	177
blca	11424	100	gbm	11452	276	laml	10964	194	ov	11536	456
brca	11535	506	hnsc	11738	306	luad	11740	230	ucec	11865	248

Computing Environment and Solver Configuration. We implemented B&C in Python 2.7.5 using Gurobi 6.5.0 and callback functions. All experiments were conducted on local nodes of a computing cluster. Each node had the following configuration: two Intel E5-2680v3 CPUs with 12 CPU cores each, amounting to 24 cores in total, 64 GB RAM and 200 GB local SSD storage. All parameters in Gurobi were left at their default values, except for the number of threads that was set to one. In this way, experiments to compare the running times among different programs are conducted with serial computation. In permutation tests, we first solved the real dataset instance using a single thread and, then, the permuted datasets in parallel, one dataset per process, using python multiprocessing module and work stealing strategy.

We run the approximation and the greedy algorithms described in Sect. 2.1 and our B&C on the datasets of Table 1 for $k = 10$, 15, and 20. For each pair (dataset, k), Table 2 shows the coverage of the best solution found by the various algorithms and the running time (median over 10 runs). We observe that for only five of the 36 pairs (dataset, k) the approximation or greedy algorithms identify solutions with coverage as high as the the optimal found by B&C. Even more interestingly, the runtime of the B&C is comparable to the runtime of the greedy algorithm, and it is above 1 min only for 11 pairs (instance, k), and only twice above 5 min. For $k = 10$, we also compared the runtime of the B&C with the runtime of an ILP formulation that models connectivity constraints as in [6],

---

[1] http://cbdm-01.zdv.uni-mainz.de/~mschaefer/hippie/.
[2] http://compbio-research.cs.brown.edu/pancancer/hotnet2/.

Table 2. Comparison of algorithms. For each pair (dataset, k), the coverage (cov) of the solution reported by the various algorithms and their running time (*time* [hh:mm:ss]) are shown. In bold: coverages of solutions from B&C that are strictly higher then coverage of solutions from approximation and greedy algorithms; runtimes of B&C that are lower than runtimes of greedy algorithm.

Dataset	k	approximation		bfs		ratio		greedy		flow		B&C	
		Cov	Time	Cov	Time	Cov	Time	Cov	Time	Cov	Time	Cov	Time
pancan	10	1804	4:00:12	1804	2:08:30	-	>12h	1469	0:00:55	1855	>18h	**1855**	**0:00:34**
	15	2072	4:30:14	2079	2:42:24	-	>12h	1648	0:01:48	2168	>18h	2168	**0:00:38**
	20	2276	5:01:51	2277	3:10:17	-	>12h	1817	0:02:33	2361	>18h	2361	**0:02:05**
blca	10	84	2:29:19	85	1:04:08	87	6:15:03	79	0:00:34	87	9:09:23	87	0:02:06
	15	94	2:44:52	93	1:19:21	96	6:44:01	86	0:01:01	97	8:35:49	**97**	0:02:40
	20	100	2:59:28	100	1:36:49	100	7:03:14	93	0:01:29	100	4:40:14	100	0:02:23
brca	10	190	2:30:25	193	1:03:45	189	7:42:21	162	0:00:34	196	6:55:42	**196**	**0:00:09**
	15	229	2:43:46	229	1:19:40	226	8:17:40	184	0:01:03	236	6:59:14	**236**	**0:00:43**
	20	258	3:00:06	258	1:32:21	256	8:42:12	233	0:01:34	270	8:16:49	270	0:01:42
coadread	10	468	3:06:52	468	1:25:02	-	>12h	454	0:00:41	472	0:14:41	**472**	**0:00:23**
	15	479	3:25:38	479	1:46:10	-	>12h	465	0:01:13	481	6:29:51	**481**	**0:00:50**
	20	485	3:48:49	484	2:04:56	-	>12h	473	0:01:49	488	7:31:48	**488**	**0:01:05**
gbm	10	173	2:27:19	170	1:03:22	172	6:38:31	142	0:00:32	176	1:33:48	**176**	**0:00:23**
	15	193	2:42:53	192	1:17:23	194	7:14:15	152	0:00:56	198	3:04:07	**198**	**0:00:38**
	20	209	2:58:39	209	1:34:27	210	7:39:26	158	0:01:23	215	3:11:56	**215**	0:01:07
hnsc	10	208	2:42:36	208	1:08:37	205	7:50:45	181	0:00:37	214	6:58:14	**214**	0:01:22
	15	241	2:56:58	241	1:23:36	240	8:36:33	206	0:01:08	248	7:01:28	**248**	**0:00:58**
	20	259	3:11:00	260	1:43:39	260	9:11:04	225	0:01:44	267	12:37:39	**267**	0:03:16
kirc	10	335	2:32:54	337	1:02:42	328	9:17:26	306	0:00:34	337	0:22:21	337	**0:00:06**
	15	352	2:48:38	353	1:19:03	350	9:49:22	321	0:01:04	359	0:32:10	**359**	**0:00:18**
	20	366	3:07:14	366	1:38:15	362	10:33:34	331	0:01:36	374	0:56:15	**374**	**0:00:34**
laml	10	96	2:07:23	97	0:55:42	93	5:10:48	79	0:00:27	98	1:56:59	**98**	**0:00:08**
	15	109	2:20:35	109	1:09:39	105	5:30:37	84	0:00:44	111	>18h	**111**	**0:00:18**
	20	118	2:35:59	119	1:23:56	113	5:52:13	89	0:00:58	122	0:04:13	**122**	**0:00:28**
luad	10	183	2:42:34	184	1:08:58	185	9:24:50	171	0:00:36	188	4:16:21	**188**	0:00:55
	15	202	2:55:32	201	1:25:29	201	8:22:01	183	0:01:06	206	7:20:46	**206**	0:02:18
	20	213	3:12:17	213	1:41:32	213	8:58:25	189	0:01:40	219	7:25:30	**219**	0:04:55
lusc	10	159	2:43:13	160	1:10:05	159	7:36:18	145	0:00:36	160	13:24:59	160	0:04:09
	15	170	2:57:22	170	1:26:39	169	8:04:27	158	0:01:07	173	>18h	**173**	0:14:07
	20	176	3:16:17	177	1:45:46	176	8:53:21	167	0:01:39	177	>18h	177	0:05:38
ov	10	259	2:33:15	258	1:05:38	263	9:51:42	253	0:00:34	264	0:45:25	**264**	**0:00:14**
	15	284	2:48:59	284	1:22:26	288	10:27:14	267	0:01:00	290	2:31:51	**290**	**0:00:25**
	20	304	3:04:25	306	1:38:59	306	11:18:41	277	0:01:28	313	2:57:04	**313**	**0:00:42**
ucec	10	209	2:53:05	209	1:18:22	210	9:03:36	194	0:00:38	211	0:21:33	**211**	**0:00:12**
	15	218	3:12:04	218	1:35:53	219	9:41:43	204	0:01:07	222	9:44:56	**222**	**0:00:35**
	20	226	3:31:24	228	1:58:31	227	10:22:08	211	0:01:39	231	10:23:55	**231**	**0:00:49**

adds forbidden solution cuts whenever a nonconnected integer solution is found, and employs a min-cut flow algorithm on fractional solutions for separation (as in [6]). The results in Table 2 (column flow) show that the B&C approach we propose is much faster than this alternative ILP approach.

To test the scalability of our B&C algorithm, we generated one larger instance by replicating the mutations in the pancan dataset three times, for a total of

Table 3. Permutation test results. For each pair (instance, k) and each combination of algorithms used on real dataset and permuted datasets, the p-value (from 100 permutations) is reported.

Instance	k	p-value: real dataset/permuted datasets			Instance	k	p-value: real dataset/permuted datasets		
		B&C/B&C	bfs/B&C	bfs/bfs			B&C/B&C	bfs/B&C	bfs/bfs
pancan	10	0.01	0.02	0.01	kirc	10	0.02	0.02	0.01
	15	0.01	0.09	*		15	0.01	0.18	0.03
blca	10	0.13	0.54	0.28	laml	10	0.02	0.06	0.02
	15	0.34	0.99	0.85		15	0.02	0.08	0.02
brca	10	0.01	0.02	0.01	luad	10	0.14	0.53	0.3
	15	0.03	0.17	0.04		15	0.32	0.85	0.49
coadread	10	0.01	0.3	0.16	lusc	10	0.77	0.77	0.6
	15	0.11	0.52	0.14		15	0.69	1	0.83
gbm	10	0.13	0.55	0.39	ov	10	0.03	0.19	0.1
	15	0.18	0.75	0.39		15	0.06	0.41	0.14
hnsc	10	0.17	0.56	0.17	ucec	10	0.01	0.03	0.02
	15	0.07	0.38	0.06		15	0.02	0.36	0.12

*denotes experiments that did not complete in <2 h.

Table 4. Weighted model results. For each dataset and value of k, the weight of the optimal solution and the median runtime over 10 runs is shown.

Dataset	k	Weight	Runtime [s]	p-value	k	Weight	Runtime [s]	p-value	k	Weight	Runtime [s]	p-value
brca	10	127.06	22.29	0.01	15	137.06	141.04	0.01	20	141.98	184.16	0.03
gbm	10	93.37	130.41	0.03	15	94.26	341.02	0.02	20	94.69	479.55	0.01

10275 patients in P. On such instance, B&C identifies the optimal solution in 257 s for $k = 10$, 270 s for $k = 15$, and 435 s for $k = 20$.

We also compared the statistical significance of the results obtained using B&C for $k = 10, 15$ with the statistical significance of the results obtained using the variant bfs of the approximation algorithm, that reported the best solution among the approximation and greedy algorithm in most cases. In particular, we used bfs to obtain the best solution on the instances from Table 1 and used bfs in the permutation test of Sect. 2.4 to compute the p-value for such solutions. We repeated the same experiment (with the same permuted datasets) using B&C instead of bfs for both the instances in Table 1 and the permuted datasets. We also used B&C to compute the p-value for the solutions obtained by bfs on real data. Results are shown in Table 3. We observe that B&C almost always identifies more statistically significant solutions compared to bfs. Moreover, in several instances we see that the solution obtained by bfs appears significant when bfs (that does not identify the optimal solution) is used for the permuted datasets, while the significance of such a solution is greatly reduced when B&C (that does identify the optimal solution) is used instead.

We considered the model with weights from Sect. 2.3 and tested it on the brca and gbm datasets. Similarly to the analysis performed in [18], the weights are obtained as $-\log_{10} q_i$, where q_i is the MutSigCV[3] [17] q-value for gene i. Weight, runtime, and p-value of optimal solutions are presented in Table 4.

---

[3] http://firebrowse.org.

While the runtime increases, as expected for these more complicated formulation, it still remains feasible to identify statistically significant large subnetworks of high weight. For `brca` and $k = 10$, our `B&C` algorithm identifies the subnetwork containing genes {BMI1, CTCF, ELAV1L, FOXA1, GATA3, MLL3, NCOR1, PTEN, RUNX1, TBX3}. While the last 6 genes were reported as significantly mutated by single gene test in the TCGA publication on the same dataset [20], CTCF and FOXA1 are known cancer genes that did not pass significance for single gene testing in [20]. Further, the polycomb group gene BMI1 is mutated with low frequency, but has been reported to be involved in various cancers [2].

4   Conclusions and Discussion

We presented a novel algorithm for the connected maximum coverage problem, previously proposed for finding frequently mutated subnetworks in cancer. Our algorithm is based on an ILP formulation solved in a branch and cut framework. Our results show that our algorithm identifies subnetworks more frequently mutated and of higher statistical significance compared to previously proposed algorithms and to greedy approaches, while maintaining a runtime lower than or comparable to the runtime of greedy approaches. We also generalised our formulation to the case of weights for each gene in each patient, and showed that using this formulation we identify networks containing cancer genes that are not identified by single gene tests. While we considered a protein-protein interaction networks as interaction graph, our approach is also applicable when a diffusion-based influence graph [25] is used. In this work we focused on CMCP, but we believe that our framework could be beneficial to other optimization problems in bioinformatics where connected subgraphs are sought.

Acknowledgments. Computation for the work described in this paper was supported by the DeiC National HPC Center, University of Southern Denmark. This work is supported, in part, by MIUR of Italy under project AMANDA and by NSF grant IIS-1247581, and has been done, in part, while FV was a research fellow at the Simons Institute for the Theory of Computing (University of California, Berkeley). The results presented in this manuscript are in whole or part based upon data generated by the TCGA Research Network: http://cancergenome.nih.gov/.

References

1. Álvarez-Miranda, E., Ljubić, I., Mutzel, P.: The maximum weight connected subgraph problem. In: Jünger, M., Reinelt, G. (eds.) Facets of Combinatorial Optimization, pp. 245–270. Springer, Heidelberg (2013)
2. Benetatos, L., Vartholomatos, G., Hatzimichael, E.: Polycomb group proteins and MYC: the cancer connection. Cell. Mol. Life Sci. **71**(2), 257–269 (2014)
3. Bruckner, S., Hüffner, F., Karp, R.M., et al.: Topology-free querying of protein interaction networks. J. Comput. Biol. **17**(3), 237–252 (2010)
4. Creixell, P., Reimand, J., Haider, S., et al.: Pathway and network analysis of cancer genomes. Nat. Methods **12**(7), 615–621 (2015)

5. Ding, L., Wendl, M.C., McMichael, J.F., et al.: Expanding the computational toolbox for mining cancer genomes. Nat. Rev. Genet. **15**(8), 556–570 (2014)
6. El-Kebir, M., Klau, G.W.: Solving the maximum-weight connected subgraph problem to optimality. CoRR abs/1409.5308 (2014)
7. Vandin, F., Upfal, E., Raphael, B.J.: Algorithms and genome sequencing: identifying driver pathways in cancer. Computer **45**(3), 39–46 (2012)
8. Fischetti, M., Leitner, M., Ljubic, I., et al.: Thinning out steiner trees: a node based model for uniform edge costs. Math. Progr. Comput. (2015, submitted)
9. Garraway, L.A., Lander, E.S.: Lessons from the cancer genome. Cell **153**(1), 17–37 (2013)
10. Hanahan, D., Weinberg, R.A.: Hallmarks of cancer: the next generation. Cell **144**(5), 646–674 (2011)
11. Hochbaum, D.S.: Approximation Algorithms for NP-hard Problems. PWS Publishing Co., Boston (1996)
12. Hudson, T.J., Anderson, W., Aretz, A., et al.: International network of cancer genome projects. Nature **464**(7291), 993–998 (2010)
13. Johnson, D.S., Koch, T., Werneck, R.F., et al.: The eleventh dimacs implementation challenge. http://dimacs11.cs.princeton.edu/home.html
14. Kim, Y.A., Cho, D.Y., Dao, P., et al.: Memcover: integrated analysis of mutual exclusivity and functional network reveals dysregulated pathways across multiple cancer types. Bioinformatics **31**(12), i284–i292 (2015)
15. Kim, Y.A., Salari, R., Wuchty, S., et al.: Module cover-a new approach to genotype-phenotype studies. Pac. Symp. Biocomput. 135–146 (2013)
16. Kuo, T.W., Lin, K.C.J., Tsai, M.J.: Maximizing submodular set function with connectivity constraint: theory and application to networks. IEEE/ACM Trans. Netw. **23**(2), 533–546 (2015)
17. Lawrence, M.S., Stojanov, P., Polak, P., et al.: Mutational heterogeneity in cancer and the search for new cancer-associated genes. Nature **499**(7457), 214–218 (2013)
18. Leiserson, M.D., Vandin, F., Wu, H.T., et al.: Pan-cancer network analysis identifies combinations of rare somatic mutations across pathways and protein complexes. Nat. Genet. **47**(2), 106–114 (2015)
19. Ljubić, I., Weiskircher, R., Pferschy, U., et al.: An algorithmic framework for the exact solution of the prize-collecting steiner tree problem. Math. Program. **105**(2–3), 427–449 (2006)
20. TCGA Network: Comprehensive molecular portraits of human breast tumours. Nature **490**(7418), 61–70 (2012)
21. Raphael, B.J., Dobson, J.R., Oesper, L., et al.: Identifying driver mutations in sequenced cancer genomes: computational approaches to enable precision medicine. Genome Med. **6**(1), 5 (2014)
22. Schaefer, M.H., Fontaine, J.F., Vinayagam, A., et al.: Hippie: integrating protein interaction networks with experiment based quality scores. PLoS One **7**(2), e31826 (2012)
23. Shrestha, R., et al.: HIT'nDRIVE: multi-driver gene prioritization based on hitting time. In: Sharan, R. (ed.) RECOMB 2014. LNCS, vol. 8394, pp. 293–306. Springer, Heidelberg (2014)
24. Ulitsky, I., Karp, R.M., Shamir, R.: Detecting disease-specific dysregulated pathways via analysis of clinical expression profiles. In: Vingron, M., Wong, L. (eds.) RECOMB 2008. LNCS (LNBI), vol. 4955, pp. 347–359. Springer, Heidelberg (2008)
25. Vandin, F., Upfal, E., Raphael, B.J.: Algorithms for detecting significantly mutated pathways in cancer. J. Comput. Biol. **18**(3), 507–522 (2011)

26. Vogelstein, B., Kinzler, K.W.: Cancer genes and the pathways they control. Nat. Med. **10**(8), 789–799 (2004)
27. Vogelstein, B., Papadopoulos, N., Velculescu, V.E.: Cancer genome landscapes. Science **339**(6127), 1546–1558 (2013)
28. Weinstein, J.N., Collisson, E.A., Mills, G.B., et al.: The cancer genome atlas pan-cancer analysis project. Nat. Genet. **45**(10), 1113–1120 (2013)
29. Wang, Y., Buchanan, A., Butenko, S.: On imposing connectivity constraints in integer programs (2015). http://www.optimization-online.org/DB_HTML/2015/02/4768.html

Isometric Gene Tree Reconciliation Revisited

Broňa Brejová[⊠], Askar Gafurov, Dana Pardubská, Michal Sabo,
and Tomáš Vinař

Faculty of Mathematics, Physics and Informatics, Comenius University,
Mlynská Dolina, 842 48 Bratislava, Slovakia
{brejova,pardubska,vinar}@fmph.uniba.sk, {gafurov1,sabo19}@uniba.sk

Abstract. Isometric gene tree reconciliation is a gene tree/species tree reconciliation problem where both the gene tree and the species tree include branch lengths, and these branch lengths must be respected by the reconciliation. The problem was introduced by Ma et al. (2008a) in the context of reconstructing evolutionary histories of genomes in the infinite sites model. In this paper, we show that the original algorithm by Ma et al. (2008a) is incorrect, and we propose a modified algorithm that addresses the problems that we discovered. Moreover, by adapting a data structure by Amir et al. (2007), we were able to improve the running time from $O(mn)$ to $O(n + m \log m)$, where n is the size of the species tree, and m is the size of the gene tree.

1 Introduction

In this paper, we revisit the problem of isometric gene tree reconciliation introduced by Ma et al. (2008a). We point out several mistakes in the original publication and provide a corrected and simplified version of the algorithm. We also improve its running time by employing appropriate data structures and suggest several related open problems.

We will consider evolution of a single gene family. The evolutionary history starts with a single ancestral gene which evolves by a series of duplications, speciations and losses, resulting in several present-day species, each carrying some number of copies of the studied gene. A particular evolutionary history of a gene family defines a gene tree G and a species tree S (see Fig. 1). The leaves of the species tree S are the present-day species, and the internal nodes correspond to speciation events. The leaves of the gene tree G are the present-day copies of the gene and the internal nodes correspond to duplications or speciations.

Species trees and gene trees can be reconstructed from sequence data by well-established methods (Felsenstein 2004). However, a pair of gene tree G and species tree S can correspond to many different histories, because it is not clear, which nodes of G correspond to speciations from S. The goal of gene tree/species tree reconciliation is to map nodes of the gene tree to the species tree, and thus to reconstruct the evolutionary history of the gene family.

Classical approaches to reconciliation consider only topologies of G and S. As the reconciliation is not unique, the goal is to find the most parsimonious

© Springer International Publishing Switzerland 2016
M. Frith and C.N.S. Pedersen (Eds.): WABI 2016, LNBI 9838, pp. 40–51, 2016.
DOI: 10.1007/978-3-319-43681-4_4

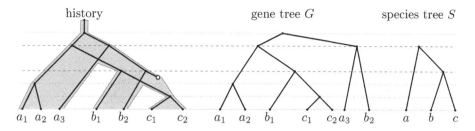

Fig. 1. An example of an evolutionary history of a gene family and its corresponding gene tree G and species tree S. Gene identifiers start with species label; thus a_1, a_2 and a_3 are three copies of the studied gene in species a. Gene loss is depicted as an empty circle. Duplications are highlighted by gray dotted horizontal lines, speciations by dashed lines.

reconciliation minimizing the number of events. This problem is studied since 1979 (Goodman et al. 1979), and multiple algorithms were developed (Guigo et al. 1996; Zhang 1997; Eulenstein 1997; Zmasek and Eddy 2001).

In this paper, we consider a different variant of the problem called *isometric gene tree reconciliation*. In this problem, branch lengths in both the gene tree and the species tree are known exactly, and the reconciliation should obey them. This problem was introduced by Ma et al. (2008a), who used this form of gene tree reconciliation as one of the steps in their polynomial-time algorithm which can reconstruct evolutionary history of several genomes in a rich model which includes duplications, two and three breakpoint rearrangements, deletions, and insertions under assumptions of the infinite sites model. This result is rather remarkable, as reconstruction of rearrangement histories is typically NP hard even in simple models (Fertin et al. 2009).

If both the gene tree and the species tree are rooted, their isometric reconciliation can be found by a straightforward algorithm: we map each leaf of S to the corresponding leaf of G and once a node v is mapped, we can map its parent p to the unique place in G determined by the length of the edge (p, v). Note that in general, a node from G can map either to one of the nodes in S or to a point on an edge of S. For example, in Fig. 2 node y from G maps to a point in the middle of edge (r, x) in S, because this is the unique point on the path from the root to leaf c, which is situated in distance 3 from c. Note that this simple algorithm maps node p using one of its children. When the other child of p suggests mapping of p to a different place, the trees cannot be reconciled. However, if the reconciliation exists, it is clearly unique.

However, Ma et al. (2008a) and (2008b) consider a more difficult problem, in which the species tree is rooted, but the gene tree is unrooted. This is needed, because in practice most of the phylogenetic reconstruction methods produce unrooted trees. While the species tree can be rooted by including an outgroup, finding an appropriate outgroup for a multi-gene family, which may harbor ancient duplications, is more problematic. Ma et al. (2008a) and (2008b) give a polynomial-time algorithm for the isometric reconciliation problem, and after

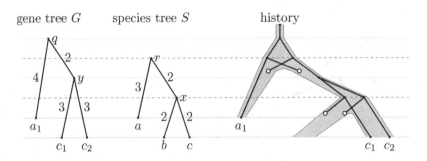

Fig. 2. Isometric reconciliation of rooted G and S.

some unspecified modifications, apply it to real data with inexact branch lengths. In this paper, we point out several mistakes in their version of the algorithm and provide a corrected version. Note that although it is not obvious, the isometric reconciliation is still unique in this case.

Reconciliation with some branch length information was also considered in several more complex models, such as probabilistic approaches considering branch lengths in S (Sennblad and Lagergren 2009; Górecki et al. 2011; Doyon et al. 2012) and models allowing horizontal gene transfer (Doyon et al. 2010; Bansal et al. 2012).

Notation. We will now introduce notation used in this paper, which slightly differs from notation in the publications by Ma et al. Given two nodes u and v belonging to the same phylogenetic tree, $d(u, v)$ denotes their distance, i.e. the sum of edge lengths on the unique simple path between u and v. If u is a node in a rooted phylogenetic tree, $\mathrm{anc}(u, d)$ denotes the point in the tree which is at a distance exactly d from u on the path towards the root. The result can be one of the ancestors of u, or a point on some edge on the path from u to the root, or even on an implicit edge of infinite length leading from the root upward. By $\mathrm{lca}(u, v)$, we denote the lowest common ancestor of nodes u and v, and by $\mathrm{lca}(X)$ the lowest common ancestor of a whole set of nodes X. We will assume that all phylogenetic trees have strictly positive edge lengths.

2 Problems in the Original Algorithm

Both Ma et al. (2008a) and (2008b) include the same algorithm for isometric gene tree reconciliation. In this section, we describe some of its details and point out mistakes in the original paper. We start with the original definition of isometric reconciliation as given by Ma et al. (2008a).

Definition 1 (Original Definition). *Any mapping Φ from a gene tree G to a species tree S that roots the gene tree is an isometric reconciliation if*

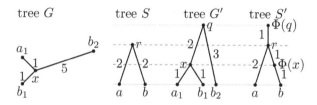

Fig. 3. A counter-example for the original definition of isometric reconciliation. G and S are input trees, G' is the rooted version of G and S' is S with duplication nodes added. Although this reconciliation satisfies Definition 1, it does not correspond to any evolutionary history.

1. *Every leaf of G maps to the leaf of the designated species in S.*
2. *Each internal node of G maps to a speciation node in S or a point on a branch in S.*
3. *The new root q of G maps to a point $\Phi(q)$ on a branch in S such that any other node x in G maps to $\Phi(x)$ below $\Phi(q)$ and $d(\Phi(x), \Phi(q)) = d(x, q)$*

As the example in Fig. 3 shows, this definition is not sufficiently stringent to characterize meaningful reconciliations. The mapping Φ shown satisfies all the above conditions, but does not correspond to any valid history. In particular, node x in the rooted version of G is a parent of leaf a_1, but node $\Phi(x)$ is not an ancestor of $\Phi(a_1)$.

This problem is very easily corrected by demanding that mapping Φ preserves distances and ancestor relationships between every pair of nodes in G, or equivalently, between every pair of adjacent nodes in G. We will also introduce a more explicit notation distinguishing the input trees and their modified versions, as follows.

Definition 2 (Modified Definition). *Let G be an unrooted gene tree and S be a rooted species tree. Isometric reconciliation of G and S is a triple (Φ, G', S'). Tree G' is a rooted version of G. Tree S' is obtained from S by potentially subdividing some edges by new nodes and potentially adding a path leading to the original root of S from above. Mapping Φ maps nodes of G' to nodes of S' so that:*

1'. *Every leaf of G' maps to the leaf of the designated species in S'.*
2'. *Each internal node of G' maps to an internal node in S'.*
3'. *If node x is the parent of y in G', then $\Phi(x)$ is an ancestor of $\Phi(y)$ in S' and $d(\Phi(x), \Phi(y)) = d(x, y)$; or in our notation, $\Phi(x) = anc(\Phi(y), d(x, y))$.*

In addition, we require that each node from S' either exists in S or some node from G' maps to it.

Overall Scheme of the Original Algorithm. The algorithm of Ma et al. (2008a) proceeds by first mapping leaves of G to corresponding leaves of S and then repeatedly choosing one unmapped node x from G which has at least two of its

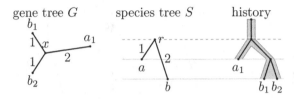

Fig. 4. An example which can be reconciled, but which the original algorithm may recognize as irreconcilable. When mapping node x, all three neighbors are already mapped, and thus the algorithm by Ma et al. can choose any two of them as nodes u and v. If it chooses b_1 and b_2, it works correctly, but if it chooses a_1 and b_1, it will reject the tree in step 7(b)iii.

three neighbours already mapped. Each such node is mapped to its corresponding point $\Phi(x)$, and if one of the adjacent edges in G contains a root, the gene tree is rooted. This process continues, until only one node remains.

The overall scheme of the algorithm reveals another minor issue: the algorithm does not work for gene trees with two leaves. The leaves can be mapped trivially, but we also need to find the position of the root on the edge connecting them in G, and since in the algorithm, rooting is done simultaneously with mapping internal nodes, it is not obvious how to find the root in this case.

Mapping One Node. The algorithm for mapping an internal node and, if appropriate, rooting the gene tree, consists of a rather extensive case analysis, with about ten different cases. After simulating the algorithm on several examples, we have discovered that it does not always work correctly. Figure 4 shows a simple input, which can be reconciled. The algorithm maps the only internal node x correctly, but sometimes fails during rooting, rejecting the input as irreconcilable. When mapping the last internal vertex, the algorithm arbitrarily chooses, which of the two neighbours of this vertex are considered first, and depending on this choice, the algorithm may fail or succeed on this input.

Speciation and Duplication Happening at the Same Time. Ma et al. (2008a) and (2008b) assume that the input trees are binary and that two events (two duplications or duplication and speciation) never happen at the same time. However, even binary input trees may lead to situations, where two events happen at the same time. A simple example is when the root of G' coincides with one of the internal vertices of G, and thus it has three children. This situation is handled by the original algorithm and rejected in case 7(b)iii. An example of such an input is shown in Fig. 5.

Figure 6 shows a similar input, with only one branch length changed. Here also duplication happens at the same time as speciation, but due to later losses the rooted gene tree G' remains binary. The original algorithm accepts this input, which seems inconsistent with handling the input from Fig. 5. Note that both of these inputs can be reconciled so that they satisfy the corrected definition of isometric reconciliation (as well as the weaker original definition).

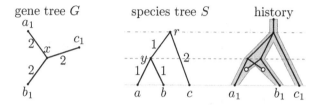

Fig. 5. An input rejected by the original algorithm where reconciliation maps duplication and speciation to the same point in the history.

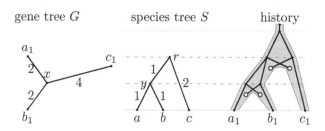

Fig. 6. An input accepted by the original algorithm, where reconciliation maps duplication and speciation to the same point in the history.

If we work in a model where no two events may happen at the same time, we would need to modify the algorithm so that it reject both of these inputs and also modify the definition of isometric reconciliation so that when a node v of G' maps to a speciation node $\Phi(v)$ in S', then indeed one of the subtrees of v contains only leaves corresponding to species in one of the subtrees of $\Phi(v)$ and the other subtree of v analogously contains only leaves corresponding to species from the other subtree of $\Phi(v)$. Alternatively, we can use a more relaxed evolutionary model, in which we allow an arbitrary combination of events happen at the same time; we will take this approach in our modified algorithm.

Summary of Issues. To summarize, the original definition of isometric reconciliation allows nonsense mappings that do not correspond to any evolutionary history and does not handle sufficiently clearly cases with simultaneous duplication

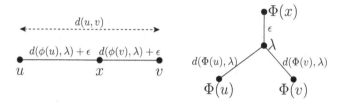

Fig. 7. Illustration of stage 3: creating a potential root x inside edge (u, v).

and speciation. In addition, the original algorithm does not handle gene trees with two leaves and sometimes fails to root valid inputs. In the next section, we present a new algorithm that corrects these problems and, at the same time, simplifies the proof of correctness by reducing the case analysis to minimum.

3   The Modified Algorithm

In this section, we describe a new version of the algorithm for isometric gene tree reconciliation. Although the overall idea is similar to the original algorithm, we have made it more modular, with several passes through the tree, each relatively simple. This allows us to avoid complicated case analysis in both the algorithm and the proof. Our algorithm works even for non-binary input trees and allows an arbitrary combination of events to happen at any given point in the evolutionary history. If desired, a reconciliation with such simultaneous events can be rejected in the verification stage. Our algorithm proceeds in five stages:

Stage 1: Map leaves of G to the corresponding leaves of S.
Stage 2: Map internal vertices of G.
  Repeatedly consider an unmapped internal node x of G with at least two mapped neighbors u and v. Let $x_u = \mathrm{anc}(\Phi(u), d(u, x))$ and $x_v = \mathrm{anc}(\Phi(v), d(v, x))$. If one of x_u and x_v is a descendant of the other, use the descendant as $\Phi(x)$ (this includes the case when $x_u = x_v$). Otherwise reject the input tree.
Stage 3: Add the root of G.
  Consider each edge (u, v) of G. We want to decide if this edge should be subdivided by a new node, which will be a *potential root*. Let $\lambda = \mathrm{lca}(\Phi(u), \Phi(v))$ and let $\epsilon = (d(u, v) - d(\Phi(u), \lambda) - d(\Phi(v), \lambda))/2$. If $\epsilon > 0$ or $\lambda \notin \{\Phi(u), \Phi(v)\}$, create a new node on the edge (u, v) in distance $d(\Phi(u), \lambda) + \epsilon$ from u and $d(\Phi(v), \lambda) + \epsilon$ from v (see Fig. 7). Map the new node to $\mathrm{anc}(\lambda, \epsilon)$.
Stage 4: Create rooted tree G'.
  Let V be the set of all nodes of G, including nodes created in stage 3, if any. Let $\Phi(V)$ be the set of points in S where nodes from V map, and let $\lambda = \mathrm{lca}(\Phi(V))$. If there is a unique node $q \in V$ that maps to λ, select it as the root of G'. Otherwise reject the input.
Stage 5: Check that the mapping satisfies the definition of isometric reconciliation: consider each edge of G' and verify condition (3') of Definition 2.

To keep the algorithm efficient, we will not explicitly construct S' with added duplication nodes. When a node x maps to a point on an edge (u, v) in S, we will keep the mapping as a pair (u, d), where u is the bottom endpoint of the edge and d is distance from u to the mapped point.

Proof of Correctness. If the algorithm does not reject the input, it will produce a correct isometric reconciliation, thanks to the verification in stage 5. Therefore it is sufficient to prove that we will never falsely reject an input if a reconciliation exists. Assume that (Φ^*, G', S') is an isometric reconciliation; we will prove that the algorithm will find it. This also proves that the isometric reconciliation is unique, because if there were two distinct reconciliations, the algorithm cannot produce both of them simultaneously.

Correctness of mapping constructed in stage 2 can be obtained by induction using Lemma 1. Note that the existence of an unmapped node with at least two mapped neighbors in each iteration of the algorithm is guaranteed by basic properties of trees.

Lemma 1. *If stage 2 of the algorithm considers an unmapped node x with two neighbors u and v which are mapped correctly to $\Phi^*(u)$ and $\Phi^*(v)$, it will map x correctly to $\Phi^*(x)$.*

Proof. We will distinguish several cases based on the location of the root q of G' in the correct reconciliation. If we remove edges (u, x) and (v, x) from G, we will get three connected components. Root q can be in one of these components or on one of the removed edges.

Case 1: q is in the component containing x (including the case $x = q$). Then u and v are children of x in G', and thus by Definition 2, $\Phi^*(x) = \mathrm{anc}(\Phi^*(u), d(u, x))$ and also $\Phi^*(x) = \mathrm{anc}(\Phi^*(v), d(v, x))$. In the algorithm, we thus get $x_u = x_v = \Phi^*(x)$, and this point is correctly selected as $\Phi(x)$.

Case 2: q is in the component containing u (including the case $u = q$). Then x is a child of u and v is a child of x. Thus by definition, $\Phi^*(u) = \mathrm{anc}(\Phi^*(x), d(u, x))$ and $\Phi^*(x) = \mathrm{anc}(\Phi^*(v), d(x, v))$. In the algorithm, x_u will be an ancestor of $\Phi^*(u)$ and x_v will be $\Phi^*(x)$. Thus x_v will be correctly selected as $\Phi(x)$, because it is a descendant of x_u.

Case 2': q is in the component containing v. This case is symmetrical to case 2.

Case 3: q is inside edge (x, u), excluding the endpoints. Vertex u is a child of q in G' and thus $\Phi^*(q) = \mathrm{anc}(\Phi^*(u), d(u, q))$. Since $d(u, q) < d(u, x)$, x_u is an ancestor of $\Phi^*(q)$. On the other hand, $\Phi^*(v)$ is a descendant of $\Phi^*(x)$, which is a descendant of $\Phi^*(q)$. Node x_v will be equal to $\Phi^*(x)$ and will be correctly selected by the algorithm as $\Phi(x)$.

Case 3': q is inside edge (x, v). This case is again symmetrical to case 3.   □

Similarly, we can prove by case analysis that the algorithm correctly finds and maps the root of the tree.

Lemma 2. *If stage 3 of the algorithm considers an edge (u, v) with its endpoints correctly mapped to S, it will subdivide this edge by a new node if and only if the correct reconciliation has a root inside this edge. The new node will be created at the correct position and mapped correctly.*

Proof. Let q be the root of G' in the correct reconciliation, and let us consider three cases regarding the position of q after removal of edge (u, v) from G.

Case 1: q is in the connected component containing u. Then $\Phi(u) = \text{anc}(\Phi(v),$ $d(u, v))$ and thus $\lambda = \Phi(u)$. In addition, $d(\Phi(u), \lambda) = 0$, $d(\phi(v), \lambda) = d(u, v)$ and thus $\epsilon = 0$. No node will be created, which is correct, as in this case the root is not inside this edge.

Case 2: q is in the connected component containing v. This case is symmetrical to case 1.

Case 3: q is inside edge (u, v). Let $d = d(u, v)$ and let $\delta = d(q, u)$. We have $0 < \delta < d$. Since q is the parent of both u and v in G', $\Phi^*(q) = \text{anc}(\Phi(u), \delta)$ and $\Phi^*(q) = \text{anc}(\Phi(v), d - \delta)$. We will consider two subcases concerning the position of $\lambda = \text{lca}(\Phi(u), \Phi(v))$. (a) If $\lambda = \Phi^*(q)$, new node will be created, because $\lambda \notin \{\Phi(u), \Phi(v)\}$. We have $\epsilon = (d - \delta - (d - \delta))/2 = 0$, so the node will be created at distance δ from u, as desired. (b) If $\Phi^*(q)$ is not λ, λ must be a descendant of $\Phi^*(q)$, with $d(\lambda, \Phi^*(q)) = \epsilon' > 0$. Note that in this case it is possible that $\lambda \in \{\Phi(u), \Phi(v)\}$. However, $\delta = d(\phi(u), \lambda) + \epsilon'$ and $d - \delta = d(\phi(v), \lambda) + \epsilon'$ and thus $\epsilon = (d - (\delta - \epsilon') - (d - \delta - \epsilon'))/2 = \epsilon' > 0$. Thus a node will be created and its distance from u will be correctly set to $d(\phi(u), \lambda) + \epsilon = \delta$.   □

After stage 3, we will thus have all nodes of G correctly mapped to S and if the root of G' is not one of the nodes of G, it was also correctly added and mapped. The definition of isometric reconciliation implies that the true root of G' will be indeed the only node mapping to $\text{lca}(\Phi(V))$, so the gene tree will be correctly rooted by the algorithm.

Running Time Analysis. Let m be the size of G and n the size of S. Ma et al. (2008b) claim that their algorithm works in $O(mn)$ time; however, we will prove that a more efficient implementation of isometric reconciliation is possible. Within the algorithm, we use several nontrivial operations on tree S:

- Finding lca of two nodes. We can use efficient data structures for solving lca queries in $O(1)$ time after $O(n)$ preprocessing of the tree (Harel and Tarjan 1984; Bender and Farach-Colton 2000).
- Determining if node u is an ancestor of v. This is equivalent to asking if $u = \text{lca}(u, v)$.
- Computing the distance between node v and its ancestor u. This can be done in $O(1)$ time by keeping the distance from the root of S in each node and subtracting these distances for u and v.
- Finding $\text{anc}(u, d)$. This operation is known as level ancestor. For unweighted trees, it can be solved in $O(1)$ time after $O(n)$ preprocessing (Berkman and Vishkin 1994) and for trees with integer weights in $O(\log \log u)$ time, where u is the maximum edge weight (Amir et al. 2007). Below, we outline a simplified version of this data structure which achieves $O(\log n)$ time per query, but works for arbitrary edge weights, as edge weights in phylogenetics are typically not expressed as integers.

Using these building blocks, the rest of the algorithm is relatively straightforward. Stage 1 of the algorithm is trivial, because identifiers of leaves in G directly indicate the correct leaf in S. For stage 2, we maintain a counter of mapped neighbors for each internal node of G and a stack of unprocessed nodes with at least two neighbors already mapped. In each step, we remove one node from the stack, map it, and increase the counters of its neighbors. If any counter reaches 2, the corresponding node is added to the stack. The overall overhead for selecting nodes for mapping in stage 2 is thus $O(m)$, and mapping each node works in $O(\log n)$. Stage 3 involves a simple loop through all edges, and each edge is processed in $O(\log n)$ time. Stages 4 and 5 work in linear time; note that $\mathrm{lca}(\Phi(V))$ can be computed by $m - 1$ applications of pairwise lca. The overall running time of the algorithm is thus $O(n + m \log n)$.

Finally, note that our algorithm does not explicitly create species tree S' with edges subdivided by duplication nodes. To do so, one would have to sort points (u, d) mapped to each edge (u, v) of S by distance d from u. This can be done in $O(n + m \log m)$ time. Typically, $m \geq n$, and so the overall running time would be $O(m \log m)$.

A Simple Level Ancestor for Arbitrary Weights. For completeness, we briefly describe a data structure for finding $\mathrm{anc}(u, d)$ in $O(\log n)$ time for arbitrary edge weights, provided that we can do addition, subtraction, and sign in constant time. We use a simplified version of the data structure by Amir et al. (2007); the simplification is possible thanks to the fact that the running time is worse than the running time achievable for integer edge weights.

Let node weight $w(u)$ be the sum of edge weights on the path from the root to node u. To compute $\mathrm{anc}(u, d)$, we are looking for the highest ancestor v of u such that $w(u) - w(v) \leq d$. If we had only a single path instead of a tree, we would be looking for a predecessor of value $x = w(u) - d$ in the sequence of node weights. Since this sequence is increasing, we can use binary search to find the desired index v.

In a general tree, we will use the heavy path decomposition (Harel and Tarjan 1984). An edge connecting node v to its parent p in a tree is called *heavy* if the size of the subtree rooted at v (the number of nodes) is at least half of the size of the subtree rooted at p. Otherwise the edge is called *light*. For each node, at most one of its children is connected to it by a heavy edge. Therefore, heavy edges form a set of vertex-disjoint paths. Vertices which are not incident to any heavy edge will be considered as heavy paths of length 0 so that each node is included in exactly one heavy path.

We create an array of node weights for each heavy path. Each vertex v also keeps the reference to the highest node on its heavy path. When searching for $\mathrm{anc}(u, d)$, we search along the path from u to the root to find the heavy path that contains the answer. Thanks to the properties of the heavy path decomposition, there are at most $O(\log n)$ light edges on any leaf-to-root path, and thus we can use linear search to iterate through heavy paths encountered on the way to the root. In constant time, we can jump to the head of the path and comparing x to the value stored in the head and in the head's parent, we can determine if this path contains the answer. Within the correct path, we then find the answer by binary search. The overall time is thus $O(\log n)$.

The data structure for integer weights by Amir et al. (2007) uses binary search over heavy paths, which requires repeated use of the unweighted level ancestor data structure. Instead of binary search within a path, they use efficient data structures for the predecessor problem with integer keys.

4   Conclusion and Open Problems

In this paper, we have corrected an algorithm for isometric gene tree reconciliation, first presented by Ma et al. (2008a) in the context of reconstruction of evolutionary histories in the infinite sites model. We have also improved running time of the algorithm from $O(nm)$ to $O(n + m \log m)$, where n is the size of the species tree, and m is the size of the gene tree.

In our problem, the gene tree is unrooted, and the species tree is rooted. This corresponds to a common practice, where gene trees are inferred computationally without outgroup genes, while species trees reflect well established phylogenetic relationships. However, it may be of interest to extend isometric gene tree reconciliation to unrooted species trees. We have shown that if the species tree is rooted, there is always at most one solution. In case of unrooted species tree, can there be multiple solutions?

In practical applications, we cannot rely on the assumption that the branch lengths are exactly correct. Algorithms that would allow for errors in branch lengths, e.g. assuming that branch lengths are correct up to some degree of tolerance, would be useful for practical applications.

Finally, it is well known that the rate of evolution varies between individual gene families. This fact could be captured if the exact branch length assumption is relaxed so that all branch lengths in the gene tree are scaled by an unknown constant factor α. The reconciliation algorithm would then simultaneously infer node mapping and scaling factor α.

Acknowledgements. This research was funded by a grant from the Slovak Research and Development Agency APVV-14-0253 and by VEGA grants 1/0684/16, 1/0719/14, and 2/0165/16.

References

Amir, A., Landau, G.M., Lewenstein, M., Sokol, D.: Dynamic text and static pattern matching. ACM Trans. Algorithms **3**(2), 19 (2007)

Bansal, M.S., Alm, E.J., Kellis, M.: Efficient algorithms for the reconciliation problem with gene duplication, horizontal transfer and loss. Bioinformatics **28**(12), i283–i291 (2012)

Bender, M.A., Farach-Colton, M.: The LCA problem revisited. In: Gonnet, G.H., Viola, A. (eds.) LATIN 2000. LNCS, vol. 1776, pp. 88–94. Springer, Heidelberg (2000)

Berkman, O., Vishkin, U.: Finding level-ancestors in trees. J. Comput. Syst. Sci. **48**(2), 214–230 (1994)

Doyon, J.-P., Hamel, S., Chauve, C.: An efficient method for exploring the space of gene tree/species tree reconciliations in a probabilistic framework. IEEE/ACM Trans. Comput. Biol. Bioinform. **9**(1), 26–39 (2012)

Doyon, J.-P., Scornavacca, C., Gorbunov, K.Y., Szöllősi, G.J., Ranwez, V., Berry, V.: An efficient algorithm for gene/species trees parsimonious reconciliation with losses, duplications and transfers. In: Tannier, E. (ed.) RECOMB-CG 2010. LNCS, vol. 6398, pp. 93–108. Springer, Heidelberg (2010)

Eulenstein, O.: A linear time algorithm for tree mapping. GMD-Forschungszentrum Informationstechnik (1997)

Felsenstein, J.: Inferring Phylogenies. Sinauer Associates, Sunderland (2004)

Fertin, G., Labarre, A., Rusu, I., Tannier, E., Vialette, S.: Combinatorics of Genome Rearrangements. MIT Press, Cambridge (2009)

Goodman, M., Czelusniak, J., Moore, G.W., Romero-Herrera, A., Matsuda, G.: Fitting the gene lineage into its species lineage, a parsimony strategy illustrated by cladograms constructed from globin sequences. Syst. Biol. **28**(2), 132–163 (1979)

Górecki, P., Burleigh, G.J., Eulenstein, O.: Maximum likelihood models and algorithms for gene tree evolution with duplications and losses. BMC Bioinform. **12**(1), 1 (2011)

Guigo, R., Muchnik, I., Smith, T.F.: Reconstruction of ancient molecular phylogeny. Mol. Phylogenet. Evol. **6**(2), 189–213 (1996)

Harel, D., Tarjan, R.E.: Fast algorithms for finding nearest common ancestors. SIAM J. Comput. **13**(2), 338–355 (1984)

Ma, J., Ratan, A., Raney, B.J., Suh, B.B., Miller, W., Haussler, D.: The infinite sites model of genome evolution. Proc. Nat. Acad. Sci. **105**(38), 14254–14261 (2008a)

Ma, J., Ratan, A., Raney, B.J., Suh, B.B., Zhang, L., Miller, W., Haussler, D.: DUP-CAR: reconstructing contiguous ancestral regions with duplications. J. Comput. Biol. **15**(8), 1007–1027 (2008b)

Sennblad, B., Lagergren, J.: Probabilistic orthology analysis. Syst. Biol. **58**(4), 411–424 (2009)

Zhang, L.: On a Mirkin-Muchnik-Smith conjecture for comparing molecular phylogenies. J. Comput. Biol. **4**(2), 177–187 (1997)

Zmasek, C.M., Eddy, S.R.: A simple algorithm to infer gene duplication and speciation events on a gene tree. Bioinformatics **17**(9), 821–828 (2001)

Further Improvement in Approximating the Maximum Duo-Preservation String Mapping Problem

Brian Brubach[(✉)]

Department of Computer Science,
University of Maryland–College Park, College Park, MD, USA
bbrubach@cs.umd.edu

Abstract. We present an improved approximation for the Maximum Duo-Preservation String Mapping Problem (MPSM). This problem was introduced in [7] as the complement to the well-studied Minimum Common String Partition problem (MCSP). Prior work also considers the k-MPSM and k-MCSP variants in which each letter occurs at most k times. The authors of [7] showed a k^2-appoximation for $k \geq 3$ and 2-approximation for $k = 2$. A 4-approximation independent of k was shown in [4]. In [4], they also showed that k-MPSM is APX-Hard and achieved approximation ratios of 8/5 for $k = 2$ and 3 for $k = 3$. In this paper, we show an algorithm which achieves a 13/4-approximation for the general MPSM problem using a new combinatorial triplet matching approach. During publication of this paper, [3] presented a local search algorithm yielding 7/2, which falls in between the previous best and this paper. The remainder of the paper has not been altered to reflect this.

Keywords: String algorithms · Polynomial-time approximation · Max Duo-Preservation String Mapping Problem · Min Common String Partition Problem

1 Introduction

String comparison is one of the most fundamental problems in many fields such as bioinformatics and data compression. In computer science, the difference between two strings is often measured by edit distance, the number of edit operations required to transform one string into the other. The most widely known definitions of edit distance include insertion, deletion, and/or substitution operations. However, the more general edit distance with moves problem studied in [10] allows an additional operation wherein an entire block of text is shifted within a string.

B. Brubach—Supported in part by NSF award CCF-1422569.

The author wishes to acknowledge internship mentor Prof. Srinivas Aluru for his support.

M. Frith and C.N.S. Pedersen (Eds.): WABI 2016, LNBI 9838, pp. 52–64, 2016.
DOI: 10.1007/978-3-319-43681-4_5

These shift operations, also known as rearrangements, are especially relevant in biology [8,18]. String comparison can be performed on DNA or protein sequences to estimate how closely related different species are. In data compression, we may want to store many similar strings as a single string along with the edits required to recover all strings. These two applications even overlap naturally in the field of bioinformatics where extremely large datasets of biological sequences are common. For example, the challenge of pan-genome storage is to store many highly similar sequences from the same clade such as a bacterial species.

One way to capture just the "moves" operation is to solve the Minimum Common String Partition problem (MCSP) which seeks to partition two strings into minimum cardinality sets of substrings that are permutations of each other. While the MCSP problem has been heavily studied, the complementary Maximum Duo-Preservation String Mapping Problem (MPSM) is a relatively new and under-explored problem in this area.

1.1   Problem Description

The Maximum Duo-Preservation String Mapping Problem (MPSM) is defined as follows. We are given two strings $A = a_1 a_2 \ldots a_n$ and $B = b_1 b_2 \ldots b_n$ of length n such that B is a permutation of A. Let a_i and b_j be the ith and j^{th} characters of their respective strings. A *proper* mapping π from A to B is a one-to-one mapping with $a_i = b_{\pi(i)}$ for all $i = 1, \ldots, n$. A *duo* is simply two consecutive characters from the same string. We say that a duo (a_i, a_{i+1}) is *preserved* if a_i is mapped some b_j and a_{i+1} is mapped to b_{j+1}. The objective is to return a proper mapping from the letters of A to the letters of B which preserves the maximum number of duos. Note that the number of duos preserved in each string is identical and by convention we count the number of duos preserved in a single string rather than the sum over both strings. Let OPT_{MPSM} denote the number of duos preserved from a single string in an optimal solution to the MPSM problem. Figure 1 shows an example of an optimal mapping which preserves the maximum possible number of duos.

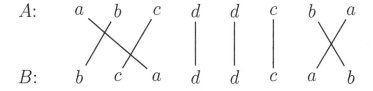

Fig. 1. Illustration of a mapping π from A to B that preserves 3 duos: bc, dd, and dc. A solution to the complementary MCSP problem on the same strings would be partitions $P_A = a, bc, ddc, b, a$ and $P_B = bc, a, ddc, a, b$ with $|P_A| = |P_B| = 5$.

The complementary Minimum Common String Partition problem (MCSP) seeks to find partitions of the strings A and B where a partition P_A of A is defined as a set of substrings whose concatenation is A. The objective is to find minimum cardinality partitions P_A of A and P_B of B such that P_B is a permutation of P_A. Let OPT_{MCSP} denote the cardinality of a partition in an optimal solution to this problem. We can see that

$$OPT_{MCSP} = |P_A| = |P_B| = n - OPT_{MPSM}$$

The variants, k-MPSM and k-MCSP, add the restriction that each letter occurs at most k times in each string. For a given algorithm, let ALG_{MPSM} be number of duos preserved by the algorithm. The approximation ratio for that algorithm is defined as

$$\frac{OPT_{MPSM}}{ALG_{MPSM}}$$

1.2   Related Work

The Maximum Duo-Preservation String Mapping Problem (MPSM) was introduced in [7] along with the related Constrained Maximum Induced Subgraph (CMIS) and Constrained Minimum Induced Subgraph (CNIS) problems. They used a linear programming and randomized rounding approach to approximate the k-CMIS problem which they show is a generalization of k-MPSM. This leads to a k^2-approximation for $k \geq 3$ and a 2-approximation for $k = 2$. This was improved by [4] to a 4-approximation independent of k as well as approximation ratios of 3 for $k = 3$ and 8/5 for $k = 2$. [4] also show that k-MPSM is APX-hard even for $k = 2$, meaning no polynomial-time approximation scheme (PTAS) exists assuming $P \neq NP$. The fixed-parameter tractability was studied in [1] and MPSM was shown to be fixed-parameter tractable when parameterized by the number of preserved duos.

The Minimum Common String Partition problem (MCSP) has been extensively studied from many angles including polynomial-time approximation [7,9, 10,14,16,17], fixed-parameter tractability [5,6,11,15], and heuristics [2,12,13]. FPT algorithms have been parameterized by maximum number of times any character occurs, minimum block size, and the size of the optimal minimum partition. Heuristic approaches range from an ant colony optimization algorithm [12] to integer linear programming (ILP) based strategies [2,13] which in some cases solve the problem optimally for strings up to $2,000$ characters in length.

The problem was shown to be NP-hard (thus implying MPSM is also NP-hard) and APX-hard even for 2-MCSP [14]. The current best approximations are an $O(\log n \log^* n)$-approximation due to [10] for general MCSP and an $O(k)$-approximation for k-MCSP due to [17]. Applications to evolutionary distance and genome rearrangement can be found in [8,18].

1.3   Our Contributions

We show a 13/4-approximation ratio for the general MPSM problem using a new combinatorial triplet matching approach. This improves the previous best approximation ratio of 4 for the general problem due to [4].

Theorem 1. *For any two strings A and B such that B is a permutation of A, there is an algorithm which finds a proper mapping from A to B that preserves at least 4/13 of the duos that the optimal algorithm preserves.*

2   Preliminaries

Let $A = a_1 a_2 \ldots a_n$ and $B = b_1 b_2 \ldots b_n$ be the two strings of length n with a_i and b_i being the ith characters of their respective strings. A *duo* $D_i^A = (a_i, a_{i+1})$ corresponds to the pair of consecutive characters a_i and a_{i+1} in the string. We use $D^A = (D_1^A, \ldots, D_{n-1}^A)$ and $D^B = (D_1^B, \ldots, D_{n-1}^B)$ to denote the sets of *duos* for A and B, respectively. We similarly define a *triplet* $T_i^A = (a_i, a_{i+1}, a_{i+2})$ as a set of three consecutive characters a_i, a_{i+1}, and a_{i+2} in the string and sets of *triplets* $T^A = (T_1^A, \ldots, T_{n-2}^A)$ and $T^B = (T_1^B, \ldots, T_{n-2}^B)$ for strings A and B, respectively. Observe that the duos D_i^A and D_{i+1}^A correspond to the first two and last two characters, respectively, of the triplet T_i^A. We refer to duos D_i^A and D_{i+1}^A as *subsets* of the triplet T_i^A.

Important note: In the first step of our algorithm, we append a special character '&' to the beginning and end of each string (indices 0 and $n+1$). We define this character to be not equal to any other character including itself (meaning $\& \neq \&$). This ensures that each duo can be a subset of exactly two triplets.

A proper mapping π from A to B is a one-to-one mapping from the letters of A to the letters of B with $a_i = b_{\pi(i)}$ for all $\forall i = 1, \ldots, n$. Recall that a duo (a_i, a_{i+1}) is preserved if and only if a_i is mapped to some b_j and a_{i+1} is mapped to b_{j+1}. We call a pair of duos (D_i^A, D_j^B) *preservable* if and only if $a_i = b_j$ and $a_{i+1} = b_{j+1}$.

For consistency, we define the concept of conflicting pairs of duos using the terminology of [4] with a small modification to accommodate our particular analysis. Two preservable pairs of duos (D_i^A, D_j^B) and (D_h^A, D_ℓ^B) are said to be *conflicting* if no proper mapping can preserve both of them. These conflicts can be of two types Type 1 and Type 2.

- Type 1: Either $i = h \wedge j \neq \ell$ or $i \neq h \wedge j = \ell$.
- Type 2: Either $i = h+1 \wedge j \neq \ell+1$ or $i \neq h+1 \wedge j = \ell+1$.

Exception: In our analysis, we also consider two pairs of consecutive preservable duos (D_i^A, D_j^B) and (D_{i+1}^A, D_{j+1}^B) and a third pair of duos (D_h^A, D_ℓ^B) which conflicts with one or both of them, potentially creating conflicts of both Type 1 and Type 2. However, we classify such conflicts simply as Type 1 conflicts.

3   Triplet Matching Approach

In this section, we introduce and analyze the triplet matching algorithm.

3.1   The Triplet Matching Algorithm

We start by finding a weighted matching on triplets that upper bounds the optimal solution, translating that to a fractional matching on duos, and rounding the fractional solution to a mapping S that preserves a number of duos that is at least $4/13$ the weight of the triplet matching.

Step 1: Construct a weighted bipartite graph G_T on the triplets.

We first append the special character '&' to the beginning and end of each string as discussed in the preliminaries. Recall that $\& \neq \&$. This ensures that each duo can be a subset of exactly two triplets.

We then construct a weighted bipartite graph $G_T = (T^A \cup T^B, E)$ with each partition being the set of triplets from a string. We add three types of edges to this graph: *full* edges, *first-half* edges, and *last-half* edges. For a given pair of triplets, (T_i^A, T_j^B), from the different strings, we can add at most one type of edge. A full edge is added if $(a_i = b_j) \wedge (a_{i+1} = b_{j+1}) \wedge (a_{i+2} = b_{j+2})$. A first-half edge is added if $(a_i = b_j) \wedge (a_{i+1} = b_{j+1}) \wedge (a_{i+2} \neq b_{j+2})$. Similarly, a last-half edge is added if $(a_i \neq b_j) \wedge (a_{i+1} = b_{j+1}) \wedge (a_{i+2} = b_{j+2})$. The full edges have weight 1 and the half edges have weight $1/2$.

In other words, if the triplets are a perfect match, the weight of the edge is 1. Otherwise, if only the first two or last two characters match, the weight is $1/2$. Finally, if the previous conditions are not met, we do not add an edge between these triplets. Figure 2 illustrates this step.

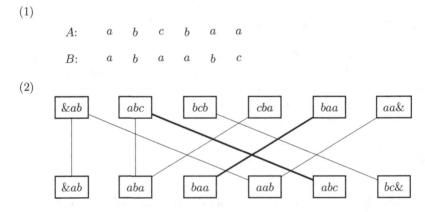

Fig. 2. Step 1 of the algorithm. (1) shows the original strings. (2) shows the bipartite triplet graph G_T.

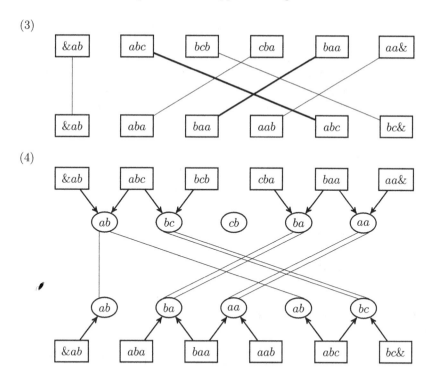

Fig. 3. Steps 2 and 3 of the algorithm. (3) shows the maximum weight matching found in Step 2. (4) shows the construction of the bipartite duo graph G_D in Step 3. Note that the double edges connecting the duos (b, c), (b, a), and (a, a) will each be collapsed into single edges of weight 1.

Step 2: Find a maximum weight matching M_T on the triplets in G_T.

We find a maximum weight matching M_T in the graph G_T. We will prove later that the weight of this matching is a valid upper bound on the optimum solution to the MPSM problem. Figure 3 illustrates this step.

Step 3: Transfer the matching to a weighted bipartite graph G_D on the duos.

We now construct a bipartite graph $G_D = (D^A \cup D^B, E)$ on the duos using the edges of the matching M_T found on G_T. For every edge $(T_i^A, T_j^B) \in M_T$, we add one or two edges to G_D. Each edge added to G_D has weight $1/2$. Since each duo from the original string is contained in two separate triplets, it can happen that we get two copies of the edge (D_i^A, D_j^B). In this case, we simply merge them into a single edge with weight 1. Edges are added according to the following simple rules:

- If (T_i^A, T_j^B) is a full edge, we add the edges (D_i^A, D_j^B) and (D_{i+1}^A, D_{j+1}^B) to G_D.
- If (T_i^A, T_j^B) is a first-half edge, we add the edge (D_i^A, D_j^B) to G_D.

– If (T_i^A, T_j^B) is a last-half edge, we add the edge (D_{i+1}^A, D_{j+1}^B) to G_D.

Recall that T_i^A and D_i^A refer to the triplet and duo, respectively, starting at letter a_i in the string A and the duos D_i^A and D_{i+1}^A are both subsets of the triplet T_i^A. If the triplet edge (T_i^A, T_j^B) causes duo edges (D_i^A, D_j^B) or (D_{i+1}^A, D_{j+1}^B), we say that the triplets *support* the duo edges. The extra '&' characters are discarded in this step since by definition, they can't be part of any pair of matched duos.

Step 4: Use G_D to find a mapping from string A to string B.

In this step, we select a subset of the edges in G_D to be the duos preserved in our final mapping solution S. This step happens in three phases, each of which may include many iterations. Each iteration of a phase removes edges from G_D corresponding to one or two pairs of duos preserved as well as any conflicting edges. The first two phases each remove all instances of a particular structure from the graph while the third phase tries to preserve as many duos as possible from the remaining graph.

Phase 1. For each edge $(D_i^A, D_j^B) \in G_D$ with weight 1. We remove (D_i^A, D_j^B) from G_D and map a_i and a_{i+1} to b_i and b_{i+1} in S. We also remove any conflicting edges from G_D.

Phase 2. Define a *pair of consecutive parallel edges* to be edges (D_i^A, D_j^B) and (D_{i+1}^A, D_{j+1}^B) in G_D such that the triplet edge (T_i^A, T_j^B) was chosen in M_T. Starting at the beginning of string A, we choose the first pair of consecutive parallel edges (D_i^A, D_j^B) and (D_{i+1}^A, D_{j+1}^B) in G_D. In other words, we find the smallest i such that (D_i^A, D_j^B) and (D_{i+1}^A, D_{j+1}^B) are a pair of consecutive parallel edges. We map a_i, a_{i+1}, and a_{i+2} to b_i, b_{i+1}, and b_{i+2} in S. We then remove the edges (D_i^A, D_j^B) and (D_{i+1}^A, D_{j+1}^B) from G_D as well as any conflicting edges. We continue this process until we reach the end of string A and no pairs of consecutive parallel edges remain in G_D.

Phase 3. Starting at the beginning of string A, we add the duos of the first edge we encounter to S and remove any conflicting edges. We repeat this step until we reach the end of A and no edges remain in G_D.

3.2   Proof of 13/4-approximation

We will first show that the weight of the maximum weight triplet matching M_T found in Step 2 (and by construction the total weight of G_D) is an upper bound on the maximum number of duos preserved. Then, we will show that the number of preserved duos added to S in each iteration of Step 4 is at least 4/13 of the total weight of edges removed from G_D in that iteration. Finally, we will show that at the end of Phase 3 of Step 4, no edges remain in G_D.

Lemma 1. *The weights of the maximum weight triplet matching M_T and the corresponding duo graph G_D are an upper bound on the maximum number of duos preserved.*

Proof. We show that any proper mapping π from A to B which preserves Δ duos implies a matching M_T of weight at least Δ in the corresponding triplet graph G_T.

For each preserved duo (D_i^A, D_j^B) in π, we add the triplet edges (T_{i-1}^A, T_{j-1}^B) and (T_i^A, T_j^B) to M_T if they have not been added already. Note that in the construction of G_T, (D_i^A, D_j^B) was responsible for adding $1/2$ to the weights of both (T_{i-1}^A, T_{j-1}^B) and (T_i^A, T_j^B) for a total contribution of 1. Thus, if we can guarantee that both triplet edges are added to the matching for each preserved duo, that ensures M_T has weight at least Δ.

Assume for the sake of contradiction that we encounter some preserved duo (D_i^A, D_j^B) and at least one of the triplet edges corresponding to (D_i^A, D_j^B) cannot be added. WLOG assume the triplet edge which cannot be added is (T_i^A, T_j^B) and it is blocked by some other edge (T_i^A, T_ℓ^B), $j \neq \ell$. The edge (T_i^A, T_ℓ^B) must have been added by either the preserved duo (D_i^A, D_ℓ^B) or $(D_{i+1}^A, D_{\ell+1}^B)$. However, both of those duos are in conflict with (D_i^A, D_j^B) and therefore could not exist in the mapping π, leading to a contradiction. It follows that both triplet edges are added to the matching for each preserved duo in π. $\square$

Lemma 2. *The number of preserved duos added to S in each iteration of Phase 1 of Step 4 is at least $1/3$ the total weight of edges removed from G_D in that iteration.*

Proof. The worst case structure for this phase is illustrated in Fig. 4.

Suppose some edge (D_i^A, D_j^B) has weight 1 in G_D. Then both triplet edges (T_{i-1}^A, T_{j-1}^B) and (T_i^A, T_j^B) containing (D_i^A, D_j^B) must have been chosen in the matching M_T. Therefore, there can be no conflicts of Type 1.

Note that the edge (D_i^A, D_j^B) can have at most four conflicts of Type 2 arising from the neighboring duos D_{i-1}^A, D_{i+1}^A, D_{j-1}^B, and D_{j+1}^B. Each of these potential conflicts is symmetric. So WLOG, we focus on the conflict with D_{i-1}^A and show that there is at most one edge (D_{i-1}^A, D_ℓ^B), $\ell \neq j - 1$, and this edge can only have weight $1/2$.

By construction, the edge (D_{i-1}^A, D_ℓ^B) can only be added by the triplet edges $(T_{i-2}^A, T_{\ell-1}^B)$ or $(T_{i-1}^A,, T_\ell^B)$. However, the latter triplet edge (T_{i-1}^A, T_ℓ^B) could not exist in M_T since we assume M_T contains the edge (T_{i-1}^A, T_{j-1}^B) and M_T is a matching. Therefore, there is at most one such edge (D_{i-1}^A, D_ℓ^B) with weight $1/2$ which must have come from a triplet edge $(T_{i-2}^A, T_{\ell-1}^B)$ chosen in the matching M_T.

Then the sum of weights of edges removed is at most the weight (D_i^A, D_j^B) plus the weight of four Type 2 conflicting edges with weight $1/2$ each:

$$1 + 4(1/2) = 3$$

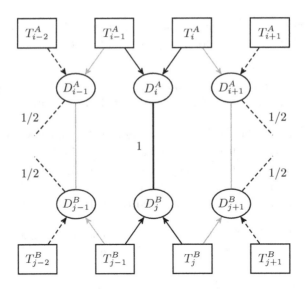

Fig. 4. Illustration of the worst case for Phase 1 of Step 4 in Lemma 2. The solid black lines correspond to the edge (D_i^A, D_j^B) and its supporting triplets. The dashed lines correspond to conflicting edges of Type 2 that must be removed. The gray lines illustrate other edges that may or may not exist, but are not conflicting.

It follows that the ratio of the number of preserved duos added to S to weight of edges removed from G_D is at least $1/3$. □

Note that after Phase 1 of Step 4, all remaining edges in G_D have weight $1/2$ since the edges with weight 1 have been removed.

Lemma 3. *The number of preserved duos added to S in each iteration of Phase 2 of Step 4 is at least $4/13$ of the total weight of edges removed from G_D in that iteration.*

Proof. The worst case structure for this phase is illustrated in Fig. 5.

Suppose we select edges (D_i^A, D_j^B) and (D_{i+1}^A, D_{j+1}^B) in Phase 2. We can upper bound the number of edges removed by identifying all triplets that could support conflicting duo edges and bounding the number of such edges they could have supported. Recall that a triplet supports a duo edge if it belongs to a triplet edge in M_T and thus caused the duo edge to be added to G_D.

First, there are four triplets at distance two from i and j that could each support at most one conflicting edge. These are triplets T_{i-2}^A, T_{i+2}^A, T_{j-2}^B, and T_{j+2}^B. Second, there are four triplets at distance one. Three of these, T_{i+1}^A, T_{j-1}^B, and T_{j+1}^B, can support two conflicting edges. However, the fourth triplet, T_{i-1}^A, can support at most one conflicting edge since we chose the smallest i such that (D_i^A, D_j^B) and (D_{i+1}^A, D_{j+1}^B) are a pair of consecutive parallel edges.

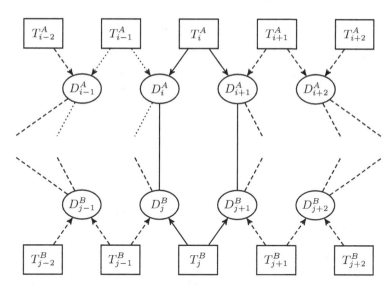

Fig. 5. Illustration of the worst case for Phase 2 of Step 4 in Lemma 3. The solid lines correspond to the edges (D_i^A, D_j^B) and (D_{i+1}^A, D_{j+1}^B) and their supporting triplets. The dashed lines correspond to conflicting edges that must be removed. The pair of parallel dotted lines originating from T_{i-1}^A represent two edges that could not both exist. This is due to the assumption that (D_i^A, D_j^B) and (D_{i+1}^A, D_{j+1}^B) is the first pair of consecutive parallel edges in the string A.

In addition to these conflicting edges, we also remove the two edges (D_i^A, D_j^B) and (D_{i+1}^A, D_{j+1}^B) leading to a total weight removed of

$$4(1/2) + 7(1/2) + 2(1/2) = 6.5$$

It follows that the ratio of the number of preserved duos added to S to weight of edges removed from G_D is at least $2/6.5 = 4/13$.   □

Lemma 4. *The number of preserved duos added to S in each iteration of Phase 3 of Step 4 is at least $1/3$ of the total weight of edges removed from G_D in that iteration.*

Proof. Suppose we select the duo edge (D_i^A, D_j^B) in some iteration of Phase 3. We can upper bound the weight of edges deleted from G_D by counting the number of triplets which could have supported duo edges that conflict with (D_i^A, D_j^B). Recall that a triplet supports a duo edge if it belongs to a triplet edge in M_T and thus caused the duo edge to be added to G_D. Because we have removed all pairs of consecutive parallel edges in Phase 2, each triplet can support at most one duo edge remaining in G_D.

There are eight triplets which could potentially support a conflicting duo edge: T_{i-2}^A, T_{i-1}^A, T_i^A, T_{i+1}^A, T_{j-2}^B, T_{j-1}^B, T_j^B, T_{j+1}^B. Notice that we have been selecting edges starting from the beginning of A and moving towards the end. Therefore any edge supported by the triplet T_{i-2}^A would have already been selected or removed prior to the current iteration. Further note that two of those triplets must support the currently selected edge. Therefore, we removed the selected duo edge (D_i^A, D_j^B) and at most five other duo edges. Each of these edges has weight at most $1/2$ since all edges of weight 1 were removed in Phase 1. Then the sum of weights of edges removed is at most

$$1/2 + 5(1/2) = 3$$

It follows that the ratio of the number of preserved duos added to S to weight of edges removed from G_D is at least $1/3$. $\qquad\square$

Lemma 5. *At the conclusion of Step 4, no edges remain in G_D.*

Proof. Phase 3 iterates through every remaining edge in G_D, thus removing all of them. $\qquad\square$

4   Conclusion and Future Directions

We have shown that a combinatorial triplet matching approach yields an improved approximation to the Maximum Duo Preservation String Mapping problem. Given the fact that triplet matching allows for an improvement over the ratio achieved by the duo matching approach in [4], a natural question is whether a 4-tuple matching could yield even better results. However, a direct extension of the work in this paper to a 4-tuple matching approach is not possible because the 4-tuple matching would not provide an upper bound on the MPSM. The issue with such an approach is that the first and last duos in a 4-tuple have no potential to be conflicting and likely should not be grouped together. On the bright side, we conjecture that the triplet matching approach can be pushed further to achieve a 3-approximation. The clear bottleneck in this paper arises from Phase 2 of Step 4, but we're hopeful this obstacle can be avoided somehow.

Other interesting future directions would be to follow the lead of the work on the MCSP problem. This could include analyzing the performance of faster algorithms such as greedy algorithms or searching for heuristics that solve smaller instances of the problem near optimally. Further, since MPSM currently appears to be "easier" than MCSP, it could be fruitful to explore more applications for this problem in fields such as bioinformatics and data compression.

References

1. Beretta, S., Castelli, M., Dondi, R.: Parameterized tractability of the maximum-duo preservation string mapping problem. CoRR abs/1512.03220 (2015). http://arxiv.org/abs/1512.03220

2. Blum, C., Lozano, J.A., Davidson, P.: Mathematical programming strategies for solving the minimum common string partition problem. Eur. J. Oper. Res. **242**(3), 769–777 (2015). http://www.sciencedirect.com/science/article/pii/S0377221714008716

3. Boria, N., Cabodi, G., Camurati, P., Palena, M., Pasini, P., Quer, S.: A 7/2-approximation algorithm for the maximum duo-preservation string mapping problem. In: Grossi, R., Lewenstein, M. (eds.) CPM 2016. LIPIcs, vol. 54, pp. 11:1–11:8. Schloss Dagstuhl–Leibniz-Zentrum fuer Informatik, Dagstuhl (2016). http://dx.doi.org/10.4230/LIPIcs.CPM.2016.11

4. Boria, N., Kurpisz, A., Leppänen, S., Mastrolilli, M.: Improved approximation for the maximum duo-preservation string mapping problem. In: Brown, D., Morgenstern, B. (eds.) WABI 2014. LNCS, vol. 8701, pp. 14–25. Springer, Heidelberg (2014)

5. Bulteau, L., Fertin, G., Komusiewicz, C., Rusu, I.: A Fixed-parameter algorithm for minimum common string partition with few duplications. In: Darling, A., Stoye, J. (eds.) WABI 2013. LNCS, vol. 8126, pp. 244–258. Springer, Heidelberg (2013)

6. Bulteau, L., Komusiewicz, C.: Minimum common string partition parameterized by partition size is fixed-parameter tractable, Chap. 8, pp. 102–121 (2014). http://epubs.siam.org/doi/abs/10.1137/1.9781611973402.8

7. Chen, W., Chen, Z., Samatova, N.F., Peng, L., Wang, J., Tang, M.: Solving the maximum duo-preservation string mapping problem with linear programming. Theoret. Comput. Sci. **530**, 1–11 (2014). http://www.sciencedirect.com/science/article/pii/S0304397514001108

8. Chen, X., Zheng, J., Fu, Z., Nan, P., Zhong, Y., Lonardi, S., Jiang, T.: Assignment of orthologous genes via genome rearrangement. IEEE/ACM Trans. Comput. Biol. Bioinform. **2**(4), 302–315 (2005)

9. Chrobak, M., Kolman, P., Sgall, J.: The greedy algorithm for the minimum common string partition problem. In: Jansen, K., Khanna, S., Rolim, J.D.P., Ron, D. (eds.) RANDOM 2004 and APPROX 2004. LNCS, vol. 3122, pp. 84–95. Springer, Heidelberg (2004)

10. Cormode, G., Muthukrishnan, S.: The string edit distance matching problem with moves. ACM Trans. Algorithms **3**(1), 2:1–2:19 (2007). http://doi.acm.org/10.1145/1186810.1186812

11. Damaschke, P.: Minimum common string partition parameterized. In: Crandall, K.A., Lagergren, J. (eds.) WABI 2008. LNCS (LNBI), vol. 5251, pp. 87–98. Springer, Heidelberg (2008)

12. Ferdous, S.M., Rahman, M.S.: Solving the minimum common string partition problem with the help of ants. In: Tan, Y., Shi, Y., Mo, H. (eds.) ICSI 2013, Part I. LNCS, vol. 7928, pp. 306–313. Springer, Heidelberg (2013)

13. Ferdous, S.M., Rahman, M.S.: An integer programming formulation of the minimum common string partition problem. PLoS ONE **10**(7), 1–16 (2015)

14. Goldstein, A., Kolman, P., Zheng, J.: Minimum common string partition problem: hardness and approximations. In: Fleischer, R., Trippen, G. (eds.) ISAAC 2004. LNCS, vol. 3341, pp. 484–495. Springer, Heidelberg (2004)

15. Jiang, H., Zhu, B., Zhu, D., Zhu, H.: Minimum common string partition revisited. J. Comb. Optim. **23**(4), 519–527 (2012). http://dx.doi.org/10.1007/s10878-010-9370-2

16. Kolman, P., Waleń, T.: Approximating reversal distance for strings with bounded number of duplicates. Disc. Appl. Math. **155**(3), 327–336 (2007). http://www.sciencedirect.com/science/article/pii/S0166218X0600309X

17. Kolman, P., Waleń, T.: Reversal distance for strings with duplicates: linear time approximation using hitting set. In: Erlebach, T., Kaklamanis, C. (eds.) WAOA 2006. LNCS, vol. 4368, pp. 279–289. Springer, Heidelberg (2007)
18. Swenson, K.M., Marron, M., Earnest-deyoung, J.V., Moret, B.M.E.: Approximating the true evolutionary distance between two genomes. In: Proceedings of 7th SIAM Workshop on Algorithm Engineering and Experiments (ALENEX 2005), p. 121. SIAM Press (2005)

SpecTrees: An Efficient Without a Priori Data Structure for MS/MS Spectra Identification

Matthieu David[1,2(✉)], Guillaume Fertin[1], and Dominique Tessier[2]

[1] LINA UMR CNRS 6241, Université de Nantes, Nantes, France
[2] INRA UR1268 Biopolymères Interactions Assemblages, 44316 Nantes, France

Abstract. Tandem Mass Spectrometry (or MS/MS) is the most common strategy used to identify unknown proteins present in a mixture. It generates thousands of MS/MS spectra per sample, each one having to be compared to a large reference database from which artificial spectra are produced. The goal is to map each experimental spectrum to an artificial one, so as to identify the proteins they come from. However, this comparison step is highly time consuming. Thus, in order to reduce computation time, most methods filter *a priori* the reference database. This tends to discard potential candidates and leads to frequent errors and lacks of identifications. We have developed an original alternate method, efficient both in terms of memory and computation time, that allows to pairwise compare spectra without any *a priori* filtering. The core of our method is SPECTREES, a data structure designed towards this goal, that stores all the input spectra without any filtering. It is designed to be easy to implement, and is also highly scalable and incremental. Once SPECTREES is built, one can run its own identification process by extracting from SPECTREES any information of interest, including pairwise spectra comparison. In this paper, we first present SPECTREES, its main features and how to implement it. We then experiment our method on two sets of experimental spectra from the ISB standard 18 proteins mixture, thereby showing its rapidity and its ability to make identifications that other software do not reach.

1 Introduction

General Context. Proteins play an essential role in many biological processes in living organisms. They are composed of long chains of *amino acids* – up to several thousands. A frequent and important task in experimental biology is to *identify* the proteins that constitute a complex mixture, i.e. determine the sequence of amino acids of each of them. *Tandem Mass Spectrometry*, also called *MS/MS*, has become the most usual technique to achieve this task. The MS/MS process can be roughly sketched as follows: first, an enzyme cuts the proteins of the mixture into smaller pieces, called *peptides*. Then, the mass spectrometer breaks each peptide p into fragments, the masses of the most intense fragments are

Supported by GRIOTE project, funded by Région Pays de la Loire.

M. Frith and C.N.S. Pedersen (Eds.): WABI 2016, LNBI 9838, pp. 65–76, 2016.
DOI: 10.1007/978-3-319-43681-4_6

measured, and from this a *mass spectrum* (that we will model here as an ordered list of the measured masses) is obtained. In a given experiment, thousands of MS/MS spectra are thus generated, and the set S_E of experimental spectra is the one from which identification of proteins is undertaken.

The most common strategy to interpret experimental spectra is through comparison to known databases of related proteins. First, from the database DB at hand, a set of peptides P is computed, by simulating the action of the enzyme on all the proteins in DB. Next, a set of artificial spectra S_P is also created *in silico* from the set of peptides P, using theoretical knowledge about the principles of MS/MS. Thus, for each peptide p in P, an artificial spectrum is created and added to S_P. Once this is done, for each spectrum s in S_E, we look for the artificial spectrum s' in S_P that is the closest to s. Most of the current popular software, such as Sequest [4], Mascot [7] or X!Tandem [3] define the above notion of "closest" in the following intuitive way: maximize some scoring function f that represents the similarity between s and s'. Although f may vary between algorithms, it is essentially based on the number of common masses $Sim(s, s')$ between s and s'. Finally, the set of peptides that have been inferred in the above described process is used to determine which proteins are likely to be in the initial experimental mixture.

Although this process seems simple and easily achievable, identification of peptides, let alone of proteins, remains a very complicated task. Because of chemical noise, chemical modifications, imperfect fragmentation, contaminations in the experiments and lack of mass resolution from the mass spectrometer, experimental spectra are far from being perfect. In addition, protein databases are still incomplete and consequently, artificial spectra corresponding to experimental spectra may be absent. These difficulties lead to partial and sometimes unsatisfactory results: typically, only around 30 % of MS/MS spectra are identified with high confidence. This is why the development of better identification methods remains a very active research area [5]. Moreover, even if the experiment was perfect, the identification process also presents computational limitations. Indeed, in a typical experiment, S_E is composed of 5,000 to 10,000 spectra, while the size of S_P is usually larger than 500,000. Therefore, comparing each $s \in S_E$ to each $s' \in S_P$ requires a huge computational effort. Most of the algorithms thus tend to avoid this complexity by filtering S_P: for each $s \in S_E$, they only consider a small *subset* $S'_P \in S_P$ of artificial spectra to compare s to. S'_P is chosen based on the assumption that the best-scoring artificial spectrum should have a total mass that is close to the total mass of s, m_s; thus, any $s' \in S_P$ whose total mass is too far from m_s is discarded. Unfortunately, this filtering is too stringent, and certainly forbids some identifications. For instance, it is unable to take into account a large number of chemical modifications, called *post-translational modifications*, that yet are very common in proteins. In order to address this problem, some unrestricted database search algorithms have been recently proposed, see e.g. [1,2,8,9]. In spite of recent progress, the development of such methods is still challenging because they remain too time-consuming.

Our Proposal. In this paper, we are mainly interested in the peptide identification process, during which each spectrum from S_E has to be compared to all spectra from S_P. More precisely, we propose a new method to compare sets of spectra of large size without *a priori* restriction (i.e. without filtering) of the search space. At the core of our method lies SPECTREES, an easy to implement data structure that allows to rapidly store the data contained both in S_E and S_P without any filtering, even for very large sets of spectra. Moreover, SPECTREES is incremental: any new set of spectra can be easily added to the currently built structure. SPECTREES is designed so as to efficiently compute $Sim(s, s')$ for any spectra $s \in S_E$ and $s' \in S_P$. As previously mentioned, this number is essential for defining the scoring function that is at the heart of any peptide identification method. Once SPECTREES is built, one can exploit it and run its own identification process, by extracting from SPECTREES any further information of interest. The paper is organized as follows: in Sect. 2, we describe the main features of SPECTREES and detail its construction algorithm. In Sect. 3, we propose a peptide identification method based on SPECTREES, that we experiment on two sets of experimental spectra from the ISB standard 18 proteins mixture [6], and confront to the *Arabidopsis thaliana* protein database. We analyze our results, and in particular compare them to those obtained by one of the most popular software, X!Tandem [3], and show the interest of our method in terms of space occupancy, execution time and richness of the results.

2   From Sets of Spectra to SpecTrees

In this section, we develop and illustrate our algorithm, that takes S_E and S_P as input, and builds our SPECTREES data structure. We then explain how to exploit SPECTREES so as to compute efficiently $Sim(s, s')$ for any $s \in S_E$ and $s' \in S_P$.

Main Features of SpecTrees. Assume that each spectrum in S is identified by a unique integer *s*-id, for *spectrum identifier*. We will suppose that $|S| = N_s$, and for simplicity we will assume that any *s*-id lies in the interval $[0; N_s - 1]$. Conceptually, SPECTREES stores all available information from S in the form of a forest, i.e. a set of trees. Each tree T of SPECTREES is rooted and directed in such a way that T is an in-tree, i.e. any child node c in T is connected to its parent p by an arc from c to p. Thus the root of T is the only node without outgoing arc. Suppose SPECTREES contains n nodes, say $v_0, v_1 \ldots v_{n-1}$, and let each $0 \le i \le n - 1$ be the *node identifier* (or *n*-id) of SPECTREES. Each node v_i in SPECTREES, $0 \le i \le n - 1$, contains two pieces of information: a spectrum identifier *s*-id $\in [0; N_s - 1]$, and a non negative integer cpt_i, which will act as a counter. It is important to note that distinct nodes in SPECTREES may contain the same value *s*-id, and thus each *s*-id may be present several times in SPECTREES. Because we want to access rapidly all the nodes storing the same value *s*-id in the in-trees, we add to SPECTREES a set of linked lists, which we call

Transverse Pathways (or *T-Path*); there will be one list per spectrum identifier *s*-id. We refer the reader to Fig. 1 for an example of a SPECTREES structure, where the dotted red lines represent the T-Paths. Although SPECTREES is conceptually a forest of in-trees to which we superimpose the T-Paths, we have chosen to implement it most simply, in the form of five integer arrays, which we describe now. Arrays $Id[]$, $Cpt[]$ and $Par[]$ are all of length n, indexed from 0 to $n-1$, and for any $0 \leq i \leq n-1$, (i) $Id[i]$ contains the *s*-id of node v_i, (ii) $Cpt[i]$ contains the counter cpt_i of node v_i and (iii) $Par[i]$ contains the *n*-id j of the node v_j which is the parent of v_i in the in-tree T that contains v_i ($Par[i]$ is set to -1 if v_i is the root of T). The T-Paths are not implemented as actual linked lists, but through two arrays $TP[]$ and $Next[]$, again for simplicity, and also for code optimization purposes. $Next[]$ is of length n, and for any $0 \leq i \leq n-1$, $Next[i]$ contains the *n*-id j of the node v_j that follows v_i in the T-Path. We set $Next[i]$ to -1 when we have reached the end of the list. In order to access the *first* node of every linked list, a fifth array called $TP[]$, of length N_s, is defined. It works as follows: for any $0 \leq i \leq N_s - 1$, $TP[i] = k$, where k is the smallest integer such that v_k stores a value *s*-id $= i$. Note that v_k is also the leftmost node in our forest of in-trees that stores *s*-id $= i$ (see Fig. 1). Now, in order to access *all nodes* that store *s*-id $= i$, it suffices to start with $TP[i]$, then to use $Next[]$ iteratively until we reach the value -1.

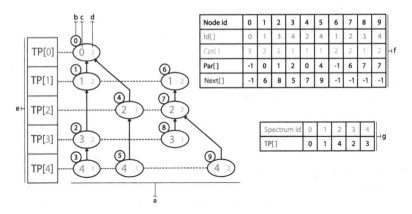

Fig. 1. The different components of SPECTREES. (a) is the forest of in-trees. Each node possesses a (black) *n*-id (b), and contains both a (blue) *s*-id (c) and a (green) counter (d). T-Paths are represented as dotted red lines (e). (f) and (g) are the arrays that encode SPECTREES. (Color figure online)

Building SpecTrees. The algorithm that builds SPECTREES from any set $\mathcal{S}$ of spectra works in two steps. First, we operate a pre-processing of $\mathcal{S}$, that we call *Bucket Clustering*. Second, we create the in-trees of SPECTREES (together with the T-Paths), with the help of the clusters (the *buckets*) that we computed during the Bucket Clustering.

Bucket Clustering. Let us denote by $\mathcal{M_S}$ the set of distinct masses that are present over all the spectra present in $\mathcal{S}$, and let $N_m = |\mathcal{M_S}|$. We can then write $\mathcal{M_S} = \{m_0, m_1 \ldots m_{N_m}\}$. The *Bucket Clustering* pre-processing step creates a set BC of buckets, where each bucket $B_i \in BC$, $0 \leq i \leq N_m - 1$, is a set which contains all the *s*-ids of the spectra that contain m_i in their set of masses. Moreover, we impose that each bucket B_i is sorted in increasing order of the *s*-ids it contains. In practice, this sorting is obtained at no extra cost, because spectra from S_E are already given in this order when they come out of the mass spectrometer. Spectra from S_P are produced *in silico*, thus can be indexed in such a way as to respect this property too. Next, the set BC of buckets is also sorted by lexicographic order, and we denote this new order as follows: $BC = \{B'_0, B'_1 \ldots B'_{N_m - 1}\}$. For instance, if $\mathcal{S} = \{s_0, \ldots, s_4\}$ with $s_0 = \{m_0, m_1, m_3\}$, $s_1 = \{m_1, m_2, m_3, m_4\}$, $s_2 = \{m_0, m_2, m_4\}$, $s_3 = \{m_1, m_3, m_4\}$ and $s_4 = \{m_0, m_1, m_2\}$, then $BC = \{B'_0, B'_1, B'_2, B'_3, B'_4\}$, with $B'_0 = \{0, 1, 3\}$, $B'_1 = \{0, 1, 3, 4\}$, $B'_2 = \{0, 2, 4\}$, $B'_3 = \{1, 2, 3\}$, $B'_4 = \{1, 2, 4\}$. This pre-processing step has a complexity of $O(\sum_{s_i \in \mathcal{S}} |s_i| + N_m \log N_m)$. If we consider that any spectrum contains a constant number of masses (which is the case in practice) then $\sum_{s_i \in \mathcal{S}} |s_i| = O(N_s)$ and $N_m = O(N_s)$ too, and the previous complexity becomes $O(N_s \log N_s)$.

Construction of SPECTREES. The second step consists in building SPECTREES, based on the bucket clustering BC we just computed. We refer to Algorithm 1 for a detailed description. We roughly sketch the algorithm here: we scan through the buckets $B'_0, B'_1 \ldots B'_{N_m - 1}$, following the order of the buckets, and inside each bucket following the (increasing) order of *s*-ids. Then, depending on the contents found, we either follow an already constructed branch of SPECTREES (and update the counters contained in its nodes), or create either a new branch or a new tree. First, bucket B'_0 is used for the initialization step (lines 1–4): starting from an empty structure, we create an in-tree T which is actually an in-path of root r, that contains as many nodes as there are elements in B'_0. The *s*-id stored in each node (from r to the unique leaf of T) is the *s*-id found in B'_0 (read from left to right). All *cpt* values of these nodes are set to 1. Then, each bucket B'_i, $1 \leq i \leq N_m - 1$ is compared to B'_{i-1}: if the first values of B'_i and B'_{i-1} differ, create a new in-tree T' the same way it has been done for B'_0. Otherwise, let $CP = \{x_1, x_2 \ldots x_c\}$, $c > 0$, be the longest common prefix of B'_i and B'_{i-1}, where any x_p, $1 \leq p \leq c$ is an *s*-id (lines 7–9). Then, follow in SPECTREES the latest produced tree, from its root (which necessarily contains x_1), and along the branch whose nodes share the same *s*-id, while incrementing by 1 the counters that have been met. Whenever x_{c+1} (the first *s*-id in B'_i that is not in CP) is encountered, create a new branch starting from the latest visited node, again in the same way as was done for B'_0 (lines 10–15). Note that *pre* is used as an index keeping track of the previously accessed node, in order to define it as a parent in $Par[]$ when we insert a new node which is not a tree root. Besides, every time a new node is created, it is added to the T-Paths, by updating $Next[]$ and/or $TP[]$ (lines 3 and 13–15). As T-Paths simulate linked lists, adding an element at the end of a list would require to go through the entire list, and thus would be too time-consuming. In order to avoid this, we use array $Last[]$ to keep track of the s-id of the last node inserted in each list. As mentioned in the previous section, SPECTREES is managed through five integer arrays, which makes Algorithm 1 very easy to implement. Moreover, the complexity of Algorithm 1 is clearly in $O(N_m)$, which, as discussed in the previous paragraph, may be assumed to be an $O(N_s)$, since N_s and N_m differ in practice by a constant factor.

Algorithm 1. Building SPECTREES from the buckets

Input : A bucket clustering $BC = \{B'_0, B'_1 \ldots B'_{N_m-1}\}$.
Output: Structure SPECTREES, i.e. integer arrays $Id[]$, $Cpt[]$, $Par[]$, $Next[]$ of size n, and $TP[]$ of size N_s.
Variables: Integer array $Last[]$ of size N_s, integers $pre = -1$, $in = 0$ and pos.

1: **for all** $id \in B'_0$ **do**   ▷ A node is inserted for each element in the first bucket
2:     $Id[in] \leftarrow id$; $Cpt[in] \leftarrow 1$; $Par[in] \leftarrow pre$;
3:     $TP[id] \leftarrow in$; $Last[id] \leftarrow in$;
4:     $pre \leftarrow in$; $in \leftarrow in + 1$;
5: **for** i from 1 to $N_m - 1$ **do**   ▷ For each bucket, parent and position are reset
6:     $pre \leftarrow -1$; $pos \leftarrow 0$;
7:     **while** $B_i[pos] = B_{i-1}[pos]$ **do**   ▷ B_i and B_{i-1} share a common prefix
8:         $Cpt[Last[B_i[pos]]] \leftarrow Cpt[Last[B_i[pos]]] + 1$;
9:         $pre \leftarrow Last[B_i[pos]]$; $pos \leftarrow pos + 1$;
10:     **while** $pos < |B_i|$ **do**   ▷ Now the rest of B_i differs from B_{i-1}
11:         $Id[in] \leftarrow B_i[pos]$; $Cpt[in] \leftarrow 1$; $Par[in] \leftarrow pre$;
12:         $Next[Last[B_i[pos]]] \leftarrow in$;
13:         $Last[B_i[pos]] \leftarrow in$;
14:         **if** $TP[B_i[pos]] = -1$ **then**
15:             $TP[B_i[pos]] \leftarrow in$;
16:         $pre \leftarrow in$; $in \leftarrow in + 1$; $pos \leftarrow pos + 1$;

Exploiting SpecTrees to Compute $Sim(s_i, s_j)$. Now that SPECTREES has been built from a set of spectra $\mathcal{S}$, it is possible to compute $Sim(s_i, s_j)$ for any $s_i, s_j \in \mathcal{S}$, using the algorithm we describe in this section. We will assume for simplicity that the s-id of s_i (resp. s_j) is equal to i (resp. j). Note that for a fixed value i, our algorithm actually computes $Sim(s_i, s_j)$ for *all* values j such that $j < i$, at no extra cost. For this, we use an integer accumulator array $Acc[]$, of length N_s, with all values initialized to -1. At the end of the process $Acc[j]$ will contain a value k which is exactly $Sim(s_i, s_j)$. The rough idea of the algorithm is the following: we find all nodes of $v_{x_1}, v_{x_2} \ldots v_{x_t}$ of SPECTREES that contain i – this can be easily done by our T-Paths. Then, for each node v_{x_p}, $1 \leq p \leq t$, we follow the path from v_{x_p} to the root of the tree it belongs to. Whenever a node with s-id $= j$ is encountered, we add cpt_{x_p} to $Acc[j]$. For instance, in the example given in Fig. 2(a) with $i = 3$, on the path from node with n-id $= 3$ to the root (n-id$=0$) in the leftmost tree, nodes with s-id$=1$ and s-id$=0$ are encountered, thus after this path is visited we have $Acc[0] = Acc[1] = 2$, since the counter value of node with n-id $= 3$ equals 2. We claim that, at the end of the process, $Acc[j]$

contains exactly $Sim(s_i, s_j)$: indeed, in SPECTREES, given a node v storing an s-id $= i$, only nodes storing an s-id $= j$ with $j < i$ appear above v in its tree – this is due to the fact that buckets in BC are sorted in increasing order. Moreover, if two nodes are in the same branch of a tree, they necessarily have at least one mass in common, and the number of common masses is given by the value of the counter of the lowest of the two nodes in this branch.

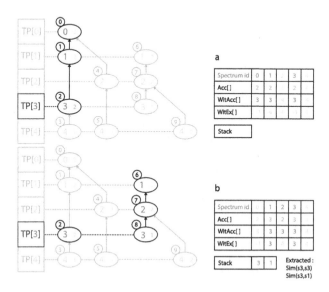

Fig. 2. Computing $Sim(s_3, s_j)$, for all $j < 3$, from SPECTREES. (a) After the first branch is visited. (b) After the second branch is visited.

Visiting all nodes in SPECTREES that store the same s-id $= i$ can be achieved easily: the leftmost node is stored in $TP[i]$, and array $Next[]$ allows to go from one node to the next in constant time. Besides, following branches up to the root is achieved using array $Par[]$. The above description allows to obtain, for a fixed i, all values of $Sim(s_i, s_j)$ with $j < i$. The algorithm that we implemented (and that is described in Algorithm 2) realizes an even bigger task: it actually computes $Sim(s_i, s_j)$ for *all i and all j* (see the **for** loop line 5). To this aim, we use three extra structures: one integer array $WitAcc[]$ ("Wit" stands for "Witness"), of length N_s, that basically avoids reinitializing $Acc[]$ when the value of i changes. We also have a stack St that stores all indices j of $Acc[]$ such that $Acc[j] \neq -1$, i.e. St contains only those j for which $Sim(s_i, s_j) \neq 0$. In practice, this speeds up computations, since the length N_s of $Acc[]$ is very large, and using St avoids scanning through the whole array $Acc[]$ to display similarities of interest. Finally, array $WitEx[]$ is used to avoid duplicates in St. At the end of each iteration for i, stack St is emptied (for instance, the result for this specific value of i may be copied elsewhere), so as to be reused for the next value of i. Note that the interest of our implementation also lies in the fact

---

Algorithm 2. Extracting similarities from SPECTREES

Input: Structure SPECTREES, i.e. integer arrays $Id[]$, $Cpt[]$, $Par[]$, $Next[]$ of size n, and $TP[]$ of size N_s.

Output: Res a set of triples of integers $(s_i, s_j, Sim(s_i, s_j))$.

Variables: Integer arrays $Acc[]$, $WitAcc[]$, $WitEx[]$ of size N_S, integer stack St, integers $start$, pos, τ and $temp$.

1: **for** i from 0 to $N_s - 1$ **do**   ▷ Initialization
2:     $Acc[i] \leftarrow -1$; $WitAcc[i] \leftarrow -1$; $WitEx[i] \leftarrow -1$;
3: $St \leftarrow \emptyset$; $Res \leftarrow \emptyset$;
4: **for** i from $N_s - 1$ to 0 **do** ▷ Iteration over the spectra indices in decreasing order
5:     $start \leftarrow TP[i]$;
6:     **while** $start \neq -1$ **do**   ▷ Until the end of the 'linked list' is reached
7:         $pos \leftarrow start$;
8:         **while** $pos \neq -1$ **do**   ▷ Until the root of the tree is reached
9:             **if** $WitAcc[Id[pos]] = i$ **then**   ▷ If the accumulator has been modified this iteration we accumulate values
10:                 $Acc[Id[pos]] = Acc[Id[pos]] + Cpt[start]$;
11:                 **if** $Acc[Id[pos]] \geq \tau$ $\&\&WitEx[Id[pos]] \neq i$ **then**   ▷ If an accumulator goes over the threshold put the index in the stack and track it
12:                     $push(St, Id[pos])$;
13:                     $WitEx[Id[pos]] = i$;
14:             **else**   ▷ Otherwise we start a new sum
15:                 $Acc[Id[pos]] = Cpt[start]$;
16:                 $WitAcc[Id[pos]] = i$;
17:                 **if** $Acc[Id[pos]] \geq \tau$ $\&\&WitEx[Id[pos]] \neq i$ **then**   ▷ If an accumulator goes over the threshold put the index in the stack and track it
18:                     $push(St, Id[pos])$;
19:                     $WitEx[Id[pos]] = i$;
20:             $pos \leftarrow Par[pos]$;
21:         $start \leftarrow Next[start]$;
22:     **while** $St \neq \emptyset$ **do**
23:         $temp \leftarrow pop(St)$;   ▷ Unstack and create similarity measures
24:         $Res \leftarrow Res \cap (s_i, s_{temp}, Acc[temp])$;
25: **return** Res;

---

that for any new value of i, no specific initialization or structure allocation is necessary. Thus we avoid these procedures, which are time-consuming due to the size of the data we consider in practice. We also added an extra feature to Algorithm 2. Indeed, in practice, for a given i, the number of spectra for which $Sim(s_i, s_j) \neq 0$ is exceedingly large, and the fact that two spectra have *at least one* mass in common is not informative enough. Hence, it is desirable to introduce a parameter τ that acts as a threshold, and to memorize only those spectra s_j such that $Sim(s_i, s_j) \geq \tau$. Thus, we added τ as a tunable parameter (see lines 11 and 17 of Algorithm 2) in order to provide such flexibility. Similarly to Bucket Clustering and Algorithm 1, we chose to implement Algorithm 2 with simple structures, namely a few arrays and one stack. Let us now turn to the complexity analysis of Algorithm 2. For a fixed value $0 \leq i \leq N_s - 1$, Algorithm 2 is linearly dependent of n_i, where n_i is the total number of nodes of SPECTREES which lie above the nodes having s-id $= i$ in the in-trees that compose SPECTREES. Theoretically, it is difficult to determine precisely n_i, and all can be said is that $n_i = O(n)$, where n is the total number of nodes in SPECTREES. Thus the overall complexity of Algorithm 2 is in $O(n \cdot N_s)$, but we will see in the following section that, in practice, its behavior is much more efficient.

3   Experiments

The tests we propose in this section have two purposes: first, show that SPEC-TREES is rapidly built, does not require too much memory space, and that computing function $Sim()$ can be done efficiently. Second, show that even with a very basic usage of $Sim()$, SPECTREES allows peptide identifications that other software are unable to discover.

Data Collection. We downloaded the ISB standard 18 proteins mixture (or 18mix) [6] from https://regis-web.systemsbiology.net/PublicDatasets/ in order to obtain two different sets S_E. The first experimental dataset datA contains about 3,000 spectra provided by the 18mix. The second dataset datB contains about 9,000 spectra generated by triplicating datA: datB is thus large enough to represent a classical MS/MS experiment. The database used for peptide identification is a set S_P of roughly 500,000 artificial spectra obtained from the addition of the 18mix proteins, the most common contaminant proteins and the *Arabidopsis thaliana* database (release 2012). Our algorithms were written in Java, and our tests were executed on a workstation (Intel Xeon E3-1245, 8 MB cache, 16 GB RAM) running the 64-bits Windows 7 operating system.

Construction of SPECTREES *and Computation of* $Sim(s_i, s_j)$. The first series of tests was dedicated to SPECTREES itself. We evaluated independently (a) the execution time and the memory requirements for the Bucket Clustering algorithm and the construction of SPECTREES (see Table 1), and (b) the execution time and volume of data generated for the pairwise computation of spectra similarities depending of the value of the threshold τ (see Table 2).

From Table 1, what we first note is that our method is extremely fast, as it takes only a few seconds to build SPECTREES from scratch, on realistic datasets,

Table 1. Time and space needed to build SPECTREES from our two datasets.

	SPECTREES		
	Bucket clustering (s)	Construction (s)	Memory occupancy (Gb)
datA $+ S_P$	1.4	1.3	1.48
datB $+ S_P$	1.4	4.3	1.49

Table 2. Execution time and volume of the data generated by Algorithm 2 from SPECTREES, depending on the threshold τ.

Threshold τ	Execution time				Volume of generated data (Mb)	
	With storage (s)		Without storage (s)			
	datA + S_P	datB + S_P	datA + S_P	datB + S_P	datA + S_P	datB + S_P
6	211	631	26	81	1324	3923
8	167	498	26	79	98	291
10	165	494	26	80	5	15
12	164	490	26	79	0.2	0.6
14	162	486	26	79	0.02	0.05
16	163	485	26	79	0.008	0.02

Table 3. Execution time for X!Tandem depending on the mass tolerance parameter.

Mass filter tolerance (Daltons)	±1	±5	±25	±50	±100	±200	±500	±1000
	Execution time (s)							
datA + S_P	6	16	65	125	251	492	1156	1987
datB + S_P	17	44	185	360	698	1414	3348	5689

Table 4. Examples of peptides identified by SPECXTRACT that may not be identified by X!Tandem.

Identified peptide	Number of occurrences	Minimum mass tolerance (Daltons)	Maximum mass tolerance (Daltons)
GAGAFGYFEVTHDITR	5	5.37×10^{-4}	437.187
GGADATEDVTAVEVDPADR	3	0.001	786.303
SVDDYQECYIAMVPSHAVVAR	11	0.002	215.135
VPQVSTPTIVEVSR	1	128.093	128.093
PDYVTDSAASATAWSTGVK	9	286.152	368.104
IQGGFVWDWVDQSIIK	1	617.051	617.051

even on a regular workstation. The lack of significant difference between the processing of datA and datB is explained by the relatively small difference between datA and datB in terms of number of spectra, compared to the size of S_P.

Table 2 shows the execution time for the computation of $Sim(s_i, s_j)$ for all pairs of spectra in SPECTREES with $s_i \in S_E$ and $s_j \in S_P$, for different values of the threshold τ (see Algorithm 2). As writing the results on the hard-drive is extremely slow, in order to assess the rapidity of the computation itself, we also provide the execution times when storage is excluded. What we can see from Table 2 is that low values of the threshold τ generate a considerable amount of similarities, the large majority of which being irrelevant, and this gives rise to an important computational overtime; however, when $\tau \geq 8$, the execution time is stabilized. Nevertheless, extracting the similarities *without* adding an intermediate writing step drastically reduces the execution time, even with a small τ. This shows that the computation of the pairwise similarities in itself is very efficient, and is only slowed down by the storage process.

Identification of Peptides Using SPECTREES. In a second series of tests, we compared a very basic peptide identification procedure, that we called SPECXTRACT, to X!Tandem (version Sledgehammer), a widely-used peptide identification software specifically designed to be fast [3]. SPECXTRACT does the following: for every experimental spectrum $s_i \in S_E$, SPECXTRACT returns the peptides in S_P that share a *maximum number* of common masses with s_i. This procedure is very basic in that it does not incorporate any additional information. Our goal here is to test whether, even with such a simple analysis, the absence of a priori filtering in SPECTREES is sufficient to provide identifications that other software cannot provide. We also want to compare SPECXTRACT to X!Tandem in terms of execution time.

Note that X!Tandem uses a notion of *mass tolerance* in order to filter the artificial spectra. We thus evaluated the execution time required by X!Tandem to analyze datA and datB according to different mass tolerances. The concept of mass tolerance is a filtering process which works as follows: let $s_i \in S_E$ and let m_i be the total mass (expressed in Daltons, or Da) of the peptide associated to s_i. Given a value $\Delta \in \mathbb{R}^+$ (the mass tolerance), the idea is to identify the peptide associated to s_i by only considering those spectra in S_P whose total mass lies in the interval $[m_i - \Delta : m_i + \Delta]$. This filtering is widely used, as discussed in Sect. 1, to reduce the number of computations of $Sim()$. We configured the X!Tandem parameters so that we could compare SPECXTRACT and X!Tandem under close conditions. Table 3 shows our results, that have to be compared to the ones presented in Table 2. It can be seen that on small datasets and with restricted filters, X!Tandem performs faster than SPECXTRACT, but the running time increases proportionally to the number of spectra and the mass tolerance. In particular, as soon as $\tau = 8$ (a reasonable threshold in practice), SPECXTRACT outperforms X!Tandem whenever $\Delta \geq 100\,Da$, and competes reasonably well when $\Delta = 50\,Da$. Moreover, as illustrated by some examples in Table 4, a large number of spectra cannot be identified with a low mass tolerance. In Table 4, each line lists a given peptide p, whose amino-acid sequence is given in the first column, and the "*Number of occurrences*" column indicates how many spectra in S_E have led to identify p. The last two columns indicate the minimum (resp. maximum) value one has to set Δ to in order to detect at least one spectrum (resp. all spectra) that led to identify p. Note that with a standard mass tolerance $\Delta = 1\,Da$, only 17 % of the spectra found a match in datA with X!Tandem. Table 4 shows that X!Tandem misses some identifications, either completely (see the last three peptides), or partially (i.e. not all occurrences of p will be found, thus the confidence that p is an actual peptide may be underestimated – see the first three peptides). However, SPECXTRACT is able to identify these peptides with the right number of occurrences, since it does not filter spectra before comparing them.

4   Conclusion

SPECTREES is a very well designed data structure that allows to provide rapidly the number of common masses between experimental and theoretical spectra in an MS/MS experiment. In a first approach, we developed a simple extraction algorithm to demonstrate the potential of SPECTREES. The promising results of SPECXTRACT call for going further in order to fit a more complete identification demand, e.g. by designing a more elaborate scoring function, and by associating some degree of confidence to our results using statistical measurements. This line of work is currently followed by the authors.

References

1. Chi, H., He, K., Yang, B., Chen, Z., Sun, R.-X., Fan, S.-B., Zhang, K., Liu, C., Yuan, Z.-F., Wang, Q.-H., Liu, S.-Q., Dong, M.-Q., He, S.-M.: pFind-Alioth: a novel unrestricted database search algorithm to improve the interpretation of high-resolution MS/MS data. J. Proteomics **125**, 89–97 (2015)
2. Cliquet, F., Fertin, G., Rusu, I., Tessier, D.: Comparison of spectra in unsequenced species. In: Guimarães, K.S., Panchenko, A., Przytycka, T.M. (eds.) BSB 2009. LNCS, vol. 5676, pp. 24–35. Springer, Heidelberg (2009)
3. Craig, R., Beavis, R.C.: TANDEM: matching proteins with tandem mass spectra. Bioinformatics **20**(9), 1466–1467 (2004). (Oxford, England)
4. Eng, J.K., McCormack, A.L., Yates, J.R.: An approach to correlate tandem mass spectral data of peptides with amino acid sequences in a protein database. J. Am. Soc. Mass Spectrom. **5**(11), 976–989 (1994)
5. Käll, L., Vitek, O.: Computational mass spectrometry-based proteomics. PLoS Comput. Biol. **7**(12), e1002277 (2011)
6. Klimek, J., Eddes, J.S., Hohmann, L., Jackson, J., Peterson, A., Letarte, S., Gafken, P.R., Katz, J.E., Mallick, P., Lee, H., Schmidt, A., Ossola, R., Eng, J.K., Aebersold, R., Martin, D.B.: The standard protein mix database: a diverse dataset to assist in the production of improved peptide and protein identification software tools. J. Proteome Res. **7**(1), 96–103 (2008)
7. Perkins, D.N., Pappin, D.J., Creasy, D.M., Cottrell, J.S.: Probability-based protein identification by searching sequence databases using mass spectrometry data. Electrophoresis **20**(18), 3551–3567 (1999)
8. Pevzner, P.A., Dancik, V., Tang, C.L.: Mutation-tolerant protein identification by mass spectrometry. J. Comput. Biol. **7**(6), 777–787 (2000)
9. Tanner, S., Payne, S.H., Dasari, S., Shen, Z., Wilmarth, P.A., David, L.L., Loomis, W.F., Briggs, S.P., Bafna, V.: Accurate annotation of peptide modifications through unrestrictive database search. J. Proteome Res. **7**(1), 170–181 (2008)

Predicting Core Columns of Protein Multiple Sequence Alignments for Improved Parameter Advising

Dan DeBlasio$^{(\boxtimes)}$ and John Kececioglu

Department of Computer Science
The University of Arizona, Tucson AZ 85721, USA
{deblasio,kece}@cs.arizona.edu

Abstract. In a computed protein multiple sequence alignment, the *coreness* of a column is the fraction of its substitutions that are in so-called core columns of the gold-standard reference alignment of its proteins. In benchmark suites of protein reference alignments, the core columns of the reference are those that can be confidently labeled as correct, usually due to all residues in the column being sufficiently close in the spatial superposition of the folded three-dimensional structures of the proteins. When computing a protein multiple sequence alignment in practice, a reference alignment is not known, so its coreness can only be predicted.

We develop for the first time a *predictor* of column coreness for protein multiple sequence alignments. This allows us to predict which columns of a computed alignment are core, and hence better estimate the alignment's accuracy. Our approach to predicting coreness is similar to nearest-neighbor classification from machine learning, except we transform nearest-neighbor distances into a coreness prediction via a regression function, and we learn an appropriate distance function through a new optimization formulation that solves a large-scale linear programming problem. We apply our coreness predictor to *parameter advising*, the task of choosing parameter values for an aligner's scoring function to obtain a more accurate alignment of a specific set of sequences. We show that for this task, our predictor strongly outperforms other column-confidence estimators from the literature, and affords a substantial boost in alignment accuracy.

1 Introduction

The accuracy of a multiple sequence alignment computed on a benchmark set of input sequences is usually measured with respect to a *reference alignment* that represents the gold-standard alignment of the sequences. For protein sequences, reference alignments are typically determined by structural superposition of the known three-dimensional structures of the proteins in the benchmark. The accuracy of a computed alignment is then defined to be the fraction of pairs of residues aligned in the so-called *core columns* of the reference alignment that are also present in columns of the computed alignment. Core columns represent those in the reference that are deemed to be reliable, and can be objectively

© Springer International Publishing Switzerland 2016
M. Frith and C.N.S. Pedersen (Eds.): WABI 2016, LNBI 9838, pp. 77–89, 2016.
DOI: 10.1007/978-3-319-43681-4_7

defined as those columns containing a residue from every input sequence such that the pairwise distances between these residues in the structural superposition of the proteins are all within some threshold (typically a few angstroms). In short, given a known reference alignment whose columns are labeled as either core or non-core, we can determine the accuracy of any other computed alignment of its proteins by evaluating the fraction of aligned residue pairs from these core columns that are recovered. For a given column in a computed alignment, we can also define the *coreness* value of the column to be the fraction of its aligned residue pairs that are in core columns of the reference alignment. (Note that column coreness is a fully objective quantity when core columns are identified through superposition of protein structures, as in `PALI` [1] benchmarks.) A coreness value of 1 means the column of the computed alignment corresponds to a core column of the reference alignment.

When aligning sequences in practice, obviously such a reference alignment is not known, and the accuracy of a computed alignment, or the coreness of its columns, can only be estimated. A good *accuracy estimator* for computed alignments is extremely useful [7]. It can be leveraged to: pick among alternate alignments of the same sequences the one of highest estimated accuracy, for example, to choose good parameter values for an aligner's scoring function as in *parameter advising* [15]; or to select the best result from a collection of different aligners, yielding a natural *ensemble aligner* that can be far more accurate than any individual aligner in the collection [5].

Similarly, a good *coreness predictor* for columns in a computed alignment can be used to: mask out unreliable regions of the alignment before computing an evolutionary tree; or to improve an alignment accuracy estimator by concentrating its evaluation function on columns of higher predicted coreness, thereby boosting the performance of parameter advising. In fact, in principle a perfect coreness predictor would itself yield an ideal accuracy estimator.

In this paper, we develop for the first time a column coreness predictor for protein multiple sequence alignments. Our approach to predicting coreness is similar in some respects to nearest-neighbor classification from machine learning, except we transform nearest-neighbor distance into a coreness prediction via a regression function, and we learn an appropriate distance function through a new optimization formulation that solves a large-scale linear programming problem. We evaluate the performance of our new coreness predictor by applying it to the task of parameter advising in multiple sequence alignment.

Related Work. To our knowledge, this is the first fully general attempt to directly predict the coreness of columns in computed protein alignments. Tools are available that assess the quality of columns in an alignment, and can be categorized into: (a) those that only identify columns as unreliable, for removal from further analysis; and (b) those that compute a column quality score, which can be thresholded. Tools that simply mask unreliable columns include `GBLOCKS` [3], `TrimAL` [2], and `ALISCORE` [16]. Popular quality-score tools are `Noisy` [8], `ZORRO` [21], `TCS` [4], and `GUIDANCE` [17].

Our experiments compare our coreness predictor to TCS and ZORRO: the most recent tools that provide quality scores, as opposed to masking columns. GUIDANCE requires four or more sequences, which excludes many benchmarks. Noisy is dominated by an earlier version of GUIDANCE, which along with ALISCORE and GBLOCKS are in turn dominated by ZORRO.

Plan of the Paper. Section 2 next describes how we learn our coreness predictor. Section 3 then explains how we use predicted coreness to improve accuracy estimation for protein alignments. Section 4 evaluates our approach to coreness prediction by applying the improved accuracy estimator to alignment parameter advising. Section 5 concludes.

2   Learning a Coreness Predictor

To describe how we learn a column coreness predictor, we first discuss our *representation* of alignment columns, and our grouping of consecutive columns into *window classes*. We then present our *regression function* for predicting coreness, which transforms the nearest-neighbor distance from a window to a class into a coreness value. Finally, we describe how we learn this window distance function by solving a large-scale *linear programming* problem.

2.1   Representing Alignment Columns

The information used by our coreness predictor, beyond the multiple sequence alignment itself, is an annotation of its protein sequences by predicted secondary structure (which can be obtained in a preprocessing step by running the sequences through a standard protein secondary structure prediction tool such as PSIPRED [12]). When inputting a column from such an annotated alignment to our coreness predictor, we need a column representation that, while capturing the association of amino acids and predicted secondary structure types, is also independent of the number of sequences in the column. This is necessary as our predictor will be trained on example alignments of particular sizes, yet the resulting predictor must apply to alignments with arbitrary numbers of sequences.

Let Σ be the 20-letter amino acid alphabet, and $\Gamma = \{\alpha, \beta, \gamma\}$ be the secondary structure alphabet, corresponding respectively to types α-*helix*, β-*strand*, and *other* (also called *coil*). We encode the association of an amino acid $c \in \Sigma$ with its predicted secondary structure type $s \in \Gamma$ by an ordered pair (c, s) that we call a *state*, from the set $Q = (\Sigma \times \Gamma) \cup \{\xi\}$. Here $\xi = (\text{-}, \text{-})$ is the *gap state*, where the dash symbol '-' $\notin \Sigma$ is the alignment *gap character*.

We represent a multiple alignment column as a distribution over the set of states Q, which we call its *profile* (mirroring standard terminology [9, p. 101]). We denote the profile C for a given column by a function $C(q)$ on states $q \in Q$ satisfying $C(q) \geq 0$ and $\sum_{q \in Q} C(q) = 1$. Most secondary structure prediction tools output a confidence value (not a true probability) that an amino acid in a protein sequence has a given secondary structure type. For a column of amino acids $(c_1 \cdots c_k)$ in a multiple alignment of k sequences, denote the *confidence* that amino acid c_i has structure type $s \in \Gamma$ by $p_i(s) \geq 0$, where $\sum_{s \in \Gamma} p_i(s) = 1$. For non-gap state $q = (a, s) \neq \xi$, profile C has value $C(q) := \sum_{i \,:\, c_i = a} p_i(s) / k$. In other words, $C(q)$ is the normalized total confidence across the column in state $q \neq \xi$. For gap state $q = \xi$, the profile value is $C(\xi) := \big|\{i \,:\, c_i = \text{`-'}\}\big| / k$, the relative frequency of gap characters in the column.

2.2   Classes of Column Windows

In protein benchmarks, a column of a reference alignment is labeled core if its residues are all sufficiently close in the superposition of the proteins' three-dimensional structures. The folded structure around a residue is a function of nearby residues in the protein. Consequently, to predict the coreness of a column in a computed alignment, we need contextual information from nearby columns. We gather this context for a column by forming a window of consecutive columns centered on it. Formally, a *window* W of width $w \geq 1$ is a sequence of $2w{+}1$ consecutive column profiles $C_{-w} \cdots C_{-1} C_0 C_{+1} \cdots C_{+w}$ centered around profile C_0.

We define the following set of *window classes* $\mathcal{C}$, depending on whether the columns in a labeled training window are known to be core or non-core in the reference alignment. We denote a column labeled core by C, and a column labeled non-core by N. For window width $w{=}1$ (which has three consecutive columns), such labeled windows correspond to strings of length 3 over alphabet $\{\text{C}, \text{N}\}$. The three classes of *core windows* are CCC, CCN, NCC; the three classes of *non-core windows* are CNN, NNC, NNN. (A window is considered core or non-core depending on the label of its center column. We exclude windows NCN and CNC, as these almost never occur in reference alignments.) Together these six classes comprise set $\mathcal{C}$. We call the five classes with at least one core column C in the window, *structured classes*; the one class with no core columns is the *unstructured class*, denoted by $\bot = \text{NNN}$.

2.3   The Coreness Regression Function

We learn a coreness predictor by fitting a regression function that measures the similarity between a column's window and training examples of windows with known coreness, and transforms this similarity into a coreness value.

The similarity of windows $V = V_{-w} \cdots V_w$ and $W = W_{-w} \cdots W_w$ is expressed in terms of the similarity of their corresponding column profiles V_i and W_i. We measure the dissimilarity of two such profiles from window class c at position i, using class- and position-specific *substitution scores* $\sigma_{c,i}(p, q)$ on pairs of states p, q. (Section 2.4 describes how we learn these scores.) Given substitution scores $\sigma_{c,i}$, the *distance* between windows V and W from class $c \in \mathcal{C} - \{\bot\}$ is,

$$d_c(V, W) := \sum_{-w \leq i \leq +w} \sum_{p,q \in Q} V_i(p) \, W_i(q) \, \sigma_{c,i}(p, q).$$

These positional $\sigma_{c,i}$ allow distance function d_c to score dissimilarity higher at positions i near the center of the window, and lower towards its edges.

The *regression function* that predicts the coreness of a column first forms a window W centered on the column, and then performs the following.

(1) (*Find distance to closest class*) Across all labeled training windows, in all structured window classes, find the training window that has smallest class-specific distance to W. Call this closest window V, its class c, and their distance $\delta = d_c(V, W)$.

(2) (*Transform distance to coreness*) If class c is a core class, return the coreness value given by transform function $f_{\text{core}}(\delta)$. Otherwise, return value $f_{\text{non}}(\delta)$.

To transform the *nearest-neighbor distance* δ from Step (1) into a coreness value in Step (2), we use *logistic functions* for f_{core} and f_{non}. We fit these logistic curves to empirically-measured average-coreness values at nearest-neighbor distances collected for either core or non-core training examples, using the curve fitting tools in SciPy [13]. As Fig. 1 in Sect. 4.1 later shows, these logistic transform functions fit actual coreness data remarkably well.

2.4 Learning the Distance Function by Linear Programming

We now describe the linear program used to learn the distance function on column windows. The linear program learns a *class-specific* distance function d_c for each window class $c \in \mathcal{C}$.

To construct the linear program, we partition the *training set* $\mathcal{T}$ of labeled windows by window class: subset $\mathcal{T}_c \subseteq \mathcal{T}$ contains all training windows of class $c \in \mathcal{C}$. We then form a smaller *training sample* $S_c \subseteq \mathcal{T}_c$ for each class c by choosing a random subset of $\mathcal{T}_c$ with a specified cardinality $|S_c|$.

The *constraints* of the linear program fall in several categories. For a sample training window $W \in S_c$, we identify other windows $V \in \mathcal{T}_c$ from the same class c in the full training set that are close to W (under a default distance $\tilde{d}_c$). We call these close windows V from the same class c, *targets*. Similarly for $W \in S_c$, we identify other windows $U \in \mathcal{T}_b$ from a different class $b \neq c$ in the full training set that are also close to W (under $\tilde{d}_b$). We call these other close windows U from a different class b, *impostors*. More formally, the *neighborhood* $\mathcal{N}_c(W, i)$ for a structured class $c \in \mathcal{C} - \{\bot\}$ denotes the set of i-nearest-neighbors to W (not including W) from training set $\mathcal{T}_c$ under the class-specific *default distance* function $\tilde{d}_c$. (The default distance function that we use in our experiments is described in Sect. 4.1.) The constraints of the linear program find distance functions that for a sample window $W \in S_c$, *pull in* targets $V \in \mathcal{N}_c(W, i)$ by making $d_c(V, W)$ small, and *push away* impostors $U \in \mathcal{N}_b(W, i)$ for $b \neq c$ by making $d_b(U, W)$ large.

The *target constraints* for each sample window $W \in S_c$ from each structured class $c \in \mathcal{C} - \{\bot\}$, and each target window $V \in \mathcal{N}_c(W, k)$, are,

$$e_{VW} \geq d_c(V, W) - \tau, \tag{1}$$

$$e_{VW} \geq 0, \tag{2}$$

where e_{VW} is a target *error variable* and τ is a *threshold variable*. In the above, quantity $d_c(V, W)$ is a linear expression in the *substitution score variables* $\sigma_{c,i}(p, q)$, so constraint (1) is a linear inequality in the variables. Intuitively, we would like condition $d_c(V, W) \leq \tau$ to hold (so W will be considered to be in its correct class c); in the solution, variable e_{VW} will equal $\max\{d_c(V, W) - \tau, 0\}$, the amount of error by which this ideal condition is violated.

The *impostor constraints* for each sample window $W \in S_c$ from each structured class $c \in \mathcal{C} - \{\perp\}$, and each impostor window $V \in \mathcal{N}_b(W, \ell)$ from each structured class $b \in \mathcal{C} - \{\perp\}$ with $b \neq c$, are,

$$f_W \geq \tau - d_b(V, W) + 1, \tag{3}$$
$$f_W \geq 0, \tag{4}$$

where f_W is an impostor error variable. Intuitively, we would like condition $d_b(V, W) > \tau$ to hold (so W will not be considered to be in the incorrect class b), which we can express by $d_b(V, W) \geq \tau + 1$ using a *margin* of 1. (Since the scale of the distance functions is arbitrary, we can always pick a unit margin without loss of generality.) In the solution to the linear program, variable f_W will equal $\max_{b \in \mathcal{C} - \{\perp\}, V \in \mathcal{N}_b(W,\ell)}\{\tau - d_b(V, W) + 1, 0\}$, the largest amount of error by which this condition is violated for W across all b and V.

We also have impostor constraints for each completely non-core window $W \in T_\perp$, and each core window $V \in \mathcal{N}_b(W, \ell)$ from each structured core class b (as we do not want W to be considered core), which are of the same form as inequalities (3) and (4) above.

The *triangle inequality constraints*, for each structured class $c \in \mathcal{C} - \{\perp\}$, each window position $-w \leq i \leq w$, and all states $p, q, r \in Q$ (including the gap state ξ), are: $\sigma_{c,i}(p, r) \leq \sigma_{c,i}(p, q) + \sigma_{c,i}(q, r)$. A consequence of these constraints is that the resulting distance functions d_c also satisfy the triangle-inequality property. (We omit the proof due to page limits.) This property allows us to use faster metric-space data structures for computing the nearest-neighbor distance δ from Sect. 2.3.

The remaining constraints, for classes c, positions i, and states p and q, are: $\sigma_{c,i}(p, q) = \sigma_{c,i}(q, p)$, $\sigma_{c,i}(p, p) \leq \sigma_{c,i}(p, q)$, $\sigma_{c,i}(p, q) \geq 0$, $\sigma_{c,i}(\xi, \xi) = 0$, and $\tau \geq 0$, which ensure the distance functions are symmetric and non-negative.

Finally, the *objective function* minimizes the average error over all training sample windows. Formally, we minimize,

$$\alpha \frac{1}{|\mathcal{C}| - 1} \sum_{c \in \mathcal{C} - \{\perp\}} \frac{1}{|S_c|} \sum_{W \in S_c} \frac{1}{k} \sum_{V \in \mathcal{N}_c(W,k)} e_{VW} + (1 - \alpha) \frac{1}{|\mathcal{C}|} \sum_{c \in \mathcal{C}} \frac{1}{|S_c|} \sum_{W \in S_c} f_W,$$

where $0 \leq \alpha \leq 1$ is a blend parameter controlling the weight on target error versus impostor error. We note that in an optimal solution to this linear program, variables $e_{VW} = \max\{d_c(V, W) - \tau, 0\}$ and $f_W = \max_{V,b}\{\tau - d_b(V, W) + 1, 0\}$, since inequalities (1)–(4) ensure the error variables are at least these values, while minimizing the above objective function ensures they will not exceed them. Thus solving the linear program finds distance functions d_c, given by substitution scores $\sigma_{c,i}(p,q)$, that minimize the average over the training windows $W \in S_c$ of the amount of violation of our ideal conditions $d_c(V, W) \leq \tau$ for targets $V \in T_c$ and $d_b(V, W) > \tau$ for impostors $V \in T_b$.

3   Applying Coreness to Accuracy Estimation

The Facet alignment accuracy estimator [15] is a linear combination of efficiently-computable feature functions that are positively correlated with true accuracy. As mentioned earlier, the accuracy of a computed alignment is measured only with respect to core columns of the reference alignment. We leverage our coreness predictor to improve the Facet estimator by: (1) creating a new feature function that attempts to directly estimate accuracy, and (2) concentrating the evaluation of existing feature functions on columns with high predicted coreness.

3.1   Creating a New Coreness Feature

Our new feature function on alignments, *Predicted Alignment Coreness*, is similar to the so-called total-column score sometimes used to measure alignment accuracy. Predicted Alignment Coreness counts the number of columns in the alignment that are predicted to be core, by taking a window W around each column, and determining whether its *predicted coreness* $\chi(W)$ exceeds a threshold κ. This count of predicted core columns is normalized by an estimate of the number of core columns in the unknown reference alignment of the sequences. Formally, for computed alignment $\mathcal{A}$ of sequences $\mathcal{S}$, the Predicted Alignment Coreness feature function is $F_{AC}(\mathcal{A}) := \big|\{W \in \mathcal{A} : \chi(W) \geq \kappa\}\big| / L(\mathcal{S})$.

Normalizer $L(\mathcal{S})$ is designed to be positively correlated with the number of core columns in the reference alignment for sequences $\mathcal{S}$. We consider functions L that are linear combinations of products of at most three factors from the following: aggregate measures of the lengths of sequences in $\mathcal{S}$ (their minimum, mean, and maximum length); ratios of the longest-common-subsequence length for pairs of sequences, divided by an aggregate length measure (a form of "percent identity"); and ratios of the difference in maximum and minimum length, divided by an aggregate length measure (a form of "percent indel"). Finally, we obtain $L(\mathcal{S})$ by solving a linear program to find coefficients of the linear combination that minimizes the L_1-norm with the true number of core columns.

3.2   Augmenting Former Features by Coreness

We also augment some of the features currently in Facet to concentrate their evaluation on columns of higher predicted coreness. A full description of all feature functions in Facet is in [15]; we use predicted coreness to augment: Secondary Structure Blockiness, Secondary Structure Identity, Amino Acid Identity, and Average Substitution Score. Each of these functions computes a feature value that in essence is a sum over substitutions in a column; in the modified feature, this is now a *weighted* sum over columns weighted by predicted coreness.

4   Assessing the Coreness Predictor

We evaluate our new approach to coreness prediction, and its use in accuracy estimation for alignment parameter advising, through experiments on a collection of protein multiple sequence alignment benchmarks. A full description of the benchmarks, and the universe of parameter choices for parameter advising, is in [15]. Briefly, the benchmarks in our experiments consist of reference alignments of protein sequences largely induced by structurally aligning their known three-dimensional structures. We use the BENCH suite of Edgar [10], supplemented by a selection from the PALI suite of Balaji et al. [1]. Our full benchmark collection consists of 861 reference alignments.

We use twelve-fold *cross-validation* to assess both column classification with our coreness predictor, and parameter advising with our augmented accuracy estimator. To correct for the overabundance of easy-to-align benchmarks when assessing parameter advising, we bin the benchmarks according to *difficulty*, measured by the true accuracy of their alignment computed by the Opal aligner [19,20] under its default parameter setting. We ensure folds are balanced in their representation of benchmarks from all difficulty bins. For each fold, we generate a *training set* and *testing set* of example alignments by running Opal on each benchmark for each parameter choice from a fixed universe of settings.

4.1   Constructing the Coreness Predictor

We first discuss learning distance functions and fitting transform functions.

Learning the Distance Function. To keep the linear program manageable, each training fold and each structured class has a training set of 4,000 window examples. When learning the distance functions and testing the accuracy of our coreness predictor, we use training and testing samples of 2,000 window examples representing all classes (including the unstructured class), drawn from our training and testing example alignments.

We form the initial sets of targets and impostors for the linear program by either: (1) using a default distance function whose positional substitution score is a convex combination of (a) the VTML200 substitution score on the states' amino acids (transformed to a dissimilarity value in the range $[0, 1]$) and (b) the identity function on the states' secondary structure types, with positions weighted so the center column has twice the weight of its flanking columns; or (2) randomly sampling example windows from the appropriate classes to form targets and impostors.

When learning the distance function we use 2 targets and 150 impostors per class for each window in the training sample. Once a distance function is learned, we *iterate* the process by using the learned distance function to recompute the sets of targets and impostors for another instance of the linear program that is in turn solved to learn a new distance function. For the receiver operating characteristic (ROC) curve, we give the *area under the curve* (AUC) measure

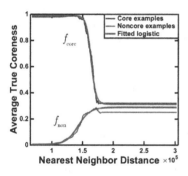 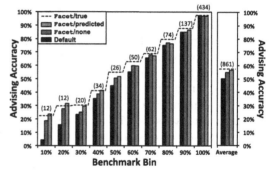

Fig. 1. Fit of the logistic transform functions to the average true coreness of training examples. (Color figure online)

Fig. 2. Average parameter advising accuracy within difficulty bins for greedy advisor sets of cardinality 7. (Color figure online)

on both training and testing data, for successive iterations of distance learning, starting from the default distance function. Across the first 5 iterations, the successive *training* AUC is $86.3, 93.9, 98.9, 99.3, 99.3$; the corresponding *testing* AUC is $83.8, 82.5, 84.9, 84.8, 85.0$. Note that the training AUC increases steadily for the first four iterations, though this translates into only a slight improvement in testing AUC; after this fifth iteration, no further improvement is seen. While iterating distance learning markedly improves our core column predictor on the training examples, it is overfitting and does not generalize well to testing examples, most likely due to the smaller training sample and training set we used to reduce the time for solving the linear program. We also found that using *random examples* for the targets and impostors led to much better generalization, namely a training and testing AUC of 85.8 and 88.7, so we use these resulting distance functions (without iterating) when evaluating results on parameter advising.

Transforming Distance to Coreness. Figure 1 shows the fitted logistic functions f_{core} and f_{non} used to transform nearest-neighbor distance to predicted coreness, superimposed on the underlying true coreness data for one fold of training examples. The horizontal axis is nearest-neighbor distance δ, while the vertical axis is the average true coreness of training examples at that distance. The blue and red curves respectively show the average true coreness of training examples for which the nearest neighbor is in either a core class or a structured non-core class. The top and bottom green curves respectively show the logistic transform functions for the core and non-core classes fitted to this training data. Note that the green logistic curves fit the data quite well.

4.2   Improving Parameter Advising

A *parameter advisor* and has two components: (1) an *accuracy estimator*, which estimates the accuracy of a computed alignment, and (2) an *advisor set*, which is a set of candidate assignments of values to the aligner's parameters. The advisor picks the choice of parameter values from the advisor set for which the aligner yields the alignment of highest estimated accuracy. In our experiments, we evaluate the true accuracy of the Opal aligner [19,20] combined with a parameter advisor using Facet (the best accuracy estimator for advising from the literature [15]), augmented by our new coreness predictor as well as by TCS and ZORRO. We compare these parameter advising results to previous results using unmodified Facet as well TCS (the next-best accuracy estimator for advising). We also compare against augmenting Facet by *true* coreness, which provides a limit for an unattainable perfect coreness predictor.

The choice of advisor set is crucial for parameter advising, as the performance of an advisor is limited by the quality of the alignments generated by this set of parameter choices. We consider two types of advisor sets [6]: estimator-independent *oracle sets*, which are optimal for a conceptual oracle advisor that uses true accuracy as its estimator; and estimator-aware *greedy sets*, which tend to perform better than oracle sets in practice, but are tuned to a specific accuracy estimator. Finding such advisor sets requires specifying a universe of possible parameter choices; we use the universe of 243 parameter choices from [6].

As mentioned earlier, we bin alignments according to difficulty to correct for the overabundance of easy-to-align benchmarks. Figure 2 lists in parentheses above the bars the number of benchmarks in each bin. When reporting advising accuracy, we give the true accuracy of the alignments chosen by the advisor, uniformly averaged over *bins* (rather than uniformly averaging over benchmarks). With this equal weighting of bins, an advisor that uses only the single optimal default parameter choice will achieve an average advising accuracy of roughly 50 % (illustrated in Fig. 3). This establishes, as a point of reference, advising accuracy 50 % as the *baseline* against which to compare advising performance.

The Augmented Facet Estimator. We use our coreness predictor to modify the Facet accuracy estimator by including the new Predicted Alignment Coreness feature of Sect. 3.1, and augmenting existing feature functions by coreness as in Sect. 3.2. We learned coefficients for these feature functions using the difference-fitting technique described in [15]. The new alignment accuracy estimator that uses our coreness predictor has non-zero coefficients for: the new feature, Predicted Alignment Coreness F_{AC}; two features augmented by predicted coreness, Amino Acid Identity F'_{AI}, and Secondary Structure Identity F'_{SI}; and four original unaugmented features, Gap Open Density F_{GO}, Secondary Structure Agreement F_{SA}, Amino Acid Identity F_{AI}, and Secondary Structure Blockiness F_{BL}. The resulting augmented accuracy estimator is: $(0.512) F_{GO} + (0.304) F'_{SI} + (0.157) F_{SA} + (0.109) F_{AI} + (0.096) F_{BL} + (0.025) F'_{AI} + (0.013) F_{AC}$.

Improvement in Advising Accuracy. We assess the parameter advising performance of our augmented Facet estimator ("Facet/predicted") by comparing it to unaugmented Facet ("Facet/none"), as well as Facet augmented by TCS

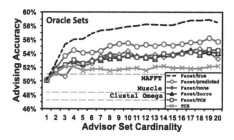
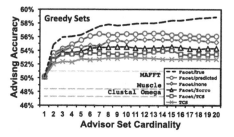

Fig. 3. Advising accuracy on *oracle sets* with modified `Facet` or TCS estimators.

Fig. 4. Advising accuracy on *greedy sets* with modified `Facet` or TCS estimators.

("Facet/TCS"), ZORRO ("Facet/ZORRO"), and true coreness ("Facet/true"). We also compare against TCS, the next-best estimator from the literature.

Parameter advising performance using oracle and greedy advisor sets is shown in Figs. 3 and 4. In both figures, the horizontal axis is advisor set cardinality, while the vertical axis is advising accuracy for testing folds, averaged across bins. The curves show performance with the `Opal` aligner [19,20]. For reference, the default alignment accuracy for three other popular aligners, `MAFFT` [14], `MUSCLE` [11], and `Clustal Omega` [18], is also shown with dashed horizontal lines.

Figure 3 shows that on *oracle* advisor sets, `Facet`/predicted compared to `Facet`/none boosts the average accuracy of parameter advising by nearly 3 %. This increase is in addition to the improvement of `Facet` over TCS.

Figure 4 shows that on *greedy* advisor sets, `Facet`/predicted at cardinality 7 boosts advising accuracy by 2 %. (Note the curves are higher for greedy sets than oracle sets.) The accuracy for `Facet`/predicted is about halfway between `Facet`/none and `Facet`/true (the perfect predictor). Interestingly, `Facet`/TCS and `Facet`/ZORRO actually have worse accuracy than `Facet`/none.

Advising accuracy within difficulty *bins* for greedy sets of cardinality 7 is shown earlier in Fig. 2. In this bar chart, for the bin at each difficulty on the horizontal axis, advising accuracy averaged over just the benchmarks in the bin is shown on the vertical axis. The final chart on the right gives accuracy averaged across all bins. Note that on the most difficult benchmarks, `Facet`/predicted boosts accuracy over `Facet`/none by more than 5 %.

For reference, advising accuracy uniformly-averaged over *benchmarks* (rather than bins), on greedy sets of cardinality 10, is: for `Facet`/none, 81.9 %; and for `Facet`/predicted, 82.1 %. On these same benchmarks, the corresponding average accuracy of other popular aligners, using their default parameter settings, is: `Clustal Omega`, 77.3 %; `MUSCLE`, 78.1 %; `MAFFT`, 79.4 %; and `Opal`, 80.5 %.

5   Conclusion

We have developed a column coreness predictor for protein multiple sequence alignments that uses a regression function on nearest neighbor distances for class distance functions learned by solving a new linear programming formulation.

When applied to alignment accuracy estimation and parameter advising, the coreness predictor strongly outperforms other column confidence estimators from the literature, and provides a substantial boost in advising accuracy.

Acknowledgement. This research was supported by NSF grant IIS-1217886 to J.K.

References

1. Balaji, S., Sujatha, S., Kumar, S., Srinivasan, N.: PALI—a database of Phylogeny and ALIgnment of homologous protein structures. NAR **29**(1), 61–65 (2001)
2. Capella-Gutierrez, S., Silla-Martinez, J.M., Gabaldón, T.: trimAl: a tool for automated alignment trimming in large-scale phylogenetic analyses. Bioinformatics **25**(15), 1972–1973 (2009)
3. Castresana, J.: Selection of conserved blocks from multiple alignments for their use in phylogenetic analysis. Mol. Biol. Evol. **17**(4), 540–552 (2000)
4. Chang, J.M., Tommaso, P.D., Notredame, C.: TCS: a new multiple sequence alignment reliability measure to estimate alignment accuracy and improve phylogenetic tree reconstruction. Mol. Biol. Evol. **31**, 1625–1637 (2014)
5. DeBlasio, D., Kececioglu, J.: Ensemble multiple sequence alignment via advising. In: Proceedings of the 6th ACM Conference on Bioinformatics, Computational Biology, and Health Informatics (BCB), pp. 452–461 (2015)
6. DeBlasio, D.F., Kececioglu, J.D.: Learning parameter sets for alignment advising. In: Proceedings of the 5th ACM Conference on Bioinformatics, Computational Biology, and Health Informatics (BCB), pp. 230–239 (2014)
7. DeBlasio, D.F., Wheeler, T.J., Kececioglu, J.D.: Estimating the accuracy of multiple alignments and its use in parameter advising. In: Chor, B. (ed.) RECOMB 2012. LNCS, vol. 7262, pp. 45–59. Springer, Heidelberg (2012)
8. Dress, A.W., Flamm, C., Fritzsch, G., Grünewald, S., Kruspe, M., Prohaska, S.J., Stadler, P.F.: Noisy: identification of problematic columns in multiple sequence alignments. Algorithms Mol. Biol. **3**(7) (2008)
9. Durbin, R., Eddy, S.R., Krogh, A., Mitchison, G.: Biological Sequence Analysis: Probablistic Models of Proteins and Nucleic Acids. Cambridge University Press, Cambridge (1998)
10. Edgar, R.C.: BENCH, January 2009. drive5.com/bench
11. Edgar, R.C.: MUSCLE: a multiple sequence alignment method with reduced time and space complexity. BMC Bioinform. **5**(113), 1–19 (2004)
12. Jones, D.T.: Protein secondary structure prediction based on position-specific scoring matrices. J. Mol. Biol. **292**(2), 195–202 (1999)
13. Jones, E., Oliphant, T., Peterson, P., et al.: SciPy: open source scientific tools for Python (2001). http://www.scipy.org
14. Katoh, K., Kuma, K.I., Toh, H., Miyata, T.: MAFFT ver. 5: improvement in accuracy of multiple sequence alignment. Nucleic Acids Res. **33**(2), 511–518 (2005)
15. Kececioglu, J., DeBlasio, D.: Accuracy estimation and parameter advising for protein multiple sequence alignment. J. Comput. Biol. **20**(4), 259–279 (2013)
16. Kück, P., Meusemann, K., Dambach, J., et al.: Parametric and non-parametric masking of randomness in sequence alignments can be improved and leads to better resolved trees. Front. Zool. **7**(10), 1–10 (2010)

17. Sela, I., Ashkenazy, H., Katoh, K., Pupko, T.: GUIDANCE2: accurate detection of unreliable alignment regions accounting for the uncertainty of multiple parameters. Nucleic Acids Res. **43**(W1), W7–W14 (2015)
18. Sievers, F., et al.: Fast, scalable generation of high-quality protein multiple sequence alignments using Clustal Omega. Mol. Syst. Biol. **7**(1), 539 (2011)
19. Wheeler, T.J., Kececioglu, J.D.: Multiple alignment by aligning alignments. Bioinformatics **23**(13), i559–i568 (2007). Proceedings of ISMB 2007
20. Wheeler, T.J., Kececioglu, J.D.: Opal: software for sum-of-pairs multiple sequence alignment, January 2012. http://opal.cs.arizona.edu
21. Wu, M., Chatterji, S., Eisen, J.A.: Accounting for alignment uncertainty in phylogenomics. PLoS One **7**(1), e30288 (2012)

Fast Compatibility Testing for Phylogenies with Nested Taxa

Yun Deng$^{(\boxtimes)}$ and David Fernández-Baca

Department of Computer Science, Iowa State University, Ames, IA 50011, USA
{yundeng,fernande}@iastate.edu

Abstract. Semi-labeled trees are phylogenies whose internal nodes may be labeled by higher-order taxa. Thus, a leaf labeled *Mus musculus* could nest within a subtree whose root node is labeled Rodentia, which itself could nest within a subtree whose root is labeled Mammalia. Suppose we are given collection $\mathcal{P}$ of semi-labeled trees over various subsets of a set of taxa. The ancestral compatibility problem asks whether there is a semi-labeled tree $\mathcal{T}$ that respects the clusterings and the ancestor/descendant relationships implied by the trees in $\mathcal{P}$. We give a $\tilde{O}(M_{\mathcal{P}})$ algorithm for the ancestral compatibility problem, where $M_{\mathcal{P}}$ is the total number of nodes and edges in the trees in $\mathcal{P}$. Unlike the best previous algorithm, the running time of our method does not depend on the degrees of the nodes in the input trees.

1 Introduction

In the *tree compatibility problem*, we are given a collection $\mathcal{P} = \{\mathcal{T}_1, \ldots, \mathcal{T}_k\}$ of rooted phylogenetic trees with partially overlapping taxon sets. $\mathcal{P}$ is called a *profile* and the trees in $\mathcal{P}$ are the *input trees*. The question is whether there exists a tree $\mathcal{T}$ whose taxon set is the union of the taxon sets of the input trees, such that $\mathcal{T}$ exhibits the clusterings implied by the input trees. That is, if two taxa are together in a subtree of some input tree, then they must also be together in some subtree of $\mathcal{T}$. The tree compatibility problem has been studied for over three decades [1,8,10,20].

In the original version of the tree compatibility problem, only the leaves of the input trees are labeled. Here we study a generalization, called *ancestral compatibility*, in which taxa may be *nested*. That is, the internal nodes may also be labeled; these labels represent *higher-order taxa*, which are, in effect, sets of taxa. Thus, for example, an input tree may contain the taxon *Glycine max* (soybean) nested within a subtree whose root is labeled Fabaceae (the legumes), itself nested within an Angiosperm subtree. Note that leaves themselves may be labeled by higher-order taxa. The question now is whether there is a tree $\mathcal{T}$ whose taxon set is the union of the taxon sets of the input trees, such that $\mathcal{T}$ exhibits

Supported in part by National Science Foundation grant CCF-1422134.

M. Frith and C.N.S. Pedersen (Eds.): WABI 2016, LNBI 9838, pp. 90–101, 2016.
DOI: 10.1007/978-3-319-43681-4_8

not only the clusterings among the taxa, but also the ancestor/descendant relationships among taxa in the input trees. Our main result is a $\tilde{O}(M_\mathcal{P})$ algorithm for the compatibility problem for trees with nested taxa, where $M_\mathcal{P}$ is the total number of nodes and edges in the trees in $\mathcal{P}$.

Background. The tree compatibility problem is a basic special case of the *supertree problem.* A supertree method is a way to synthesize a collection of phylogenetic trees with partially overlapping taxon sets into a single supertree that represents the information in the input trees. The supertree approach, proposed in the early 90s [2,15], has been used to build large-scale phylogenies [4].

The original supertree methods were limited to input trees where only the leaves are labeled. Page [13] was among the first to note the need to handle phylogenies where internal nodes are labeled, and taxa are nested. A major motivation is the desire to incorporate *taxonomies* as input trees in large-scale supertree analyses, as way to circumvent one of the obstacles to building comprehensive phylogenies: the limited taxonomic overlap among different phylogenetic studies [16]. Taxonomies group organisms according to a system of taxonomic rank (e.g., family, genus, and species); two examples are the NCBI taxonomy [17] and the Angiosperm taxonomy [21]. Taxonomies spanning a broad range of taxa provide structure and completeness that might be hard to obtain otherwise. A recent example of the utility of taxonomies is the Open Tree of Life, a draft phylogeny for over 2.3 million species [11].

Taxonomies are not, strictly speaking, phylogenies. In particular, their internal nodes and some of their leaves are labeled with higher-order taxa. Nevertheless, taxonomies have many of the same mathematical characteristics as phylogenies. Indeed, both phylogenies and taxonomies are *semi-labeled trees* [5,18]. We will use this term throughout the rest of the paper to refer to trees with nested taxa.

The fastest previous algorithm for testing ancestral compatibility, based on earlier work by Daniel and Semple [7], is due to Berry and Semple [3]. Their algorithm runs in $O(\tau_\mathcal{P} \cdot \log^2 n)$ time using $O(\tau_\mathcal{P})$ space. Here, n is the number of distinct taxa in $\mathcal{P}$ and $\tau_\mathcal{P} = \sum_{i=1}^k \sum_{v \in I(\mathcal{T}_i)} d(v)^2$, where $I(\mathcal{T}_i)$ is the set of internal nodes of $\mathcal{T}_i$, for each $i \in \{1, \ldots, k\}$, and $d(v)$ is the degree of node v. While the algorithm is polynomial, its dependence on node degrees is problematic: semi-labeled trees can be highly unresolved (i.e., contain nodes of high degree), especially if they are taxonomies.

Our Contributions. The $\tilde{O}(M_\mathcal{P})$ running time of our ancestral compatibility algorithm is independent of the degrees of the nodes of the input trees, a valuable characteristic for large datasets that include taxonomies. To achieve this time bound, we extend ideas from our recent algorithm for testing the compatibility of ordinary phylogenetic trees [8]. As in that algorithm, a central notion in the current paper is the *display graph* of profile $\mathcal{P}$, denoted $H_\mathcal{P}$. This is the graph obtained from the disjoint union of the trees in $\mathcal{P}$ by identifying nodes that have the same label (see Sect. 4). The term "display graph" was introduced by Bryant and Lagergren [6], but similar ideas have been used elsewhere. In particular, the display graph is closely related to Berry and Semple's *restricted descendancy graph* [3], a mixed graph whose directed edges correspond to the (undirected) edges of $H_\mathcal{P}$ and whose undirected edges have no correspondence in $H_\mathcal{P}$. The second kind of edges are the major component of the $\tau_\mathcal{P}$ term in the time and space complexity of Berry and Semple's algorithm. The absence of such edges makes $H_\mathcal{P}$ significantly smaller than the restricted descendancy graph. Display graphs also bear some relation to *tree alignment graphs* [19].

Here, we exploit the display graph more extensively and more directly than our previous work. Although the display graph of a collection of semi-labeled trees is more complex than that of a collection of ordinary phylogenies, we are able to extend several of the key ideas — notably, that of a semi-universal label — to the general setting of semi-labeled trees. As in [8], the implementation relies on a dynamic graph data structure, but it requires a more careful amortized analysis based on a weighing scheme.

Contents. Section 2 presents basic definitions regarding semi-labeled trees and ancestral compatibility. Section 3 introduces the display graph and discusses its properties. Section 4 presents `BuildNT`, our algorithm for testing ancestral compatibility. Section 5 gives the implementation details for `BuildNT`. Section 6 gives some concluding remarks.

2   Preliminaries

For each positive integer r, $[r]$ denotes the set $\{1, \ldots, r\}$.

Let G be a graph. $V(G)$ and $E(G)$ denote the node and edge sets of G. The *degree* of a node $v \in V(G)$ is the number of edges incident on v. A *tree* is an acyclic connected graph. In this paper, all trees are assumed to be rooted. For a tree T, $r(T)$ denotes the root of T. Suppose $u, v \in V(T)$. Then, u is an *ancestor* of v in T, denoted $u \leq_T v$, if u lies on the path from v to $r(T)$ in T. If $u \leq_T v$, then v is a *descendant* of u. Node u is a *proper descendant* of v if u is a descendant of v and $v \neq u$. If $\{u, v\} \in E(T)$ and $u \leq_T v$, then u is the *parent* of v and v is a *child* of u. If neither $u \leq_T v$ nor $v \leq_T u$ hold, then we write $u \parallel_T v$ and say that u and v are *not comparable* in T.

Semi-labeled Trees. A *semi-labeled tree* is a pair $\mathcal{T} = (T, \phi)$ where T is a tree and ϕ is a mapping from a set $L(\mathcal{T})$ to $V(T)$ such that, for every node $v \in V(T)$ of degree at most two, $v \in \phi(L(\mathcal{T}))$. $L(\mathcal{T})$ is the *label set* of $\mathcal{T}$ and ϕ is the *labeling function* of $\mathcal{T}$.

For every node $v \in V(T)$, $\phi^{-1}(v)$ denotes the (possibly empty) subset of $L(\mathcal{T})$ whose elements map into v; these elements as the *labels of v* (thus, each label is a taxon). If $\phi^{-1}(v) \neq \emptyset$, then v is *labeled*; otherwise, v is *unlabeled*. Note that, by definition, every leaf in a semi-labeled tree is labeled. Further, any node, including the root, that has a single child must be labeled. Nodes with two or more children may be labeled or unlabeled. A semi-labeled tree $\mathcal{T} = (T, \phi)$ is *singularly labeled* if every node in T has at most one label; $\mathcal{T}$ is *fully labeled* if every node in T is labeled.

Semi-labeled trees, also known as X-*trees*, generalize ordinary phylogenetic trees, also known as *phylogenetic X-trees* [18]. An ordinary phylogenetic tree is a semi-labeled tree $T = (T, \phi)$ where $r(T)$ has degree at least two and ϕ is a bijection from $L(T)$ into leaf set of T (thus, internal nodes are not labeled).

Let $T = (T, \phi)$ be a semi-labeled tree and let ℓ and ℓ' be two labels in $L(T)$. If $\phi(\ell) \leq_T \phi(\ell')$, then we write $\ell \leq_T \ell'$, and say that ℓ' is a *descendant* of ℓ in T and that ℓ is an *ancestor* of ℓ'. We write $\ell <_T \ell'$ if $\phi(\ell')$ is a proper descendant of $\phi(\ell)$. If $\phi(\ell) \parallel_T \phi(\ell')$, then we write $\ell \parallel_T \ell'$ and say that ℓ and ℓ' are *not comparable* in T. If T is fully labeled and $\phi(\ell)$ is the parent of $\phi(\ell')$ in T, then ℓ is the *parent* of ℓ' in T and ℓ' is a *child* of ℓ in T; two labels with the same parent are *siblings*.

Two semi-labelled trees $T = (T, \phi)$ and $T' = (T', \phi')$ are *isomorphic* if there exists a bijection $\psi : V(T) \to V(T')$ such that $\phi' = \psi \circ \phi$ and, for any two nodes $u, v \in V(T)$, $(u, v) \in E(T)$ if and only $(\psi(u), \psi(v)) \in E(T')$.

Let $T = (T, \phi)$ be a semi-labeled tree. For each $u \in V(T)$, $X(u)$ denotes the set of all labels in the subtree of T rooted at u; that is, $X(u) = \bigcup_{v:u \leq_T v} \phi^{-1}(v)$. $X(u)$ is called a *cluster* of T. $\mathrm{Cl}(T)$ denotes the set of all clusters of T. It is well known [18, Theorem 3.5.2] that a semi-labeled tree T is completely determined by $\mathrm{Cl}(T)$. That is, if $\mathrm{Cl}(T) = \mathrm{Cl}(T')$ for some other semi-labeled tree T', then T is isomorphic to T'.

Suppose $A \subseteq L(T)$ for a semi-labeled tree $T = (T, \phi)$. The *restriction* of T to A, denoted $T|A$, is the semi-labeled tree whose cluster set is $\mathrm{Cl}(T|A) = \{X \cap A : X \in \mathrm{Cl}(T) \text{ and } X \cap A \neq \emptyset\}$. Intuitively, $T|A$ is obtained from the minimal rooted subtree of T that connects the nodes in $\phi(A)$ by suppressing all vertices of degree two that are not in $\phi(A)$.

Let $T = (T, \phi)$ and $T' = (T', \phi')$ be semi-labeled trees such that $L(T') \subseteq L(T)$. T *ancestrally displays* T' if $\mathrm{Cl}(T') \subseteq \mathrm{Cl}(T|L(T'))$. Equivalently, T ancestrally displays T' if T' can be obtained from $T|L(T')$ by contracting edges, and, for any $\ell_1, \ell_2 \in L(T')$, (i) if $\ell_1 <_{T'} \ell_2$, then $\ell_1 <_T \ell_2$, and (ii) if $\ell_1 \parallel_{T'} \ell_2$, then $\ell_1 \parallel_T \ell_2$. The notion of "ancestrally displays" for semi-labeled trees generalizes the well-known notion of "displays" for ordinary phylogenetic trees [18].

For a semi-labelled tree T, let $D(T) = \{(\ell, \ell') : \ell, \ell' \in L(T) \text{ and } \ell <_T \ell'\}$ and $N(T) = \{\{\ell, \ell'\} : \ell, \ell' \in L(T) \text{ and } \ell \parallel_T \ell'\}$. Note that $D(T)$ consists of *ordered* pairs, while $N(T)$ consists of *unordered* pairs.

Lemma 1 (Bordewich et al. [5]). *Let T and T' be semi-labelled trees such that $L(T') \subseteq L(T)$. Then T ancestrally displays T' if and only if $D(T') \subseteq D(T)$ and $N(T') \subseteq N(T)$.*

Profiles and Ancestral Compatibility. Throughout the rest of this paper $\mathcal{P} = \{T_1, \ldots, T_k\}$ denotes a set where, for each $i \in [k]$, $T_i = (T_i, \phi_i)$ is a semi-labeled tree. We refer to $\mathcal{P}$ as a *profile*, and write $L(\mathcal{P})$ to denote $\bigcup_{i \in [k]} L(T_i)$, the *label set* of $\mathcal{P}$. Figure 1 shows a profile where $L(\mathcal{P}) = \{a, b, c, d, e, f, g, h, i\}$. We write $V(\mathcal{P})$ for $\bigcup_{i \in [k]} V(T_i)$ and $E(\mathcal{P})$ for $\bigcup_{i \in [k]} E(T_i)$, The *size* of $\mathcal{P}$ is $M_{\mathcal{P}} = |V(\mathcal{P})| + |E(\mathcal{P})|$.

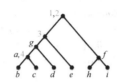
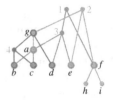

Fig. 1. A profile $\mathcal{P} = \{T_1, T_2, T_3\}$ — trees are ordered left-to-right. The letters are the original labels; grey numbers are labels added to make the trees fully labeled. (Adapted from [3].)

Fig. 2. A tree T that ancestrally displays the profile of Fig. 1. (Adapted from [3].)

Fig. 3. The display graph $H_\mathcal{P}$ for the profile of Fig. 1.

$\mathcal{P}$ is *ancestrally compatible* if there is a rooted semi-labeled tree T that ancestrally displays each of the trees in $\mathcal{P}$. If T exists, we say that T *ancestrally displays* $\mathcal{P}$ (see Fig. 2).

Given a subset X of $L(\mathcal{P})$, the *restriction* of $\mathcal{P}$ to X, denoted $\mathcal{P}|X$, is the profile $\{T_1|X \cap L(T_1), \ldots, T_k|X \cap L(T_k)\}$. The proof of the following lemma is straightforward.

Lemma 2. *Suppose $\mathcal{P}$ is ancestrally compatible and let T be a tree that ancestrally displays $\mathcal{P}$. Then, for any $X \subseteq L(\mathcal{P})$, $T|X$ ancestrally displays $\mathcal{P}|X$.*

A semi-labeled tree $T = (T, \phi)$ is *fully labeled* if every node in T is labeled. Any profile $\mathcal{P}$ that contains trees that are not fully labeled can be converted into an equivalent profile $\mathcal{P}'$ of fully-labeled trees by introducing distinct new labels for the unlabeled nodes of $\mathcal{P}$, as illustrated in Fig. 1.

Lemma 3 (Daniel and Semple [7]). *Let $\mathcal{P}'$ be a profile obtained by adding distinct new labels to $\mathcal{P}$. Then, $\mathcal{P}$ is ancestrally compatible if and only if $\mathcal{P}'$ is ancestrally compatible. Further, if T is a semi-labeled phylogenetic tree that ancestrally displays $\mathcal{P}'$, then T ancestrally displays $\mathcal{P}$.*

From this point forward, we shall assume that, for each $i \in [k]$, T_i is fully and singularly labeled. By Lemma 3, no generality is lost in assuming that all trees in $\mathcal{P}$ are fully labeled. The assumption that the trees are singularly labeled is only for clarity; even with this assumption, a tree that ancestrally displays $\mathcal{P}$ is not necessarily singularly labeled. Figure 2 illustrates this fact.

3   The Display Graph

The *display graph* of a profile $\mathcal{P}$, denoted $H_\mathcal{P}$, is the graph obtained from the disjoint union of the underlying trees $T_1, \ldots, T_k$ by identifying nodes that have the same label. Multiple edges between the same pair of nodes are replaced by a single edge. See Fig. 3.

$H_{\mathcal{P}}$ has $O(M_{\mathcal{P}})$ nodes and edges, and can be constructed in $O(M_{\mathcal{P}})$ time. By our assumption that all the trees in $\mathcal{P}$ are fully and singularly labeled, there is a bijection between the labels in $L(\mathcal{P})$ and the nodes of $H_{\mathcal{P}}$. Thus, from this point forward, we refer to the nodes of $H_{\mathcal{P}}$ by their labels. It is easy to see that if $H_{\mathcal{P}}$ is not connected, then $\mathcal{P}$ decomposes into label-disjoint sub-profiles, and that $\mathcal{P}$ is compatible if and only if each sub-profile is compatible. Thus, we shall assume, without loss of generality, that $H_{\mathcal{P}}$ is connected.

Positions. A *position* (for $\mathcal{P}$) is a vector $U = (U(1), \ldots, U(k))$, where $U(i) \subseteq L(T_i)$, for each $i \in [k]$. Since labels may be shared among trees, we may have $U(i) \cap U(j) \neq \emptyset$, for $i, j \in [k]$ with $i \neq j$. For each $i \in [k]$, let $\mathrm{Desc}_i(U) = \{\ell : \ell' \leq_{T_i} \ell, \text{ for some } \ell' \in U(i)\}$, and let $\mathrm{Desc}_{\mathcal{P}}(U) = \bigcup_{i \in [k]} \mathrm{Desc}_i(U)$.

A position U is *valid* if, for each $i \in [k]$,

(V1) if $|U(i)| \geq 2$, then the elements of $U(i)$ are siblings in T_i and
(V2) $\mathrm{Desc}_i(U) = \mathrm{Desc}_{\mathcal{P}}(U) \cap L(T_i)$.

Lemma 4. *For any valid position U, $\mathcal{P}|\mathrm{Desc}_{\mathcal{P}}(U) = \{T_1|\mathrm{Desc}_1(U), \ldots, T_k|\mathrm{Desc}_k(U)\}$.*

Let U be a valid position. $H_{\mathcal{P}}(U)$ denotes the subgraph of $H_{\mathcal{P}}$ induced by $\mathrm{Desc}_{\mathcal{P}}(U)$.

Observation 1. *For any valid position U, $H_{\mathcal{P}}(U)$ is the subgraph of $H_{\mathcal{P}}$ obtained by deleting all labels in $V(H_{\mathcal{P}}) \setminus \mathrm{Desc}_{\mathcal{P}}(U)$, along with all incident edges.*

A valid position of special interest is U_{root}, where $U_{\mathrm{root}}(i) = \phi_i^{-1}(r(T_i))$, for each $i \in [k]$. That is, $U_{\mathrm{root}}(i)$ is a singleton containing only the label of $r(T_i)$. In Fig. 3, $(U_{\mathrm{root}}(1), U_{\mathrm{root}}(2), U_{\mathrm{root}}(3)) = (\{1\}, \{2\}, \{g\})$. It is straightforward to verify that U_{root} is indeed valid, that $\mathrm{Desc}_{\mathcal{P}}(U_{\mathrm{root}}) = L(\mathcal{P})$, and that $H_{\mathcal{P}}(U_{\mathrm{root}}) = H_{\mathcal{P}}$.

Semi-universal Labels. Let U be a valid position, and let ℓ be a label in U. Then, ℓ is *semi-universal in U* if $U(i) = \{\ell\}$, for every $i \in [k]$ such that $\ell \in L(T_i)$. It can be verified that in Fig. 3, labels 1 and 2 are semi-universal in U_{root}, but g is not, since g is in both $L(T_2)$ and $L(T_3)$, but $U_{\mathrm{root}}(2) \neq \{g\}$.

The term "semi-universal", borrowed from Pe'er et al. [14], derives from the following fact. Suppose that $\mathcal{P}$ is ancestrally compatible, that T is a tree that ancestrally displays $\mathcal{P}$, and that ℓ is a semi-universal label for some valid position U. Then, as we shall see, for every i such that $\ell \in L(T_i)$, ℓ must label the root u_ℓ of a subtree of T that contains all the descendants of ℓ in T_i. The qualifier "semi" is because this subtree may also contain labels that do not descend from ℓ in any input tree, but descend from some other semi-universal label ℓ' in U instead. In this case, ℓ' also labels u_ℓ. This property of semi-universal labels is exploited in our ancestral compatibility algorithm (see Sect. 4).

For each label $\ell \in L(\mathcal{P})$, let k_ℓ denote the number of input trees that contain label ℓ. We can obtain k_ℓ for every $\ell \in L(\mathcal{P})$ in $O(M_{\mathcal{P}})$ time during the construction of $H_{\mathcal{P}}$.

Lemma 5. *Let $U = (U(1), \ldots, U(k))$ be a valid position. Then, label ℓ is semi-universal in U if $|\{i \in [k] : U(i) = \{\ell\}\}| = k_\ell$.*

Successor Positions. For every $i \in [k]$ and every $\ell \in L(\mathcal{T}_i)$, let $\mathrm{Ch}_i(\ell)$ denote the set of children of ℓ in $L(\mathcal{T}_i)$. Let U be a valid position, and S be the set of semi-universal labels in U. The *successor of U with respect to S* is the position U' defined as follows. For each $\ell \in S$ and each $i \in [k]$, if $U(i) = \{\ell\}$, then $U'(i) = \mathrm{Ch}_i(\ell)$; otherwise, $U'(i) = U(i)$.

In Fig. 3, the set of semi-universal labels in U_{root} is $S = \{1, 2\}$. Since $\mathrm{Ch}_1(1) = \{3, f\}$ and $\mathrm{Ch}_2(2) = \{e, f, g\}$, the successor of U_{root} is $U' = (\{3, f\}, \{e, f, g\}, \{g\})$.

Observation 2. *Let U be a valid position, and let U' be the successor of U with respect to the set S of semi-universal labels in U. Then, $H_{\mathcal{P}}(U')$ can be obtained from $H_{\mathcal{P}}(U)$ by doing the following for each $\ell \in S$: (1) for each $i \in [k]$ such that $U(i) = \{\ell\}$, delete all edges between ℓ and $\mathrm{Ch}_i(\ell)$; (2) delete ℓ.*

Let U be a valid position, and W be a subset of $\mathrm{Desc}_{\mathcal{P}}(U)$. Then, $U|W$ denotes the position $(U(1) \cap W, U(2) \cap W, \ldots, U(k) \cap W)$. In Fig. 3, the components of $H_{\mathcal{P}}(U')$, where U' is the successor of U_{root}, are $W_1 = \{3, 4, a, b, c, d, e, g\}$ and $W_2 = \{f, h, i\}$. Thus, $U'|W_1 = (\{3\}, \{e, g\}, \{g\})$ and $U'|W_2 = (\{f\}, \{f\}, \emptyset)$.

Lemma 6. *Let U be a valid position, and S be the set of all semi-universal labels in U. Let U' be the successor of U with respect to S, and let $W_1, W_2, \ldots, W_p$ be the label sets of the connected components of $H_{\mathcal{P}}(U')$. Then, $U'|W_j$ is a valid position, for each $j \in [p]$.*

4 Testing Ancestral Compatibility

BuildNT (Algorithm 1) is our algorithm for testing compatibility of semi-labeled trees. Its argument, U, is a valid position in $\mathcal{P}$ such that $H_{\mathcal{P}}(U)$ is connected. Line 1 computes the set S of semi-universal labels in U. If S is empty, then, as argued in Theorem 1 below, $\mathcal{P}|\mathrm{Desc}_{\mathcal{P}}(U)$ is incompatible, and, thus, so is $\mathcal{P}$. This fact is reported in Line 3. Line 4 checks if S contains exactly one label ℓ, with no proper descendants. If so, by the connectivity assumption, ℓ must be the only element in $\mathrm{Desc}_{\mathcal{P}}(U)$. Therefore, Line 5 simply returns the tree with a single node, labeled ℓ. Line 6 updates U, replacing it by its successor with respect to S. Let $W_1, \ldots, W_p$ be the connected components of $H_{\mathcal{P}}(U)$ after updating U. By Lemma 6, $U|W_j$ is a valid position, for each $j \in [p]$. Lines 7–11 recursively invoke BuildNT on $U|W_j$ for each $j \in [p]$, to determine if there is a tree t_j that ancestrally displays $\mathcal{P}|\mathrm{Desc}_{\mathcal{P}}(U \cap W_j)$. If any subproblem is incompatible, Line 11 reports that $\mathcal{P}$ is incompatible. Otherwise, Lines 12 and 13 assemble the t_js into a single tree that displays $\mathcal{P}|\mathrm{Desc}_{\mathcal{P}}(U)$, whose root is labeled by the semi-universal labels in the set S of Line 1.

Next, we argue the correctness of BuildNT.

Algorithm 1. BuildNT(U)

Input: A valid position U for $\mathcal{P}$ such that $H_\mathcal{P}(U)$ is connected.
Output: A semi-labeled tree that ancestrally displays $\mathcal{P}' = \mathcal{P}|\text{Desc}_\mathcal{P}(U)$, if $\mathcal{P}'$ is ancestrally compatible; incompatible otherwise.

1 Let $S = \{\ell \in U : \ell$ is semi-universal in $U\}$
2 **if** $S = \emptyset$ **then**
3 $\quad$ **return** incompatible
4 **if** $|S| = 1$ *and the single element,* ℓ, *of* S *has no proper descendants* **then**
5 $\quad$ **return** the tree consisting of exactly one node, whose label set is $\{\ell\}$
6 Replace U by the successor of U with respect to S.
7 Let $W_1, W_2, \ldots, W_p$ be the connected components of $H_\mathcal{P}(U)$
8 **foreach** $j \in [p]$ **do**
9 $\quad$ Let $t_j = $ BuildNT($U|W_j$)
10 $\quad$ **if** t_j *is not a tree* **then**
11 $\quad$ $\quad$ **return** incompatible
12 Create a node r_U, whose label set is S
13 **return** the tree with root r_U and subtrees $t_1, \ldots, t_p$

Theorem 1. *Let* $\mathcal{P} = \{\mathcal{T}_1, \ldots, \mathcal{T}_k\}$ *be a profile and let* $U_{\text{root}} = (U_{\text{root}}(1), \ldots, U_{\text{root}}(k))$, *where, for each* $i \in [k]$, $U_{\text{root}}(i) = \phi_i^{-1}(r(\mathcal{T}_i))$. *Then,* BuildNT($U_{\text{root}}$) *returns either (i) a semi-labeled tree* $\mathcal{T}$ *that ancestrally displays* $\mathcal{P}$, *if* $\mathcal{P}$ *is ancestrally compatible, or (ii)* incompatible *otherwise.*

Proof. (i) Suppose that BuildNT(U_{root}) outputs a semi-labeled tree $\mathcal{T}$. We prove that $\mathcal{T}$ ancestrally displays $\mathcal{P}$. By Lemma 1, it suffices to show that $D(\mathcal{T}_i) \subseteq D(\mathcal{T})$ and $N(\mathcal{T}_i) \subseteq N(\mathcal{T})$, for each $i \in [k]$.

Consider any $(\ell, \ell') \in D(\mathcal{T}_i)$. Then, ℓ has a child ℓ'' in $\mathcal{T}_i$ such that $\ell'' \leq_{\mathcal{T}_i} \ell'$. There must be a recursive call to BuildNT(U), for some valid position U, where ℓ is the set S of semi-universal labels obtained in Line 1. By Observation 2, label ℓ'', and thus ℓ', both lie in one of the connected components of the graph obtained by deleting all labels in S, including ℓ, and their incident edges from $H_\mathcal{P}(U)$. It now follows from the construction of $\mathcal{T}$ that $(\ell, \ell') \in D(\mathcal{T})$. Thus, $D(\mathcal{T}_i) \subseteq D(\mathcal{T})$.

Now, consider any $\{\ell, \ell'\} \in N(\mathcal{T}_i)$. Let v be the lowest common ancestor of $\phi_i(\ell)$ and $\phi_i(\ell')$ in $\mathcal{T}_i$ and let ℓ_v be the label of v. Then, ℓ_v has a pair of children, ℓ_1 and ℓ_2 say, in $\mathcal{T}_i$ such that $\ell_1 \leq_{\mathcal{T}_i} \ell$, and $\ell_2 \leq_{\mathcal{T}_i} \ell'$. Because BuildNT($U_{\text{root}}$) returns a tree, there are recursive calls BuildNT(U_1) and BuildNT(U_2) for valid positions U_1 and U_2 such that ℓ_1 is semi-universal for U_1 and ℓ_2 is semi-universal for U_2. We must have $U_1 \neq U_2$; otherwise, $|U_1(i)| = |U_2(i)| \geq 2$, and, thus, neither ℓ_1 nor ℓ_2 is semi-universal, a contradiction. Further, it follows from the construction of $\mathcal{T}$ that we must have $\text{Desc}_\mathcal{P}(U_1) \cap \text{Desc}_\mathcal{P}(U_2) = \emptyset$. Hence, $\ell \parallel_\mathcal{T} \ell'$, and, therefore, $\{\ell, \ell'\} \in N(\mathcal{T})$.

(ii) Asssume, by way of contradiction, that BuildNT(U_{root}) returns incompatible, but that $\mathcal{P}$ is ancestrally compatible. By assumption, there exists a semi-labeled tree $\mathcal{T}$ that ancestrally displays $\mathcal{P}$. Since BuildNT(U_{root}) returns `incompatible`, there is a recursive call to `BuildNT`(U) for some valid position U such that U has no semi-universal label, and the set S of Line 1 is empty.

By Lemma 2, $T|\text{Desc}_{\mathcal{P}}(U)$ ancestrally displays $\mathcal{P}|\text{Desc}_{\mathcal{P}}(U)$. Thus, by Lemma 4, $T|\text{Desc}_{\mathcal{P}}(U)$ ancestrally displays $T_i|\text{Desc}_i(U)$, for every $i \in [k]$. Let ℓ be any label in the label set of the root of $T|\text{Desc}_{\mathcal{P}}(U)$. Then, for each $i \in [k]$ such that $\ell \in L(T_i)$, ℓ must be the label of the root of $T_i|\text{Desc}_i(U)$. Thus, for each such i, $U(i) = \{\ell\}$. Hence, ℓ is semi-universal in U, a contradiction.   □

5   Implementation

We focus on two key aspects of the implementation of `BuildNT`: finding semi-universal labels in Line 1, and updating U and $H_{\mathcal{P}}(U)$ in Lines 6 and 7.

By Observation 1, each recursive call to `BuildNT` deals with a graph obtained from $H_{\mathcal{P}}$ through edge and node deletions. To handle these deletions efficiently, we represent $H_{\mathcal{P}}$ using the dynamic graph connectivity data structure of Holm et al. [12], which we refer to as *HDT*. HDT maintains the list of nodes in each component, as well as the number of these nodes so that, starting with no edges in a graph with N nodes, the amortized cost of each update is $O(\log^2 N)$. Since $H_{\mathcal{P}}$ has $O(M_{\mathcal{P}})$ nodes, each update takes $O(\log^2 M_{\mathcal{P}})$ time. The total number of edge and node deletions performed by `BuildNT`(U_{root}) — including all deletions in the recursive calls — is at most the total number of edges and nodes in $H_{\mathcal{P}}$, which is $O(M_{\mathcal{P}})$. HDT allows us to maintain connectivity information throughout the entire algorithm in $O(M_{\mathcal{P}} \log^2 M_{\mathcal{P}})$ time.

As deletions are performed on $H_{\mathcal{P}}$, `BuildNT` maintains three data fields for each connected component Y that is created: $Y.\texttt{weight}$, $Y.\texttt{map}$, and $Y.\texttt{semiU}$. It also maintains a field $\ell.\texttt{count}$, for each $\ell \in L(\mathcal{P})$.

(i) $Y.\texttt{weight}$ equals $\sum_{\ell \in Y} k_\ell$.
(ii) $Y.\texttt{map}$ is a map from a set $J_Y \subseteq [k]$ to a set of nonempty subsets of $Y \cap L(T_i)$. For each $i \in J_Y$, $Y.\texttt{map}(i)$ denotes the set associated with i.
(iii) $\ell.\texttt{count}$ equals the cardinality of the set $\{i \in [k] : Y.\texttt{map}(i)$ is defined and $Y.\texttt{map}(i) = \{\ell\}\}$.
(iv) $Y.\texttt{semiU}$ is a set containing all labels $\ell \in Y$ such that $\ell.\texttt{count} = k_\ell$.

Informally, each set $Y.\texttt{map}(i)$ corresponds to a non-empty $U(i)$; $Y.\texttt{semiU}$ corresponds to the semi-universal labels in Y. Next, we formalize these ideas.

At the start of the execution of `BuildNT`(U) for any valid position U, $H_{\mathcal{P}}(U)$ has a single connected component, $Y_U = \text{Desc}_{\mathcal{P}}(U)$. Our implementation maintains the following invariant.

INV: At the beginning of the execution of `BuildNT`(U), $Y_U.\texttt{map}(i) = U(i)$ for each $i \in [k]$ such that $U(i) \neq \emptyset$, and $Y_U.\texttt{map}(i)$ is undefined for each $i \in [k]$ such that $U(i) = \emptyset$.

Thus, $\ell.\texttt{count}$ equals the number of indices $i \in [k]$ such that $U(i) = \{\ell\}$. Along with Lemma 5, INV implies that, at the beginning of the execution of

BuildNT(U), Y_U.semiU contains precisely the semi-universal labels of U. Thus, the set S of line 1 of BuildNT(U) can be retrieved in $O(1)$ time.

We establish INV for the initial valid position U_{root} as follows. By assumption, $H_{\mathcal{P}}(U_{\text{root}})$ has a single connected component, $Y_{\text{root}} = L(\mathcal{P})$. Since $H_{\mathcal{P}}(U_{\text{root}})$ equals $H_{\mathcal{P}}$, we initialize data fields (i)–(iv) for Y_{root} during the construction of $H_{\mathcal{P}}$. Y_{root}.weight is simply $\sum_{\ell \in L(\mathcal{P})} k_\ell$. For each $i \in [k]$, Y_{root}.map(i) is $\{\ell\}$, where ℓ is the label of the root of T_i. We initialize the count fields as follows. First, set ℓ.count to 0 for all $\ell \in L(\mathcal{P})$. Then, iterate through each $i \in [k]$, incrementing ℓ.count by one if Y_{root}.map(i) = $\{\ell\}$. Finally, Y_{root}.semiU consists of all $\ell \in U_{\text{root}}$ such that ℓ.count = k_ℓ. All data fields can be initialized in $O(M_{\mathcal{P}})$ time.

We now focus on Lines 6 and 7 of BuildNT. We update U and $H_{\mathcal{P}}(U)$ jointly as follows. Let G_{BNT} be a temporary variable, such that, initially, $G_{\text{BNT}} = H_{\mathcal{P}}(U)$. Now, successively consider each label $\ell \in S$, and perform two steps: (a) initialize data fields (i)–(iv) in preparation for the deletion of ℓ and (b) delete from G_{BNT} the edges incident on ℓ and then ℓ itself, updating data fields (i)–(iv) as necessary, to maintain INV. By Observation 2, after these steps are executed, G_{BNT} must equal $H_{\mathcal{P}}(U)$ for the new set U created by Line 6.

To initialize the data fields prior to deleting label ℓ, first consult HDT to identify the connected component Y that contains ℓ. Then, since ℓ will cease to be semi-universal, remove ℓ from Y.semiU. Next, for each $i \in [k]$ such that $\ell \in L(T_i)$, if $\text{Ch}_i(\ell) \neq \emptyset$, replace Y.map(i) by $\text{Ch}_i(\ell)$; otherwise, Y.map(i) is undefined. These operations may create new singleton sets Y.map(i), so we may need to update certain count and semiU fields. Since each label is considered at most once, the total number of operations on map fields of the various sets Y considered over the entire execution of BuildNT(U_{root}) is $O(\sum_{i \in [k]:\ell \in L(T_i)} |\text{Ch}_i(\ell)|)$, which is $O(M_{\mathcal{P}})$. The same bound holds for updates to count and semiU fields.

To delete a label ℓ, we begin by successively deleting each edge between ℓ and a child α of ℓ, updating the appropriate data fields for the resulting connected components. This is done as follows.

1. Query HDT to determine the connected component Y containing ℓ.
2. Delete (ℓ, α), querying HDT to determine whether this disconnects Y.
3. If Y remains connected, skip the next steps and proceed directly to the next child of ℓ. Otherwise, Y is split into two components, Y_1 and Y_2.
4. Identify which of Y_1 and Y_2 has the smaller weight field. Without loss of generality, assume that Y_1.weight $\leq Y_2$.weight.
5. Initialize Y_1.map and Y_1.semiU to null and Y_2.map and Y_2.semiU to the corresponding fields of Y.
6. Iterate through each label β in Y_1. For every i such that $\beta \in L(T_i)$, move β from Y_2.map(i) to Y_1.map(i). As we do this, record in a set J the indices i such that Y_1.map(i) and Y_2.map(i) are modified.
7. For each $i \in J$, (i) if Y_1.map(i) is empty, delete it, and (ii) if Y_1.map(i) consists of a single label γ, increment γ.count by one and, if γ.count = k_γ, add γ to Y_1.semiU. Proceed similarly for Y_2.

After all edges incident on ℓ are deleted, ℓ itself is deleted.

Let us track the number of operations on `map` fields in step 5 that can be attributed to some specific label $\beta \in L(\mathcal{P})$ over the entire execution of `BuildNT`(U_{root}). Each execution of step 5 for β performs k_β operations on `map` fields. Let $w_r(\beta)$ be the weight of the connected component containing β at the beginning of step 5, at the rth time that β is considered in that step; thus, $w_0(\beta) \leq \sum_{\ell \in L(\mathcal{P})} k_\ell$. We claim that $w_r(\beta) \leq w_0(\beta)/2^r$. The reason is that we only consider β if (a) β is contained in one of the two components that result from deleting an edge in step 5 and (b) the component containing β has the smaller weight. Thus, the number of times β is considered in step 5 over the entire execution of `BuildNT`(U_{root}) is $O(\log w_0(\beta))$, which is $O(\log M_\mathcal{P})$, since $w_0(\beta) = O(M_\mathcal{P})$. Therefore, the total number of updates of `map` fields over all labels is $O(\log M_\mathcal{P} \cdot \sum_{\ell \in L(\mathcal{P})} k_\ell)$, which is $O(M_\mathcal{P} \log M_\mathcal{P})$. We remark that this analysis, in effect, invokes a weighted version of the well-known technique of scanning the smaller component [9].

The `weight` fields of the connected components are maintained using the original, unweighted, version of technique of scanning the smaller component. It can be shown that total time to update these fields over all edge deletions performed by `BuildNT`(U_{root}) is $O(M_\mathcal{P} \log M_\mathcal{P})$. The number of updates to `count` and `semiU` fields in step 5 can also be shown to be $O(M_\mathcal{P} \log M_\mathcal{P})$; each update takes $O(1)$ time. Due to space limitations, we omit further details.

To summarize, the work done by `BuildNT`(U_{root}) consists of three parts: (i) initialization, (ii) maintaining connected components, and (iii) maintaining the data fields for each connected component and each label. Part (i) takes $O(M_\mathcal{P})$ time. Part (ii) involves $O(M_\mathcal{P})$ edge and node deletions on the HDT data structure, at an amortized cost of $O(\log^2 M_\mathcal{P})$ per deletion. Part (iii) requires a total of $O(M_\mathcal{P} \log M_\mathcal{P})$ updates to the various fields. Using data structures that take logarithmic time per update, leads to our main result.

Theorem 2. `BuildNT` *can be implemented so that* `BuildNT`(U_{root}) *runs in* $O(M_\mathcal{P} \log^2 M_\mathcal{P})$ *time.*

6   Discussion

Like our earlier algorithm for compatibility of ordinary phylogenetic trees, the more general algorithm presented here, `BuildNT`, is a polylogarithmic factor away from optimality (a trivial lower bound is $\Omega(M_\mathcal{P})$, the time to read the input). `BuildNT` has a linear-space implementation, using the results of Thorup [22]. A question to be investigated next is the performance of the algorithm on real data. Another important issue is integrating our algorithm into a synthesis method that deals with incompatible profiles.

References

1. Aho, A.V., Sagiv, Y., Szymanski, T.G., Ullman, J.D.: Inferring a tree from lowest common ancestors with an application to the optimization of relational expressions. SIAM J. Comput. **10**(3), 405–421 (1981)

2. Baum, B.R.: Combining trees as a way of combining data sets for phylogenetic inference, and the desirability of combining gene trees. Taxon **41**, 3–10 (1992)
3. Berry, V., Semple, C.: Fast computation of supertrees for compatible phylogenies with nested taxa. Syst. Biol. **55**(2), 270–288 (2006)
4. Bininda-Emonds, O.R.P., et al.: The delayed rise of present-day mammals. Nature **446**, 507–512 (2007)
5. Bordewich, M., Evans, G., Semple, C.: Extending the limits of supertree methods. Ann. Comb. **10**, 31–51 (2006)
6. Bryant, D., Lagergren, J.: Compatibility of unrooted phylogenetic trees is FPT. Theoret. Comput. Sci. **351**, 296–302 (2006)
7. Daniel, P., Semple, C.: Supertree algorithms for nested taxa. In: Bininda-Emonds, O.R.P. (ed.) Phylogenetic Supertrees: Combining Information to Reveal the Tree of Life, pp. 151–171. Kluwer, Dordrecht (2004)
8. Deng, Y., Fernández-Baca, D.: Fast compatibility testing for rooted phylogenetic trees. In: Proceedings of the 27th Annual Symposium on Combinatorial Pattern Matching (to appear)
9. Even, S., Shiloach, Y.: An on-line edge-deletion problem. J. ACM **28**(1), 1–4 (1981)
10. Henzinger, M.R., King, V., Warnow, T.: Constructing a tree from homeomorphic subtrees, with applications to computational evolutionary biology. Algorithmica **24**, 1–13 (1999)
11. Hinchliff, C.E., et al.: Synthesis of phylogeny and taxonomy into a comprehensive tree of life. Proc. Nat. Acad. Sci. **112**(41), 12764–12769 (2015)
12. Holm, J., de Lichtenberg, K., Thorup, M.: Poly-logarithmic deterministic fully-dynamic algorithms for connectivity, minimum spanning tree, 2-edge, and biconnectivity. J. ACM **48**(4), 723–760 (2001)
13. Page, R.M.: Taxonomy, supertrees, and the tree of life. In: Bininda-Emonds, O.R.P. (ed.) Phylogenetic Supertrees: Combining Information to Reveal the Tree of Life, pp. 247–265. Kluwer, Dordrecht (2004)
14. Pe'er, I., Pupko, T., Shamir, R., Sharan, R.: Incomplete directed perfect phylogeny. SIAM J. Comput. **33**(3), 590–607 (2004)
15. Ragan, M.A.: Phylogenetic inference based on matrix representation of trees. Mol. Phylogenet. Evol. **1**, 53–58 (1992)
16. Sanderson, M.J.: Phylogenetic signal in the eukaryotic tree of life. Science **321**(5885), 121–123 (2008)
17. Sayers, E.W., et al.: Database resources of the National Center for Biotechnology Information. Nucleic Acids Res. **37**(Database issue), D5–D15 (2009)
18. Semple, C., Steel, M.: Phylogenetics. Oxford Lecture Series in Mathematics. Oxford University Press, Oxford (2003)
19. Smith, S.A., Brown, J.W., Hinchliff, C.E.: Analyzing and synthesizing phylogenies using tree alignment graphs. PLoS Comput. Biol. **9**(9), e1003223 (2013)
20. Steel, M.A.: The complexity of reconstructing trees from qualitative characters and subtrees. J. Classif. **9**, 91–116 (1992)
21. The Angiosperm Phylogeny Group: An update of the Angiosperm Phylogeny Group classification for the orders and families of flowering plants: APG IV. Bot. J. Linn. Soc. **181**, 1–20 (2016)
22. Thorup, M.: Near-optimal fully-dynamic graph connectivity. In: Proceedings of the 32nd Annual ACM Symposium on Theory of Computing, pp. 343–350. ACM (2000)

The Gene Family-Free Median of Three

Daniel Doerr[1]([✉]), Pedro Feijão[2], Metin Balaban[1], and Cedric Chauve[3]([✉])

[1] School of Computer and Communication Sciences, EPFL,
1015 Lausanne, Switzerland
daniel.doerr@epfl.ch
[2] Faculty of Technology, Bielefeld University, 33615 Bielefeld, Germany
[3] Department of Mathematics, Simon Fraser University, Burnaby, BC, Canada
cedric.chauve@sfu.ca

Abstract. The gene family-free framework for comparative genomics aims at developing methods for gene order analysis that do not require prior gene family assignment, but work directly on a sequence similarity graph. We present a model for constructing a median of three genomes in this family-free setting, based on maximizing an objective function that generalizes the classical breakpoint distance by integrating sequence similarity in the score of a gene adjacency. We show that the corresponding computational problem is MAX SNP-hard and we present a 0–1 linear program for its exact solution. The result of this program is a median genome with median genes associated to extant genes, in which median adjacencies are assumed to define positional orthologs. We demonstrate through simulations and comparison with the OMA orthology database that the herein presented method is able compute accurate medians and positional orthologs for genomes comparable in size of bacterial genomes.

1 Introduction

The prediction of evolutionary relationships between genomic sequences is a long-standing problem in computational biology. According to Fitch [8], two genomic sequences are called *homologous* if they descended from a common ancestral sequence. Furthermore, Fitch identifies different events that give rise to a branching point in the phylogeny of homologous sequences, leading to the well-established concepts of orthologous genes (who descend from their last common ancestor through a speciation) and paralogous genes (descending from their last common ancestor through a duplication) [9]. Until quite recently, orthology and paralogy relationships were mostly inferred from sequence similarity. However it is now well accepted that the syntenic context can carry valuable evolutionary information, which has lead to the notion of *positional orthologs* [5], which are orthologs whose syntenic context was not changed in a duplication event. In the present work, we describe a method to compute groups of likely orthologous genes for a group of three genomes, through a new problem we introduce, the *gene family-free median of three*.

Most methods for detecting potential orthologous groups require a prior clustering of the genes of the considered genomes into *homologous gene families*,

© Springer International Publishing Switzerland 2016
M. Frith and C.N.S. Pedersen (Eds.): WABI 2016, LNBI 9838, pp. 102–120, 2016.
DOI: 10.1007/978-3-319-43681-4_9

defined as groups of genes assumed to originate from a single ancestral gene. Yet clustering protein sequences into families is already in itself a difficult problem.

Here, we follow the matching-based approach, framed within the gene family-free principle, that embodies the idea to perform gene order analysis without the prerequisite of gene family or homology assignments. Instead, we are given all-against-all *gene similarities* through a symmetric and reflexive *similarity measure* $\sigma : \Sigma \times \Sigma \to \mathbb{R}_{\geq 0}$ over the universe of genes Σ [3]. We use sequence similarity but other similarity measures can fit the previous definition. This leads to the formalization of the *gene similarity graph* [3], i.e. a graph where each vertex corresponds to a gene of the dataset and where each pair of vertices associated with genes of distinct genomes are connected by a strictly positively weighted edge according to gene similarity measure σ. Gene family or homology assignments represent a particular subgroup of gene similarity functions that require transitivity. Independent of the particular similarity measure σ, relations between genes imposed by σ are considered as candidates for homology assignments. A gene family-free research program was outlined in [3] (see also [7]) and has so far been developed for the pairwise comparison of genomes [6,10,13] and shown to be effective for orthology analysis [11].

In Sect. 2 we introduce a new genome median problem in the family-free framework, that generalizes the traditional breakpoint median problem [16]. For a group of three genomes, the input of the family-free median problem is a tripartite similarity graph of pairwise gene similarities. Informally, a median of three is defined as a genome, and as such is composed of a set of median genes that are associated to the genes of the input genomes and that give rise to one or more linear or circular gene order sequences. The matching of median genes to input genes as well as their ordering in the median genome is subject to an optimization problem. Hereby, our optimization criterion fully integrates both sequence similarity and gene order conservation.

In Sect. 3 we study its the computational complexity and give an exact algorithm for its solution. We show that our method can be used for positional ortholog prediction in simulated and real data sets of bacterial genomes in Sect. 4.

2   The Gene Family-Free Median of Three

Extant genomes, genes and adjacencies. In this work, a genome G is entirely represented by a tuple $G \equiv (\mathcal{C}, \mathcal{A})$, where $\mathcal{C}$ denotes a non-empty set of unique genes, and $\mathcal{A}$ is a set of *adjacencies*. Genes are represented by their *extremities*, i.e., a gene $g \equiv (g^{\mathrm{t}}, g^{\mathrm{h}})$, $g \in \mathcal{C}$, consists of a *head* g^{h} and a *tail* g^{t}. Telomeres are modeled explicitly, as special genes of $\mathcal{C}(G)$ with a single extremity, denoted by "$\circ$". Extremities g_1^a, g_2^b, $a, b \in \{\mathrm{h}, \mathrm{t}\}$ of any two genes g_1, g_2 form an *adjacency* $\{g_1^a, g_2^b\}$ if they are immediate neighbors in their genome sequence. In the following, we will conveniently use the notation $\mathcal{C}(G)$ and $\mathcal{A}(G)$ to denote the set of genes and the set of adjacencies of genome G, respectively. We indicate the presence of an adjacency $\{x_1^a, x_2^b\}$ in an extant genome X by

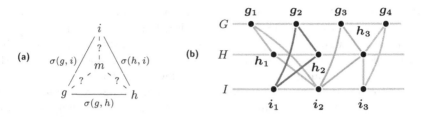

Fig. 1. (a) Illustration of the score of a candidate median gene. (b) Gene similarity graph of three genomes G, H, and I. Colored components indicate candidate median genes $m_1 = (g_1, h_1, i_2)$, $m_2 = (g_2, h_2, i_1)$, $m_3 = (g_3, h_3, i_2)$, and $m_4 = (g_4, h_3, i_3)$. Median gene pairs m_1, m_3 and m_3, m_4 are conflicting.

$$\mathbb{I}_X(x_1^a, x_2^b) = \begin{cases} 1 & \text{if } \{x_1^a, x_2^b\} \in \mathcal{A}(X) \\ 0 & \text{otherwise.} \end{cases} \tag{1}$$

Given two genomes G and H and gene similarity measure σ, two adjacencies, $\{g_1^a, g_2^b\} \in \mathcal{A}(G)$ and $\{h_1^a, h_2^b\} \in \mathcal{A}(H)$ with $a, b \in \{h, t\}$ are *conserved* iff $\sigma(g_1, h_1) > 0$ and $\sigma(g_2, h_2) > 0$. We subsequently define the *adjacency score* of any four extremities g^a, h^b, i^c, j^d, where $a, b, c, d \in \{h, t\}$ and $g, h, i, j \in \Sigma$ as the geometric mean of their corresponding gene similarities [3]:

$$s(g^a, h^b, i^c, j^d) \equiv \sqrt{\sigma(g, h) \cdot \sigma(i, j)} \tag{2}$$

Median genome, genes and adjacencies. Informally, the family-free median problem asks for a fourth genome M that maximizes the sum of pairwise adjacency scores to three given extant genomes G, H, and I. In doing so, the gene content of the requested median M must first be defined: each gene $m \in \mathcal{C}(M)$ must be unambiguously associated with a triple of extant genes (g, h, i), $g \in \mathcal{C}(G)$, $h \in \mathcal{C}(H)$, and $i \in \mathcal{C}(I)$. Moreover, we want to associate to a median gene m a sequence similarity score (g, h, i) relatively to the three extant genes it is related to. As the sequence of the median gene is obviously not available, we define this score as the geometric mean of their pairwise similarities (see Fig. 1 (a)):

$$\sigma(g, m) = \sigma(h, m) = \sigma(i, m) \equiv \sqrt[3]{\sigma(g, h) \cdot \sigma(g, i) \cdot \sigma(h, i)} \tag{3}$$

In the following we make use of mapping $\pi_G(m) \equiv g$, $\pi_H(m) \equiv h$, and $\pi_I(m) \equiv i$ to relate gene m with its extant counterparts. Two candidate median genes or telomeres m_1 and m_2 are *conflicting* if $m_1 \neq m_2$ and the intersection between associated gene sets $\{\pi_G(m_1), \pi_H(m_1), \pi_I(m_1)\}$ and $\{\pi_G(m_2), \pi_H(m_2), \pi_I(m_2)\}$ is non-empty (see Fig. 1 (b) for example). A set of candidate median genes or telomeres $\mathcal{C}$ is called *conflict-free* if no two of its members $m_1, m_2 \in \mathcal{C}$ are conflicting. This definition trivially extends to the notion of a *conflict-free* median.

Problem 1 (FF-Median). Given three genomes G, H, and I, and gene similarity measure σ, find a conflict-free median M, which maximizes the following formula:

$$\mathcal{F}_\lambda(M) = \sum_{\{m_1^a, m_2^b\} \in \mathcal{A}(M)} \sum_{\substack{X \in \{G, H, I\}, \\ \{\pi_X(m_1)^a, \pi_X(m_2)^b\} \in \mathcal{A}(X)}} s(m_1^a, \pi_X(m_1)^a, m_2^b, \pi_X(m_2)^b),$$

(4)

where $a, b \in \{h, t\}$ and $s(\cdot)$ is the adjacency score as defined by Eq. (2).

Remark 1. The adjacency score for a median adjacency $\{m_1^a, m_2^b\}$ with respect to the corresponding potential extant adjacency $\{\pi_X(m_1)^a, \pi_X(m_2)^b\}$, where $\{m_1^a, m_2^b\} \in \mathcal{A}(M)$ and $X \in \{G, H, I\}$, can be entirely expressed in terms of pairwise similarities between genes of extant genomes using Eq. (3):

$$s(m_1^a, \pi_X(m_1)^a, m_2^b, \pi_X(m_2)^b) = \sqrt[6]{\prod_{\{Y,Z\} \subset \{G,H,I\}} \sigma(\pi_Y(m_1), \pi_Z(m_1)) \cdot \sigma(\pi_Y(m_2), \pi_Z(m_2))}$$

In the following, a median gene m and its extant counterparts (g, h, i) are treated as equivalent. We denote the set of all *candidate median genes* by

$$\Sigma_\lambda = \{(g, h, i) \mid g \in \mathcal{C}(G), h \in \mathcal{C}(H), i \in \mathcal{C}(I) : \sigma(g, h) \cdot \sigma(g, i) \cdot \sigma(h, i) > 0\}. \quad (5)$$

Each pair of median genes $(g_1, h_1, i_1), (g_2, h_2, i_2) \in \Sigma_\lambda$ and extremities $a, b \in \{h, t\}$ give rise to a *candidate median adjacency* $\{(g_1^a, h_1^a, i_1^a), (g_2^b, h_2^b, i_2^b)\}$ if $(g_1^a, h_1^a, i_1^a) \neq (g_2^b, h_2^b, i_2^b)$, and (g_1^a, h_1^a, i_1^a) and (g_2^b, h_2^b, i_2^b) are non-conflicting. We denote the set of all candidate median adjacencies and the set of all *conserved* (*i.e.* present in at least one extant genome) candidate median adjacencies by $\mathcal{A}_\lambda = \{\{m_1^a, m_2^b\} \mid m_1, m_2 \in \Sigma_\lambda, a, b \in \{h, t\}\}$ and $\mathcal{A}_\lambda^C = \{\{m_1^a, m_2^b\} \in \mathcal{A}_\lambda \mid \sum_{X \in \{G, H, I\}} \mathbb{I}_X(\pi_X(m_1)^a, \pi_X(m_2)^b) \geq 1\}$, respectively.

Remark 2. A median gene can only belong to a median adjacency with non-zero adjacency score if all pairwise similarities of its corresponding extant genes g, h, i are non-zero. Thus, the search for median genes can be limited to 3-cliques (triangles) in the tripartite similarity graph.

Remark 3. The right-hand side of the above formula for the weight of an adjacency is independent of genome X. From Eq. (4), an adjacency in median M has only an impact in a solution to problem FF-Median if it participates in a gene adjacency in at least one extant genome. So including in a median genome median genes that do not belong to a candidate median adjacency in $\mathcal{A}_\lambda^C$ do not increase the objective function.

Related problems. The FF-median problem relates to previously studied gene order evolution problems. It is a generalization of the tractable mixed multi-chromosomal median problem introduced in [16], that can indeed be defined as an FF-median problem with a similarity graph composed of disjoint 3-cliques and edges having all the same weight. The FF-median problem also bears similarity with methods aimed at detecting groups of orthologous genes based on gene order evolution, especially the MultiMSOAR [15] algorithm, although other method integrate synteny and sequence conservation for inferring orthogroups, see [5].

Our approach differs first and foremost in its family-free principle (all other methods require a prior gene family assignment). Compared to MultiMSOAR, the only other method that can handle more than two genomes with an optimization criterion that considers gene order evolution, both MultiMSOAR (for three genomes) and FF-median aim at computing a maximum weight tripartite matching. However we differ fundamentally from MultiMSOAR by the full integration of sequence and synteny conservation into the objective function, while MultiMSOAR proceeds first by computing pairwise orthology assignments to define a multipartite graph.

3   Algorithmic and Complexity Results

We now describe our theoretical results: a NP-hardness proof, an exact Integer Linear Program (ILP), and an algorithm to detect local optimal structures.

Theorem 1. *Problem FF-Median is MAX SNP-hard.*

We describe the full hardness proof in Appendix A. It is based on a reduction from the Maximum Independent Set for Graphs of Bounded Degree 3.

An exact ILP algorithm to problem FF-Median. We now present program FF-Median, described by Algorithm 1, that exploits the specific properties of problem FF-Median to design an ILP using $\mathcal{O}(n^5)$ variables and statements. Program FF-Median makes use of two types of binary variables **a** and **b** as declared in domain specifications (D.01) and (D.02), that defines the set of median genes Σ_λ and of candidate conserved median adjacencies $\mathcal{A}_\lambda^C$ (Remark 3). The former variable type indicates the presence or absence of candidate genes in an optimal median M. The latter, variable type **b**, specifies if an adjacency between two gene extremities or telomeres is established in M. Constraint (C.01) ensures that M is conflict-free, by demanding that each extant gene (or telomere) can be associated with at most one median gene (or telomere). Further, constraint (C.02) dictates that a median adjacency can only be established between genes that both are part of the median. Lastly, constraint (C.03) guarantees that each gene extremity and telomere of the median participates in at most one adjacency.

Property 1. The size (i.e. number of variables and statements) of any ILP returned by program FF-Median is limited by $\mathcal{O}(n^5)$ where $n = \max(|\mathcal{C}(G)|, |\mathcal{C}(H)|, |\mathcal{C}(I)|)$.

Remark 4. The output of the algorithm FF-Median is a set of adjacencies between median genes that define a set of linear and/or circular orders, called CARs (Contiguous Ancestral Regions), where linear segments are not capped by telomeres. So formally the computed median might not be a valid genome. However, as adding adjacencies that do not belong to $\mathcal{A}_\lambda^C$ do not modify the score of a given median, a set of median adjacencies can always be completed into

---

Algorithm 1. Program `FF-Median` for three genomes (G, H, I)

Objective: Maximize

$$\sum_{\substack{\{m_1, m_2\} \in \mathcal{A}_\lambda^C, \\ m_1 = (g_1, h_1, i_1), \\ m_2 = (g_2, h_2, i_2), \\ a, b \in \{h, t\}}} \mathbf{b}(g_1^a, g_2^b, h_1^a, h_2^b, i_1^a, i_2^b) \qquad \sum_{\substack{X \in \{G, H, I\}, \\ \{\pi_X(m_1)^a, \pi_X(m_2)^b\} \in \mathcal{A}(X)}} s(m_1^a, \pi_X(m_1)^a, m_2^b, \pi_X(m_2)^b)$$

Constraints:

(C.01) $\forall\, g' \in \mathcal{C}(G)$: $\displaystyle\sum_{(g', h, i) \in \Sigma_\lambda} \mathbf{a}(g', h, i) \leq 1$

$\forall\, h' \in \mathcal{C}(H)$: $\displaystyle\sum_{(g, h', i) \in \Sigma_\lambda} \mathbf{a}(g, h', i) \leq 1$

$\forall\, i' \in \mathcal{C}(I)$: $\displaystyle\sum_{(g, h, i') \in \Sigma_\lambda} \mathbf{a}(g, h, i') \leq 1$

(C.02) $\forall\, \{(g_1, h_1, i_1), (g_2, h_2, i_2)\} \in \mathcal{A}_\lambda^C$ and $\forall\, a, b \in \{h, t\}$:
$$2 \cdot \mathbf{b}(g_1^a, g_2^b, h_1^a, h_2^b, i_1^a, i_2^b) \leq \mathbf{a}(g_1, h_1, i_1) + \mathbf{a}(g_2, h_2, i_2)$$

(C.03) $\forall\, (g_1, h_1, i_1) \in \Sigma_\lambda$ and $\forall\, a \in \{h, t\}$:
$$\sum_{(g_2, h_2, i_2) \in \Sigma_\lambda, \ b \in \{h, t\}} \mathbf{b}(g_1^a, g_2^b, h_1^a, h_2^b, i_1^a, i_2^b) \leq 1$$

Domains:

(D.01) $\forall\, (g, h, i) \in \Sigma_\lambda$: $\mathbf{a}(g, h, i) \in \{0, 1\}$

(D.02) $\forall\, \{(g_1, h_1, i_1), (g_2, h_2, i_2)\} \in \mathcal{A}_\lambda^C$ and $\forall\, a, b \in \{h, t\}$:
$$\mathbf{b}(g_1^a, g_2^b, h_1^a, h_2^b, i_1^a, i_2^b) \in \{0, 1\}$$

---

a valid genome by such adjacencies that join the linear segments together and add telomeres. These extra adjacencies would not be supported by any extant genome and thus can be considered as dubious, and in our implementation, we only return the median adjacencies computed by the ILP, *i.e.* a subset of $\mathcal{A}_\lambda^C$.

Remark 5. Following Remark 2, preprocessing the input extant genomes requires to handle the extant genes that do not belong to at least one 3-clique in the similarity graph. Such genes can not be part of any median. So one could decide to leave them in the input, and the ILP can handle them and ensures they are never part of the output solution. However, discarding them from the extant genomes can help recover adjacencies that have been disrupted by the insertion of a mobile element for example, so in our implementation we follow this approach.

As discussed at the end of Sect. 2, the FF-median problem is a generalization of the mixed multichromosomal breakpoint median [16]. However, it was shown in [16] that this breakpoint median problem can be solved in polynomial time by a Maximum-Weight Matching (MWM) algorithm. This motivates the results presented in the next paragraph that use a MWM algorithm to identify optimal median substructures by focusing on conflict-free sets of median genes.

Finding local optimal segments. Tannier *et al.* [16] solve the mixed multichromo-somal breakpoint median problem by transforming it into an MWM problem, that we outline now. A graph is defined in which each extremity of a candidate median gene and each telomere gives rise to a vertex. Any two vertices are connected by an edge, weighted according to the number of observed adjacencies between the two gene extremities in extant genomes. Edges corresponding to adjacencies between a gene extremity and telomeres are weighted only by half as much. An MWM in this graph induces a set of adjacencies that defines an optimal median.

We first describe how this approach applies to our problem. We define a graph $\Gamma(\Sigma_\lambda)$ constructed from an FF-Median instance (G, H, I, σ) that is similar to that of Tannier *et al.*, deviating by defining vertices as candidate median gene extremities and weighting an edge between two vertices $m_1^a, m_2^b, a, b \in \{h, t\}$, by

$$w(\{m_1^a, m_2^b\}) = \sum_{X \in \{G, H, I\}} \mathbb{I}_X(\pi_X(m_1)^a, \pi_X(m_2)^b) \cdot s(m_1^a, \pi_X(m_1)^a, m_2^b, \pi_X(m_2)^b).$$

$$(6)$$

We make first the following observation, where a conflict-free matching is a matching that does not contain two conflicting vertices (candidate median genes):

Observation 2. *Any conflict-free matching in graph $\Gamma(\Sigma_\lambda)$ of maximum weight defines an optimal median.*

We show now that we can define notions of sub-instances – of a full FF-median instance – that contains no internal conflicts, for which applying the MWM can allow to detect if the set of median genes defining the sub-instance is part of at least one optimal FF-median. Let $\mathcal{S}$ be a set of candidate median genes. An *internal conflict* is a conflict between two genes from $\mathcal{S}$; an *external conflict* is a conflict between a gene from $\mathcal{S}$ and a candidate median gene not in $\mathcal{S}$. We say that $\mathcal{S}$ is *contiguous* in extant genome X if the set $\pi_X(\mathcal{S})$ forms a unique, contiguous, segment in X. We say that $\mathcal{S}$ is an *internal-conflict free segment* (IC-free segment) if it contains no internal conflict and is contiguous in all three extant genomes; this can be seen as the family-free equivalent of the notion of *common interval in permutations* [2]. An IC-free segment is a *run* if the order of the extant genes is conserved in all three extant genomes, up to a full reversal of the segment.

Intuitively, one can find an optimal solution to the sub-instance defined by an IC-free segment, but it might not be part of an optimal median for the whole instance due to side effects of the rest of the instance. So we need to adapt the graph to which we apply an MWM algorithm to account for such side effects. To do so, we define the *potential* of a candidate median gene m as

$$\Delta(m) = \max_{\{m_1^a, m^b\}, \{m^a, m_2^b\} \in \mathcal{A}_\lambda} \left(w(\{m_1^a, m^b\}) + w(\{m^a, m_2^b\}) \right).$$

We then extend graph $\Gamma(\mathcal{S}) =: (V, E)$ to graph $\Gamma'(\mathcal{S}) := (V, E')$ by adding edges between the extremities of each candidate median gene of an IC-free segment $\mathcal{S}$, i.e. $E' = E \cup \{\{m^h, m^t\} \mid m \in \mathcal{S}\}$ (note that when $|\mathcal{S}| > 1$, $w(\{m^h, m^t\}) = 0$ since $\mathcal{S}$ is contiguous in all three extant genomes). In the following we refer to these edges as *conflict edges*. Let $C(m)$ be the set of candidate median genes that are involved in an (external) conflict with a given candidate median gene m of $\mathcal{S}$, then the conflict edge $\{m^h, m^t\} \in E'$ is weighted by the maximum potential of a non-conflicting subset of $C(m)$,

$$w'(\{m^h, m^t\}) = \max(\{ \sum_{m' \in C'} \Delta(m') \mid C' \subseteq C(m) : C' \text{ is conflict-free}\}).$$

A conflict-free matching in $\Gamma'(\mathcal{S})$ is a matching without a conflict edge.

Lemma 1. *Given an internal conflict-free segment $\mathcal{S}$, any maximum weight matching in graph $\Gamma'(\mathcal{S})$ that is conflict-free defines a set of median genes and adjacencies that belong to at least one optimal FF-median of the whole instance.*

A proof is presented in Appendix B. Lemma 1 leads to a procedure (Algorithm 2) that iteratively identifies and tests IC-free segments in the FF-Median instance. For each identified IC-free segment S an adjacency graph $\Gamma'(S)$ is constructed and a maximum weight matching is computed (Line 2–3). If the resulting matching is conflict free (Line 4), adjacencies of IC-free segment S are reported and S is removed from an FF-Median instance by masking its internal adjacencies and removing all candidate median genes (and consequently their associated candidate median adjacencies) corresponding to external conflicts (Line 5–6). It then follows immediately from Lemma 1 that the set median genes returned by Algorithm 2 belongs to at least one optimal solution to the FF-median problem.

Algorithm 2. Algorithm ICF–SEG

Input: FF-Median instance (G, H, I, σ)
Output: Set of adjacencies ADJ_M that is part of a median M of (G, H, I, σ).

1: **while** there exists an unobserved IC-free conserved segment S in (G, H, I, σ) **do**
2:     Construct adjacency graph $\Gamma'(S)$ of S
3:     Find maximum weight matching $\mathcal{M} \subseteq E(\Gamma'(S))$
4:     **if** $A(S) = \mathcal{M}$ **then**
5:         Add $A(S)$ to ADJ_M
6:         Remove S including external conflicts from (G, H, I, σ)
7:     **end if**
8: **end while**

In the experiments, IC-free runs are used instead of segments. Step 1 is performed efficiently by first identifying maximal IC-free runs, then breaking it down into smaller runs whenever the condition in Step 4 is not satisfied.

4   Experimental Results and Discussion

Our algorithms have been implemented in Python and require CPLEX[1]; they are freely available as part of the family-free genome comparison tool FFGC downloadable at http://bibiserv.cebitec.uni-bielefeld.de/ffgc.

---

[1] http://www.ibm.com/software/integration/optimization/cplex-optimizer/.

In subsequent analyses, gene similarities are based on local alignment hits identified with BLASTP on protein sequences using an e-value threshold of 10^{-5}. In gene similarity graphs, we discard spurious edges by applying a *stringency filter* proposed by Lechner *et al.* [12] that utilizes a local threshold parameter $f \in [0, 1]$ and BLAST bit-scores: a BLAST hit from a gene g to h is only retained if it is has a higher or equal score than f times the best BLAST hit from h to any gene g' that is member of the same genome as g. In all our experiments, we set f to 0.5. Edge weights of the gene similarity graph are then calculated according to the *relative reciprocal BLAST score* (RRBS) [14]. Finally we applied Algorithm ICF-SEG with conserved segments defined as runs.

For solving the FF-Median problem, we granted CPLEX two CPU cores, 4 GB memory, and a time limit of 3 h per dataset.

In our experiments, we compare ourselves against the orthology prediction tool MultiMSOAR [15]. This tool requires precomputed gene families, which we constructed by following the workflow described in [15].

Evaluation on simulated data. We first evaluate our algorithms on simulated data sets obtained by ALF [4]. The ALF simulator covers many aspects of genome evolution from point mutations to global modifications. The latter includes two types of genome rearrangements, as well as various options to customize the process of gene family evolution. In our simulations, we mainly use standard parameters suggested by the authors of ALF and we focus on three parameters that primarily influence the outcome of gene family-free genome analysis: (i) the rate of sequence evolution, (ii) the rate of genome rearrangements, and (iii) the rate of gene duplications and losses. We keep all three rates constant, only varying the evolutionary distance between the generated extant genomes. We confine our simulations to protein coding sequences. A comprehensive list of parameter settings used in our simulations is shown in Table 2 in Appendix C. As root genome in the simulations, we used the genomic sequence of an *E. coli* K-12 strain (Accession no: NC_000913.2) which comprises 4,320 protein coding genes. We then generated 7×10 data sets with increasing evolutionary distance ranging from 10 to 130 *percent accepted mutations* (PAM). Details about the generated data sets are shown in Table 1 in Appendix C. Figure 2(a) shows the outcome of our analysis with respect to precision and recall[2] of inferring positional orthologs. In all simulations, FF-Median generated no or very few false positives, leading to perfect or near-perfect precision score, consistently outperforming MultiMSOAR. However, since the objective of FF-Median only takes median genes into account that are conserved by synteny, the increase in mutational changes over evolutionary time causes a growing loss of syntenic context which results in a lower recall. Therefore, MultiMSOAR retains a better recall for larger evolutionary distances, while FF-Median provides better results for more closely related genomes.

Evaluation on real data. We study 15 γ-proteobacterial genomes that span a large taxonomic spectrum and are contained in the OMA database [1]. A complete list

---

[2] precision: #true positives/(#true positives + #false positives), recall: #true positives/(#true positives + #false negatives).

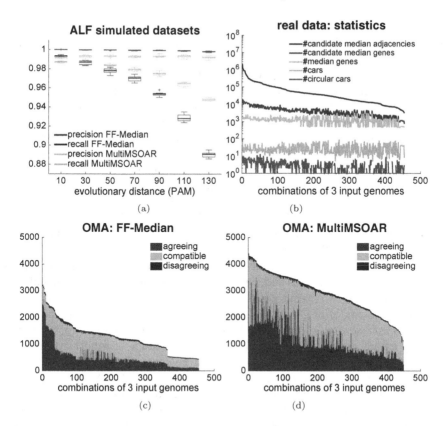

Fig. 2. Top: (a) Precision and recall of FF-Median and MultiMSOAR in simulations; (b) statistical assessment of CARs and median genes on real datasets. Bottom: agreement, compatibility and disagreement of positional orthologs inferred by (c) FF-Median and (d) MultiMSOAR with OMA database. (Color figure online)

of species names is given in Appendix D. We obtained the genomic sequences from the NCBI database and constructed for each combination of three genomes a gene similarity graph following the same procedure as in the simulated dataset. In 9 out of the 455 combinations of genomes the time limit prohibited CPLEX from finding an optimal solution. However, in those cases CPLEX was still able to find integer feasible suboptimal solutions. Figure 2(b) displays statistics of the real dataset. The number of candidate median genes and adjacencies ranges from 442 to 18,043 and 3,164 to 2,261,716, respectively, giving rise to up to 3,227 median genes that are distributed on 5 to 91 CARs per median. Some CARs are circular, indicating dubious conformations mostly arising from tandem duplications, but the number of such cases were low (mean: 2.78, max: 13).

We observed that the gene families in the OMA database are clustered tightly and therefore missing many true orthologies in the considered triples of genomes. As a result, many of the orthologous groups inferred by FF-Median and MultiMSOAR fall into more than one gene family inferred by OMA. We therefore evaluate our results by classifying the inferred orthologous groups into three categories: An orthologous group *agrees* with OMA if its three genes are in the same OMA group. It *disagrees* with OMA if extant genes x and y (of genomes X and Y respectively) are in different OMA groups but the OMA group of x contains another gene from genome Y. It is *compatible* with OMA if it neither agrees nor disagrees with OMA. We measure the number of median genes as well as the number orthologous groups of MultiMSOAR in each of the three categories. Figure 2(c) and (d) shows the outcome this analysis. MultiMSOAR is generally able to find more orthology relations in the dataset. This comes at no surprise, as it is clear from the objective of problem FF-Median and from the results of the simulated datasets that our method does not retain candidate median genes which have lost their syntenic context, which happens in triples of highly divergent genomes. The number of disagreeing orthologous groups that disagree with OMA is comparably low for both FF-Median (mean: 35.16, var: 348) and MultiMSOAR (mean: 48.61, var: 348).

We then performed another analysis to assess the *robustness* of the positional orthology predictions. To this end, we look at orthologous groups across multiple datasets that share two extant genomes, but vary in the third. Given two genes, x of genome X and y of genome Y, an orthologous group that contains x and y is called *robust* if x and y occur in the same orthologous group, whatever the third extant genome is. We computed the percentage of robust orthologous groups for all gene pairs of randomly-chosen genomes *E. coli K-12 MG 1655* and *S. enterica subsp. enterica serovar Typhimurium str. 14028s* in our dataset. The results indicate that orthologous groups inferred by FF-Median are slightly more robust (95.61 %) than robust those by MultiMSOAR (91.77 %). This is likely due to the strict constraint of defining median adjacencies only from genes that participate in at least one observed adjacency (Remark 4).

Overall, we can observe that FF-Median performed better than MultiM-SOAR only for triples of closely related genomes – which is consistent with our observation on simulated data – while being slightly more robust in general. This suggests FF-Median is an interesting alternative to identify higher confidence positional orthologs, at the expense of a higher recall rate.

Future work. We first aim to investigate alternative methods to reduce the computational load of Program FF-Median by identifying further strictly suboptimal and optimal substructures, which might require a better understanding of the impact of internal conflicts within substructures defined by intervals in the extant genomes. Without the need to modify drastically either the FF-median problem definition or the ILP, one can think about more complex weighting schemes for adjacencies that could account for known divergence time between genomes or relaxed notion of adjacencies that would address the high recall rate we observe in FF-Median. Within that regard, it would probably be interesting to combine this with the use of common intervals instead of runs to define conflict-free sub-instances. Finally, ideal family-free analysis should take into account the effects of gene family evolution. However, the presented family-free median model can only resolve certain cases of gene duplication. It is generally susceptible to gene losses that occurred along the evolutionary paths between the three extant genomes and their common ancestor. The definition of a family-free median model that tolerates events of gene family evolution at a reasonable computational cost is likely an interesting research avenue.

Acknowledgments. The authors are grateful to the anonymous reviewers for their valuable comments. DD and MB wish to thank Bernard M.E. Moret for helpful discussions. CC acknowledges funding from the NSERC Discovery Grant.

References

1. Altenhoff, A.M., Skunca, N., Glover, N., Train, C.-M., Sueki, A., Pilizota, I., Gori, K., Tomiczek, B., Müller, S., Redestig, H., Gonnet, G.H., Dessimoz, C.: The OMA orthology database in 2015: function predictions, better plant support, synteny view and other improvements. Nucleic Acids Res. **43**, 240–249 (2015)
2. Bergeron, A., Chauve, C., Gingras, Y.: Formal models of gene clusters. In: Bioinformatics Algorithms: Techniques and Applications, pp. 177–202. Wiley, New York (2008)
3. Braga, M.D.V., Chauve, C., Doerr, D., Jahn, K., Stoye, J., Thévenin, A., Wittler, R.: The potential of family-free genome comparison. In: Models and Algorithms for Genome Evolution, pp. 287–323. Springer, London (2013)
4. Dalquen, D.A., Anisimova, M., Gonnet, G.H., Dessimoz, C.: Alf - a simulation framework for genome evolution. Mol. Biol. Evol. **29**(4), 1115–1123 (2012)
5. Dewey, C.N.: Positional orthology: putting genomic evolutionary relationships into context. Brief Bioinform. **12**(5), 401–412 (2011)
6. Doerr, D., Thévenin, A., Stoye, J.: Gene family assignment-free comparative genomics. BMC Bioinform. **13**(Suppl. 19), S3 (2012)
7. Dörr, D.: Gene family-free genome comparison. Ph.D. thesis, Universität Bielefeld, Bielefeld, Germany (2016)
8. Fitch, W.M.: Homology a personal view on some of the problems. Trends Genet. **16**, 227–231 (2000)
9. Gabaldón, T., Koonin, E.V.: Functional and evolutionary implications of gene orthology. Nat. Rev. Genet. **14**, 360–366 (2013)
10. Kowada, L.A.B., Doerr, D., Dantas, S., Stoye, J.: New genome similarity measures based on conserved gene adjacencies. In: RECOMB **2016**, pp. 204–224 (2016)
11. Lechner, M., Hernandez-Rosales, M., Doerr, D., Wieseke, N., Thévenin, A., Stoye, J., Hartmann, R.K., Prohaska, S.J., Stadler, P.F.: Orthology detection combining clustering and synteny for very large datasets. PLoS ONE **9**(8), e105015 (2014)
12. Lechner, M., Findeiß, S., Steiner, L., Marz, M., Stadler, P.F., Prohaska, S.J.: Proteinortho: detection of (co-)orthologs in large-scale analysis. BMC Bioinform. **12**, 124 (2011)
13. Martinez, F.V., Feijão, P., Braga, M.D.V., Stoye, J.: On the family-free DCJ distance and similarity. Algorithms Mol. Biol. **10**, 13 (2015)
14. Pesquita, C., Faria, D., Bastos, H., Ferreira, A.E.N., Falcão, A.O., Couto, F.M.: Metrics for GO based protein semantic similarity: a systematic evaluation. BMC Bioinform. **9**(Suppl. 5), S4 (2008)
15. Shi, G., Peng, M.-C., Jiang, T.: Multimsoar 2.0: an accurate tool to identify ortholog groups among multiple genomes. PLoS ONE **6**(6), e20892 (2011)
16. Tannier, E., Zheng, C., Sankoff, D.: Multichromosomal median and halving problems under different genomic distances. BMC Bioinform. **10**, 120 (2009)

A   Hardness Proof

A.1   Reduction

The *maximum independent set problem for graphs bounded by node degree* 3, denoted as MAX IS-3 is MAX SNP-hard [3]. The corresponding decision problem can be informally stated as follows: Given a graph Λ bounded by degree 3 and some number $l \geq 1$, does there exists a set of vertices $V' \subseteq V$ of size $|V'| = l$ whose induced subgraph is unconnected? In the following, we present a transformation scheme $\mathbf{R}$ to phrase Λ as FF-median instance $\mathbf{R}(\Lambda) = (G, H, I, \sigma)$ such that the value $\mathcal{F}_\lambda(M)$ of a median M of $\mathbf{R}(\Lambda)$ is limited by $\mathcal{F}_\lambda(M) \leq 2 \cdot l + 3$. In doing so, we associate vertices of V with genes of extant genomes G, H and I. In order to keep track of associated genes, we denote by function $\xi(x)$ the list of vertices associated with gene x. We further introduce two types of unassociated genes, "$\emptyset$" and "$*$", whose members are identified by subscript notation.

Transformation $\mathbf{R}$:

1. Construct genome G such that for each vertex $v \in V$ there exists two associated genes $g_v, \bar{g}_v \in \mathcal{C}(G)$, i.e. $\xi(g_v) = \xi(\bar{g}_v) = v$. Further, let each gene pair $g_v, \bar{g}_v$ form a circular chromosome, giving rise to adjacency set $\mathcal{A}(G) = \{\{\bar{g}_v^h, g_v^t\}, \{\bar{g}_v^h, g_v^t\} \mid v \in V, g_v, \bar{g}_v \in \mathcal{C}(G)\}$.
2. For each edge $(u, v) \in E$ construct a circular chromosome $\mathcal{X}_{uv}$ hosting two genes $x_{uv}, x_\emptyset \in \mathcal{C}(\mathcal{X}_{uv})$, with gene x_{uv} being associated with both vertices u and v and gene $x_\emptyset$ being unassociated. Further, let both genes form a circular chromosome, giving rise to adjacency set $\mathcal{A}(\mathcal{X}_{uv}) = \{\{x_{uv}^h, x_\emptyset^t\}, \{x_\emptyset^h, x_{uv}^t\}\}$.
3. Assign each chromosome constructed in the previous step either to genome H or to genome I such that each vertex $v \in V$ is associated with at most two genes per genome.
4. Complete genomes H and I with additional circular chromosomes $\mathcal{X}_v$ where $\mathcal{C}(\mathcal{X}_v) = \{x_v, x_\emptyset\}$ and $\mathcal{A}(\mathcal{X}_v) = \{\{x_v^h, x_\emptyset^t\}, \{x_\emptyset^h, x_v^t\}\}$ such that each vertex in V is associated with exactly two genes per genome.
5. For each vertex $v \in V$, let $g, \bar{g} \in \mathcal{C}(G)$, $h, \bar{h} \in \mathcal{C}(H)$, and $i, \bar{i} \in \mathcal{C}(I)$ be the pairs of genes associated with v, i.e. $\xi(g) = \xi(\bar{g}) = \xi(h) \cap \xi(i) = \xi(\bar{h}) \cap \xi(\bar{i}) = v$. Assign gene similarities $\sigma(g, h) = \sigma(g, i) = \sigma(h, i) = 1$ and $\sigma(\bar{g}, \bar{h}) = \sigma(\bar{g}, \bar{i}) = \sigma(\bar{h}, \bar{i}) = 1$.
6. Add a copy of circular chromosome $\mathcal{X}_*$ to each genome G, H, and I, where $\mathcal{C}(\mathcal{X}_*) = \{x_*, \bar{x}_*\}$ and $\mathcal{A}(\mathcal{X}_*) = \{\{x_*^h, \bar{x}_*^t\}, \{\bar{x}_*^h, x_*^t\}\}$. Let $g_*, \bar{g}_* \in \mathcal{C}(G), h_*, \bar{h}_* \in \mathcal{C}(H)$, and $i_*, \bar{i}_* \in \mathcal{C}(I)$, set the gene similarity score between all pairs of genes in $\{g_*, h_*, i_*\}$ and $\{\bar{g}_*, \bar{h}_*, \bar{i}_*\}$ respectively, to 1. Lastly, set the gene similarity score of all pairs of unassociated genes of type "$\emptyset$" including genes $g_*, \bar{g}_*$ to $\frac{1}{4}$.

Except for step 3, none of the instructions of transformation scheme $\mathbf{R}$ are computationally challenging. Note that in step 3 the demanded partitioning of chromosomes into genomes H and I is always possible as consequence of Vizing's Theorem [4], by which every graph with maximum node degree d is

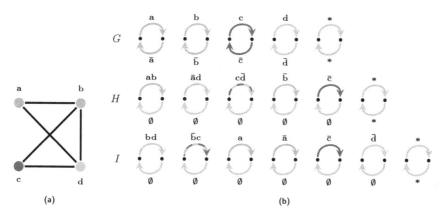

(a)

(b)

Fig. 3. (a) A simple graph bounded by degree three and (b) a corresponding FF-Median instance constructed with transformation scheme **R**.

edge-colorable using at most d or $d+1$ colors. Hence, using colors $\chi_1, \chi_2, \chi_3, \chi_4$ with $\chi_1 = \chi_2 \equiv I$, $\chi_3 = \chi_4 \equiv H$ and Misra and Gries' algorithm [2], edges of graph $\Lambda = (E, V)$ can be partitioned into two groups in $\mathcal{O}(|E||V|)$ time implying an assignment to genomes H and I.

Example 1. Fig. 3(b) shows a FF-Median instance constructed with transformation scheme **R** from the simple graph depicted in Fig. 3(a). Gene similarities between genes are not shown, but can be derived from the genes' labeling.

We structure our proof that the presented transformation is in fact a valid mapping of an MAX IS-3 instance to an instance of FF-Median into three different lemmas:

Lemma 2. *Given a median M of FF-Median instance $\mathbf{R}(\Lambda) = (G, H, I, \sigma)$, (1) for each median gene $(g, h, i) \in \mathcal{C}(M)$ where g, h, or i are associated with vertices in $V(\Lambda)$ holds $\xi(g) = \xi(h) \cap \xi(i) = v$, $v \in V(\Lambda)$; (2) there exist at most two median genes whose corresponding extant genes are not associated to any vertex in $V(\Lambda)$.*

Proof. Assume for contradiction that claim (1) does not hold. Then either $\xi(g) \neq \xi(h) \cap \xi(i)$, or $\xi(h) \cap \xi(i) = \emptyset$, both of which violate the constraint of establishing gene similarities between associated genes that is given in step 5. For claim (2), observe that the only unassociated genes in genome G are gene g_* and $\bar{g}_*$ introduced in step 6, limiting the overall number of unassociated genes in any median M. $\square$

Lemma 3. *The conserved adjacency set of any median M of FF-Median instance $\mathbf{R}(\Lambda) = (G, H, I, \sigma)$ is of the form $\mathcal{A}(M) \cap \mathcal{A}_{\Lambda}^{C} = \mathcal{A}_{\Lambda}^{G}(M) \cup \{\{m_*^h, \overline{m}_*^t\}, \{\overline{m}_*^h, m_*^t\}\}$, where the extant genes corresponding to m_* and $\overline{m}_*$ are all unassociated genes of type "*" and $\mathcal{A}(M)_{\Lambda}^{G} \subseteq \{\{m_1^h, m_2^t\} \in \mathcal{A}_{\Lambda}^{C} \mid \xi(\pi_G(m_1)) = \xi(\pi_G(m_2))\}$.*

Proof. Observe that both candidate median adjacencies $a_* = \{m_*^{\mathrm{h}}, \overline{m}_*^{\mathrm{t}}\}$ and $\bar{a}_* = \{\overline{m}_*^{\mathrm{h}}, m_*^{\mathrm{t}}\}$ are conserved in all three genomes, whereas all other conserved candidate adjacencies between associated and unassociated genes can be at most conserved in H and I. Establishing adjacencies $a_*, \bar{a}_*$ gives rise to a cumulative adjacency score of 6. Conversely, up to 4 non-conflicting adjacencies between associated and unassociated genes can be established that are conserved in both genomes H and I. However, since such adjacencies are only conserved between unassociated genes of type "$\emptyset$" whose gene similarities are set to $\frac{1}{4}$, the best cumulative adjacency score can not exceed 4. Thus, adjacencies $a_*, \bar{a}_*$ must be contained in any median. Further, because of this and the fact that in both genomes H and I, each gene associated with vertices of $V(\Lambda)$ is only adjacent to an unassociated gene, M cannot contain adjacencies that are conserved in extant genomes other than genome G, which are the adjacencies of each gene pair $(g_v, \bar{g}_v)$ associated with the same vertex $v \in V(\Lambda)$. $\qquad\square$

Lemma 4. *Given FF-median instance* $\mathbf{R}(\Lambda) = (G, H, I, \sigma)$, *let* m_u, m_v *be any pair of candidate median adjacencies of* $\mathcal{A}_\lambda$ *whose corresponding extant genes are associated to vertices* $u, v \in V(\Lambda)$, *then* m_u, m_v *are conflicting if and only if* $(u, v) \in E$.

Proof. By construction in step 5 of transformation scheme $\mathbf{R}$, each vertex $v \in V$ is associated with exactly two candidate median genes $m_v = (g, h, i), \overline{m}_v = (\bar{g}, \bar{h}, \bar{i})$, $m_v, \overline{m}_v \in \Sigma_\lambda$, such that $\xi(g) = \xi(h) \cap \xi(i) = v$ and $\xi(\bar{g}) = \xi(\bar{h}) \cap \xi(\bar{i}) = v$. Further, let u be another vertex of $V(\Lambda)$, such that $(u, v) \in E(\Lambda)$, and $m_u, \overline{m}_u$ are its two corresponding candidate median genes. Then, by construction in step 2, there exists exactly one extant gene x with $\xi(x) = uv$ (which, by assignment in step 3, is either contained in genome H or I). Consequently, either m_u is in conflict with m_v, or $\overline{m}_u$ with $\overline{m}_v$, or $\overline{m}_u$ with m_v, or m_u with $\overline{m}_v$. Recall that by construction in step 2 in $\mathbf{R}$ and by Lemma 3, $m_u, \overline{m}_u$ and $m_v, \overline{m}_v$ form conserved candidate adjacencies $\{m_u^{\mathrm{h}}, \overline{m}_u^{\mathrm{t}}\}$, $\{\overline{m}_u^{\mathrm{h}}, m_u^{\mathrm{t}}\}$ and $\{m_v^{\mathrm{h}}, \overline{m}_v^{\mathrm{t}}\}$, $\{\overline{m}_v^{\mathrm{h}}, m_v^{\mathrm{t}}\}$, respectively. Clearly, independent of which of the candidate median gene pairs of u and v are in conflict, both pairs of candidate median adjacencies are in conflict with each other.

Now, let u, v be two vertices of $V(\Lambda)$ such that edge $(u, v) \notin E(\Lambda)$, then there exists no gene x in extant genomes H and I with $\xi(x) = uv$. Even more, due to Lemma 2, there cannot exist a candidate median gene (g, h, i) with $\{u, v\} \subseteq \xi(g) \cup \xi(h) \cup \xi(i)$. Thus, the candidate median genes of u and v are not conflicting and neither are their corresponding candidate median adjacencies. $\qquad\square$

We proceed to show that the given transformation scheme gives rise to an approximation preserving reduction known as *L-reduction*. An L-reduction reduces a problem P to a problem Q by means of two polynomial-time computable transformation functions: A function $f : P \to Q' \subseteq Q$ that maps each instance of P onto an instance of Q, herein represented by transformation scheme $\mathbf{R}$, and a function $g : Q' \to P$ to transform any feasible solution of an instance in Q' to a feasible solution of an instance of P. Here, a *feasible* solution means any – not necessarily *optimal* – solution that obeys the problem's constraints. A feasible solution of FF-Median instance (G, H, I, σ) is an *ancestral genome* X where $\mathcal{C}(X) \subseteq \Sigma_\lambda$ and $\mathcal{A}(X) \subseteq \mathcal{A}_\lambda$ such that $\mathcal{A}(X)$ is conflict-free. We give the following transformation scheme to map ancestral genomes of an FF-Median instance to solutions of an MAX IS-3 instance:

Transformation **S**: Given any ancestral genome X of $\mathbf{R}(\Lambda)$, return $\{\xi(\pi_G(m_1))|$ $\{m_1^a, m_2^b\} \in \mathcal{A}(X) : \mathbb{I}_G(\pi_G(m_1)^a, \pi_G(m_2)^b) = 1$ and $\xi(\pi_G(m_1)) \neq \emptyset\}$.

We define score function $s_\lambda(X) \equiv \frac{1}{2}\mathcal{F}_\lambda(X) - 3$ of an ancestral genome X. For $(\mathbf{R}, \mathbf{S})$ to be an L-reduction the following two properties must hold for any given MAX IS-3 instance (Λ, l): (1) There is some constant α such that for any median M of the transformed FF-Median instance $\mathbf{R}(\Lambda)$ holds $s_\lambda(M) \leq \alpha \cdot l$; (2) There is some constant β such that for any ancestral genome X of $\mathbf{R}(\Lambda)$ holds $l - |\mathbf{S}(X)| \leq \beta \cdot |s_\lambda(M) - s_\lambda(X)|$. We proceed to proof the following lemma:

Lemma 5. $(\mathbf{R}, \mathbf{S})$ *is an L-reduction of problem MAX IS-3 to problem FF-Median with* $\alpha = \beta = 1$.

Proof. For any median M of FF-Median instance $\mathbf{R}(\Lambda)$, the number of conserved median adjacencies with correspondence to the same vertex of Λ is two, giving rise a cumulative adjacency score of two. From Lemmata 3 and 4 immediately follows that any ancestral genome of $\mathbf{R}(\Lambda)$ that maximizes the number of conserved adjacencies also maximizes the number of independent vertices in Λ. Recall that the two conserved adjacencies between unassociated genes of type "$*$" (which are part of all medians) give rise to a cumulative adjacency score of 6, we conclude that $|\mathcal{A}(M) \cap \mathcal{A}_\lambda^C| - 2 = \frac{1}{2}\mathcal{F}_\lambda(M) - 3 = s_\lambda(M) = l$, thus $\alpha = 1$.

Because $l = s_\lambda(M)$, it remains to show that $l - |\mathbf{S}(X)| \leq \beta|l - s_\lambda(X)|$. In a *suboptimal* ancestral genome of $\mathbf{R}(\Lambda)$, median genes with no association to vertices of Λ can also contain extant genes of type "$\emptyset$". These unassociated median genes can form "mixed" conserved adjacencies with genes that are associated with vertices of Λ. Such mixed conserved adjacencies have no correspondence to vertices in Λ and do not contribute to the transformed solution $\mathbf{S}(X)$ of an ancestral genome X. Yet, as mentioned earlier, the cumulative adjacency score of all mixed conserved adjacencies can not not exceed 4. Therefore it holds that $|\mathbf{S}(X)| \geq s_\lambda(X)$ and we conclude $\beta = 1$. □

B   Speeding up the Search for a Median

Proof of Lemma 1:

Proof. Given an IC-free segment $\mathcal{S} = \{m_1, \ldots, m_k\}$ of an FF-Median instance (G, H, I, σ). Let M be a conflict-free matching in graph $\Gamma'(\mathcal{S})$. Because M is conflict-free and $\mathcal{S}$ contiguous in all three extant genomes, M must contain all candidate median genes of $\mathcal{S}$. Now, let M' be a median such that $\mathcal{S} \not\subseteq \mathcal{C}(M')$.

Further, let $C(m)$ be the set of candidate median genes that are involved in a conflict with with a given median gene m of $\mathcal{S}$ and $X = \mathcal{C}(M') \cap (\bigcup_{m \in \mathcal{S}} C(m) \cup \mathcal{S})$. Clearly, $X \neq \emptyset$ and for the contribution $\mathcal{F}_\lambda(X)$ must hold $\mathcal{F}_\lambda(X) \geq \mathcal{F}_\lambda(\mathcal{S})$, otherwise M' is not optimal since it is straightforward to construct a median higher score which includes $\mathcal{S}$. Clearly, the contribution $\mathcal{F}(X)$ to the median is bounded by $\max(\{\sum_{m' \in C'} \Delta(m') \mid C' \subseteq C(m) : C' \text{is conflict-free}\}) + \mathcal{F}_\lambda(\mathcal{S})$. But since $\mathcal{S}$ gives rise to a conflict-free matching with maximum score, also median M'' with $\mathcal{C}(M'') = (\mathcal{C}(M') \setminus X) \cup \mathcal{C}(\mathcal{S})$ and $\mathcal{A}(M'') = (\mathcal{A}(M') \setminus \mathcal{A}(X)) \cup \mathcal{A}(\mathcal{S}))$ must be an (optimal) median. $\qquad\square$

C Simulated Sequence Evolution with ALF

Table 1. Average benchmark data of seven evolutionary distances, each comprising ten genomic datasets generated by ALF [1].

PAM	Genome	Inversions	Transpositions	Duplications	Losses
10	G	8.7	6.1	7.3	6.9
	H	7.3	4.5	6.3	5.4
	I	8.5	6.6	10.4	5.6
30	G	24.5	16.9	21.0	22.7
	H	23.4	19.8	20.6	18.4
	I	25.5	17.2	17.5	20.9
50	G	39.9	27.8	32.4	36.7
	H	41.8	31.8	31.0	31.7
	I	43.2	30.0	28.7	39.7
70	G	58.6	42.3	41.1	39.2
	H	57.0	43.6	46.3	45.1
	I	60.4	41.4	40.7	39.1
90	G	75.0	54.5	53.1	64.2
	H	69.9	50.5	54.1	65.0
	I	75.2	55.5	60.3	58.5
110	G	96.3	69.4	67.0	74.6
	H	90.6	64.2	62.5	70.9
	I	90.2	68.5	62.6	61.2
130	G	105.7	76.3	74.4	81.0
	H	108.7	78.2	79.6	82.8
	I	110.8	73.6	73.9	77.3

Table 2. Parameter settings for simulations generated by ALF [1].

Parameter name	Value	
Sequence evolution		
Substitution model	WAG (amino acid substitution model)	
Insertion and deletion	Zipfian distribution	exponent $c = 1.8214$
	Insertion rate	0.0003
	Maximum insertion length	50
Rate variation among sites	Γ-distribution	shape parameter $a = 1$
	Number of classes	5
	Rate of invariable sites	0.01
Genome rearrangement		
Inversion	rate	0.0004
	Maximum inversion length	100
Transposition	rate	0.0002
	Maximum transposition length	100
	Rate of inverted transposition	0.1
Gene family evolution		
Gene duplication	Rate	0.0001
	Max. no. of genes involved in one dupl.	5
	Probability of transposition after dupl.	0.5
	Fission/fusion after duplication	0.1
	Probability of rate change	0.2
	Rate change factor	0.9
	Probability of temporary rate change (duplicate)	0.5
	Temporary rate change factor (duplicate)	1.5
	Life of rate change (duplicate)	10 PAM
	Probability of temporary rate change (orig+duplicate)	0.3
	Temporary rate change factor (orig+duplicate)	1.2
	Life of rate change (orig+duplicate)	10 PAM
Gene loss	Rate	0.0001
	Maximum length of gene loss	5
Gene fission/fusion	rate	0.0
	Maximum number of fused genes	–

D   Real Genomes Dataset

See Table 3.

Table 3. Dataset of genomes used in comparison with the OMA database.

Genbank ID	Name
U00096.3	Escherichia coli str. K-12 substr. MG1655, complete genome
AE004439.1	Pasteurella multocida subsp. multocida str. Pm70, complete genome
AE016853.1	Pseudomonas syringae pv. tomato str. DC3000, complete genome
AM039952.1	Xanthomonas campestris pv. vesicatoria complete genome
CP000266.1	Shigella flexneri 5 str. 8401, complete genome
CP000305.1	Yersinia pestis Nepal516, complete genome
CP000569.1	Actinobacillus pleuropneumoniae L20 serotype 5b complete genome
CP000744.1	Pseudomonas aeruginosa PA7, complete genome
CP000766.3	Rickettsia rickettsii str. Iowa, complete genome
CP000950.1	Yersinia pseudotuberculosis YPIII, complete genome
CP001120.1	Salmonella enterica subsp. enterica serovar Heidelberg str. SL476, complete genome
CP001172.1	Acinetobacter baumannii AB307-0294, complete genome
CP001363.1	Salmonella enterica subsp. enterica serovar Typhimurium str. 14028S, complete genome
FM180568.1	Escherichia coli 0127:H6 E2348/69 complete genome, strain E2348/69
CP002086.1	Nitrosococcus watsoni C-113, complete genome

References

1. Dalquen, D.A., Anisimova, M., Gonnet, G.H., Dessimoz, C.: Alf - a simulation framework for genome evolution. Mol. Biol. Evol. **29**(4), 1115–1123 (2012)
2. Misra, J., Gries, D.: A constructive proof of Vizing's theorem. Inform. Process. Lett. **41**(3), 131–133 (1992)
3. Papadimitriou, C.H., Yannakakis, M.: Optimization, approximation, and complexity classes. J. Comput. Syst. Sci. **43**(3), 425–440 (1991)
4. Vizing, V.G.: On an estimate of the chromatic class of a p-graph. Diskret. Analiz **3**, 25–30 (1964)

Correction of Weighted Orthology and Paralogy Relations - Complexity and Algorithmic Results

Riccardo Dondi[1]([✉]), Nadia El-Mabrouk[2], and Manuel Lafond[2]

[1] Dipartimento di Lettere, Filosofia e Comunicazione,
Università degli Studi di Bergamo, Bergamo, Italy
riccardo.dondi@unibg.it
[2] Department of Computer Science,
Université de Montréal, Montréal, QC, Canada
mabrouk@iro.umontreal.ca, lafonman@iro.umontreal.ca

Abstract. A relation graph for a gene family is a graph with vertices representing the genes, edges connecting pairs of orthologous genes and "missing" edges representing paralogs. While a gene tree directly leads to a set of orthology and paralogy relations, the converse is not always true. Indeed a relation graph cannot necessarily be inferred from any tree, and even if it is "satisfiable" by a tree, this tree is not necessarily "consistent", i.e. does not necessarily reflect a valid history for the genes, in agreement with a species tree. Here, we consider the problems of minimally correcting a relation graph for satisfiability and consistency, when a degree of confidence is assigned to each orthology or paralogy relation, leading to a weighted relation graph. We provide complexity and algorithmic results for minimizing corrections on a weighted graph, and also for the maximization variant of the problems for unweighted graphs.

1 Introduction

As genes are the basic molecular units of heredity, key for understanding genetic diversity, a first step of most genomic studies is to group genes into families. Gene families are usually inferred from sequence similarity, the underlying idea being that similar sequences reflect *homologous* genes that have diverged from a common ancestral sequence.

Given a gene family, it is important to discriminate between two types of homologs: *orthologs* being gene copies originating from a speciation event, and *paralogs* originating from a duplication. For this purpose, tree-based methods consist in first constructing a phylogenetic tree for the gene family, and then, given a species tree, applying a reconciliation approach for inferring speciation and duplication nodes [8]. On the other hand, tree-free methods are based on gene clustering according to sequence similarity (c.f. for example [3,17,18,22]), synteny [15,16] or functional annotation of genes [5]. Results of these methods are pairwise orthology relations, or groups of orthologs, that can be represented as relation graphs, where vertices are genes and edges are orthology relations. Assuming a full inference of pairwise orthology relations, "missing" edges of the

© Springer International Publishing Switzerland 2016
M. Frith and C.N.S. Pedersen (Eds.): WABI 2016, LNBI 9838, pp. 121–136, 2016.
DOI: 10.1007/978-3-319-43681-4_10

relation graph represent paralogy relations. In addition, as different inference methods may lead to different predictions, instead of a yes or no orthology assignment, existing methods can rather motivate a way of assigning a score to a given relation [14], leading to a weighted relation graph. For example, orthology predictions with OrthoMCL [18] are based on a weighted graph, where edge weights are related to the sequence similarity score of the adjacent genes, while InParanoid [3] provides a confidence value that shows how closely related a paralog is to its "seed ortholog". Surprisingly, as far as we know, weighted orthology/paralogy relation graphs have not been formally considered in the literature.

While a gene tree induces a set of relations between genes, the converse is not true, as a set of relations may or may not represent a valid history for the gene family. Two underlying questions are: (1) is the set of relations "satisfiable" i.e. is there a tree, with internal nodes labeled as duplication or speciation, containing them all? (2) is the set of relations "S-consistent" with the known species tree S, i.e. is there a tree containing the relations that is a "valid" gene tree "in agreement" with S? Polynomial-time algorithms are known to exist for deciding satisfiability and S-consistency for full [9–11] or partial [14] pairwise gene relations.

In this paper, we address the problem of correcting a full relation graph R, and more specifically a full weighted relation graph, so that it represents a satisfiable and S-consistent set of relations. The related minimization problems consist in editing, i.e. adding or removing, edges of minimum total weight. In the unweighted case, the satisfiability correction problem reduces to editing a minimum number of edges of R in order to make it P_4-free, which is known to be NP-hard [19]. In [10], an integer linear programming formulation is used to correct relation graphs of small size, which is also applicable to weighted graphs. In [20], the authors propose an approximation algorithm of factor 4Δ, where Δ is the maximum degree of the input graph. The algorithm, however, offers no guarantees in the case of weighted graphs, as there are weighted instances on which it is arbitrarily far from optimal. It is shown in [1] that the minimum edge editing problem cannot be approximated within an "additive" factor of $n^{2-\epsilon}$, for any $\epsilon > 0$. Yet, the authors give a class of polynomial time algorithms that are approximable within an additive factor of ϵn^2, for any $\epsilon > 0$. This implies a constant factor algorithm for graphs with an edit distance of $\Omega(n^2)$, but offers no guarantee in the other cases. Moreover, this algorithm only applies to unweighted graphs, and does not consider that two genes from the same species must remain paralogs. Finally in [19], parameterized versions of the algorithm are explored. As for the S-consistency correction problem, we proved in a previous paper [13] that it is NP-hard, which is the only result so far.

We show in, Sect. 3, that the weighted satisfiability and S-consistency problems are not approximable within a constant factor, assuming the Unique Games Conjecture. In Sect. 4, we then show that they can be approximated within a factor of n (the number of vertices of the relation graph), and provide n-approximation algorithms for both the satisfiability and S-consistency problems. We end this paper by giving, in Sect. 5, a few results on the maximization variants of the problems for the unweighted case, which consists in maximizing the number of preserved relations. We first introduce the concepts and optimization problems.

2   Trees and Orthology Relations

A graph H is denoted $H = (V_H, E_H)$, where V_H is its set of vertices (or *nodes* if H is a tree) and E_H its set of edges. If H is a tree, degree one nodes are *leaves*.

2.1   Trees

All considered trees are rooted and binary. Given a set X, a *tree T for X* is a tree whose leafset $\mathcal{L}(T)$ is in bijection with X. Given an internal node u of T, the subtree rooted at u is denoted T_u and we call the leafset $\mathcal{L}(T_u)$ the *clade of* u. A node u is an *ancestor* of v if u is on the (inclusive) path between v and the root. The *lowest common ancestor* (lca) of u and v, denoted $lca_T(u, v)$, is the ancestor common to both nodes that is the most distant from the root. We define $lca_T(U)$ analogously for a set $U \subseteq V(T)$.

A *species tree S* for a species set Σ represents an ordered set of speciation events that have led to Σ: an internal node is an ancestral species at the moment of a speciation event, and its children are the new descendant species. For simplicity, we will assume that species trees are binary.

A *gene family Γ* is a set of genes accompanied with a function $s : \Gamma \to \Sigma$ mapping each gene to its corresponding species. The evolutionary history of Γ can be represented as a *node-labeled gene tree* for Γ, where each internal node refers to an ancestral gene at the moment of an event (either speciation or duplication), and is labeled as a speciation (*Spec*) or duplication (*Dup*) accordingly. Formally, we call a *DS-tree* for Γ a pair (G, ev), where G is a tree with $\mathcal{L}(G) = \Gamma$, and $ev : V_G \setminus \mathcal{L}(G) \to \{Dup, Spec\}$ is a function labeling each internal node of G as a duplication or a speciation. For example, in Fig. 1, G_1 and G_2 are two DS-trees.

According to the Fitch [7] terminology, we say that two genes x, y of Γ are *orthologous* in G if $ev(lca_G(x, y)) = Spec$, and *paralogous* in G if $ev(lca_G(x, y)) = Dup$.

A DS-tree G for Γ does not necessarily represent a valid history. For this to hold, any speciation node of G should reflect a clustering of species "in agreement" with S [14]. Formally G should be *S-consistent*, as defined below, where s_G is the *LCA-mapping* function, mapping each gene, ancestral or extant, to a species as follows: if $g \in \mathcal{L}(G)$, then $s_G(g) = s(g)$; otherwise, $s_G(g) = lca_S(\{s(g') : g' \in \mathcal{L}(G_g)\})$.

Definition 1. *Let S be a species tree and G be a DS-tree. Let v be an internal node of G such that $ev(v) = Spec$. Then the speciation node v, with children v_1 and v_2, is S-consistent iff none of $s_G(v_1)$ and $s_G(v_2)$ is an ancestor of the other. We say that G is S-consistent iff every speciation node of G is S-consistent.*

For example, in Fig. 1, G_1 is not *S*-consistent as the root of G_1 is not *S*-consistent.

2.2   Relation Graphs

For a graph $H = (V_H, E_H)$, we denote the complementary set of E_H by $\overline{E_H} = \{\{u, v\} : u, v \in V_H, \{u, v\} \notin E_H\}$. Let V' be a subset of V_H. The *subgraph of H induced by V'*, denoted $H[V']$, is the subgraph of H with vertex-set V' having every edge $\{u, v\}$ of H such that $u, v \in V'$. If I is another graph, we say H is *I-free* if there is no $V' \subseteq V_H$ such that $H[V']$ is isomorphic to I.

A *relation graph* R on a gene family Γ is a graph with vertex set $V_R = \Gamma$, in which we interpret each edge $\{u, v\}$ of E_R as an orthology relation between u and v, and each "missing" edge $\{u, v\} \in \overline{E_R}$, also called *non-edge*, as a paralogy relation. Notice that if $s(u) = s(v)$, then $\{u, v\}$ must be a non-edge (u and v are paralogous). We denote $n = |V_R|$.

A *DS-tree* G leads to a relation graph, denoted $R(G)$, with vertex set $\mathcal{L}(G)$ and edge set corresponding to all gene pairs that are orthologous in G. Conversely, a relation graph R does not necessarily lead to a *DS-tree*. If this is the case, i.e. if there exists a *DS-tree* G such that $R(G) = R$, then R is said *satisfiable*. As shown in [9], a relation graph R is satisfiable if and only if R is P_4-free, meaning that no four vertices of R induce a path of length 3 (number of edges). The P_4-free graphs are sometimes called *cographs*. See Fig. 1 for an example.

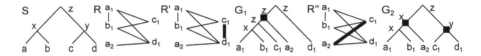

Fig. 1. S is the species tree for $\Sigma = \{a, b, c, d\}$. The internal nodes, representing ancestral species, are labeled by x, y and z. R is a relation graph on gene set $\Gamma = \{a_1, a_2, b_1, c_1, d_1\}$. A gene name corresponds to the species it belongs to (e.g. $s(a_1) = a$). R is not satisfiable as the set of vertices $\{c_1, b_1, d_1, a_2\}$ induces a P_4. R' is a satisfiable relation graph obtained from R by inserting the edge $\{c_1, d_1\}$, and G_1 is a *DS-tree* displaying every relation of R' (each internal node v is labeled by $s_{G_1}(v)$). However, G_1 is not consistent with the species tree S. R'' is another correction of R that is S-consistent, as the tree G_2 displays the relations in R'' and is S-consistent. *Dup* nodes in *DS-trees* are marked by a square; all other nodes are speciation nodes.

As a *DS-tree* does not necessarily represent a true history for Γ, satisfiability of a relation graph does not ensure a possible translation in terms of a history for Γ. For this to hold, R should also be *consistent* with the species tree, according to the following definition.

Definition 2. *Let S be a species tree. A relation graph R for Γ is S-consistent if and only if R is satisfiable by a DS-tree G which is itself S-consistent.*

2.3   Problem Statements

We call a *weight* for a relation graph $R = (V_R, E_R)$ a function $w : \binom{V_R}{2} \to \mathbb{R}^+$ on its vertex pairs. Notice that w assigns a weight to both edges (orthologies) and non-edges (paralogies). We shall assume that if $s(u) = s(v)$ for two genes u and v, then $\{u, v\} \in \overline{E_R}$ and $w(\{u, v\}) = \infty$. The weight function w is extended to any $I_R \subseteq \binom{V_R}{2}$ by defining $w(I_R) = \sum_{\{x,y\} \in I_R} w(\{x, y\})$.

Given a relation graph $R = (V_R, E_R)$, an *edge-editing* of R is a pair $E_R^* = (E_R^+, E_R^-)$ with $E_R^+ \subseteq \overline{E_R}$ and $E_R^- \subseteq E_R$. We denote by $R(E_R^*)$ the graph $R(E_R^*) = (V_R, (E_R \cup E_R^+) \backslash E_R^-, w)$. In other words, E_R^+ (respectively E_R^-) denotes inserted (respec. removed) edges. Given a relation graph $R' = (V_{R'}, E_{R'})$ computed from R by edge insertion and removal, the set of removed edges is $E_R^- = E_R \backslash E_{R'}$, and the set of inserted edges is $E_R^+ = E_{R'} \backslash E_R$. For example, for the graph R' of Fig. 1, $E_R^- = \{\{c_1, d_1\}\}$ and $E_R^+ = \emptyset$. An *edge-editing* E_R^* is said P_4-*free* if $R(E_R^*)$ is itself P_4-free.

The problems considered in Sects. 3 and 4 are the following (corresponding maximization problems are introduced in Sect. 5). The first problem asks for a satisfiable relation graph, hence no species tree is considered, while the second asks for an S-consistent relation graph, hence the input contains also a species tree.

Minimum Weighted Editing for Satisfiability (MinWES):
Input: A relation graph $R = (V_R, E_R)$ and a weight function w;
Output: A satisfiable relation graph $R' = (V_R, E_{R'})$, obtained from R by an edge-editing $E_R^* = (E_R^+, E_R^-)$ that minimizes $w(E_R^+) + w(E_R^-)$.

Minimum Weighted Editing for Consistency (MinWEC):
Input: A relation graph $R = (V_R, E_R)$, a weight function w and a species tree S for Σ (the set of species containing the genes represented by R);
Output: An S-consistent relation graph $R' = (V_R, E_{R'})$, obtained from R by an edge-editing $E_R^* = (E_R^+, E_R^-)$ that minimizes $w(E_R^+) + w(E_R^-)$.

3   Hardness of Approximation of Minimum Weighted Editing for Satisfiability and Consistency

We show that MinWES is unlikely to be approximable within a constant factor, by presenting a gap-preserving reduction from Minimum Multi-Cut. First, we consider the variant of MinWES, called Minimum Weighted Removal for Satisfiability (MinWRS), where only edge removal is allowed, then we easily extend the result to MinWES.

Given a graph $H = (V_H, E_H)$, and a set $X \subseteq \binom{V_H}{2}$ (i.e. a set of pairs), Minimum Multi-Cut asks for a set E_H' of minimum cardinality such that each pair $\{v_i, v_j\} \in X$ is disconnected in $H' = (V_H, E_H \backslash E_H')$.

Given an instance $H = (V_H, E_H, X)$ of Minimum Multi-Cut, we construct an instance $R = (V_R, E_R, w)$ of MinWRS as follows. The vertex set V_R includes, for each $v_i \in V_H$, two vertices $v_{i,R}$ and $v'_{i,R}$. For any distinct $x, y \in V_R$, we set $s(x) \neq s(y)$, and hence there are no "forced" paralogs. As for E_R, it is defined as follows, where $q = |V_H|^5 + 1$.

- For each $v \in V_H$, define an edge $\{v_{i,R}, v'_{i,R}\}$ in E_R of weight $q' = q|E_H| + 2\left(\binom{|V_H|}{2} - |E_H|\right)$;
- For each $\{v_i, v_j\} \in X$, define an edge $\{v_{i,R}, v_{j,R}\}$ in E_R with weight q if $\{v_i, v_j\} \in E_H$, and with weight 1 if $\{v_i, v_j\} \notin E_H$;
- For each $\{v_i, v_j\} \notin X$, define the edges $\{v_{i,R}, v'_{j,R}\}$ and $\{v'_{i,R}, v_{j,R}\}$ in E_R, each with weight $q/2$ if $\{v_i, v_j\} \in E_H$, and with weight 1 if $\{v_i, v_j\} \notin E_H$.

For each $\{u_R, v_R\} \in \overline{E_R}$, $\{u_R, v_R\}$ has weight q'. Notice however, that, since edge insertion is not allowed in MinWRS, the weight of $\{u_R, v_R\}$ never contributes to the cost of a solution of MinWRS.

We first show (in the Appendix) that there is a correspondance between solutions to the two problems on our constructed instances.

Lemma 1. *Let $H = (V_H, E_H, X)$ be an instance of Minimum Multi-Cut and let $R = (V_R, E_R, w)$ be the corresponding instance of MinWRS. Given a solution E'_H of Minimum Multi-Cut, we can compute in polynomial time a solution of MinWRS of weight at most $q|E'_H| + 2\left(\binom{|V_H|}{2} - |E_H|\right)$.*

Lemma 2. *Let $H = (V_H, E_H, X)$ be an instance of Minimum Multi-Cut and let $R = (V_R, E_R, w)$ be the corresponding instance of MinWRS. Given a solution R' of MinWRS of weight at most $qW + 2\left(\binom{|V_H|}{2} - |E_H|\right)$ for some integer W, we can compute in polynomial time a multicut E'_H of H of size at most W.*

Assuming the Unique Games Conjecture, the inapproximability of MinWRS is deduced from the inapproximability of Minimum Multi-Cut [4].

Theorem 1. *MinWRS is not approximable within a constant factor assuming the Unique Games Conjecture.*

The result of Theorem 1 can be easily extended to MinWES.

Corollary 1. *MinWES is not approximable within a constant factor assuming the Unique Games Conjecture.*

Proof. The result follows by a gap-preserving reduction similar to that for MinWRS. Recall that for each pair $\{u_R, v_R\} \in \overline{E_R}$, a weight of q' is associated with $\{u_R, v_R\}$. Consider a solution R' of MinWES on instance R that has cost not greater than $qW + \left(\binom{|V_H|}{2} - |E_H|\right) + \binom{|V_H|}{2}$. It is easy to see that R' is obtained without any edge insertion. $\square$

The inapproximability result for MinWES is easily extended to MinWEC. This is achieved by defining a species tree S on V_R such that the root of S is connected to two subtrees, one with leafset $\{v_{i,R} : v_i \in V_H\}$, one with leafset $\{v'_{i,R} : v_i \in V_H\}$, and showing that any solution to our instance of MinWRS must agree with this species tree.

Corollary 2. *MinWEC is not approximable within a constant factor assuming the Unique Games Conjecture.*

4 A Bounded Approximation Algorithm for Minimum Weighted Editing for Satisfiability and Consistency

While MinWES and MinWEC are not approximable within a constant factor, we show here that they can be approximated within factor $n = |V(R)|$, and we give the corresponding algorithms. Despite being a large approximation factor, this is the best known bound so far and shows that the problems have polynomially bounded approximability. We first describe the approximation algorithm for MinWES.

Denote by $\overline{R} = (V_R, \overline{E_R})$ the *complement* of the graph $R = (V_R, E_R)$. A well-known property of cographs is given by the following lemma.

Lemma 3. [6] *A graph R is P_4-free if and only if for any $X \subseteq V_R$, one of $R[X]$ or $\overline{R[X]}$ is disconnected.*

This motivates a greedy min-cut approach for MinWES, performing an edge-editing of minimum weight disconnecting the graph or its complement, and iterating recursively on the resulting components. This is the main idea of Algorithm MinCut-Cograph-Editing below. Note that assuming forced paralogs have infinite weight, this algorithm will never make two genes from the same species orthologs.

More formally, let $R = (V_R, E_R)$ be a relation graph accompanied with a weight function w. Define a *cut* $C = \{X, Y\}$ as a partition of V_R with X and Y being non-empty sets, and denote $E_R(C) = \{\{x, y\} \in E_R : x \in X, y \in Y\}$. The weight of C is $w(C) = w(E_R(C))$. The cut C is a *minimum cut* or *MinCut* if no other cut has a smaller weight $w(C)$. *Applying a cut C to R consists in removing all edges of $E_R(C)$ from R.*

ALGORITHM MINCUT-COGRAPH-EDITING(R):
  IF R has at most 2 vertices THEN RETURN;
  Find a MinCut $C = \{X, Y\}$ for R;
  Find a MinCut $\overline{C} = \{\overline{X}, \overline{Y}\}$ of $\overline{R}$;
  IF $w(C) < w(\overline{C})$ THEN
      Remove all edges between X and Y in R;
      MINCUT-COGRAPH-EDITING$(R[X])$;
      MINCUT-COGRAPH-EDITING$(R[Y])$;
  ELSE
      Add all possible edges between $\overline{X}$ and $\overline{Y}$ in R;
      MINCUT-COGRAPH-EDITING$(R[\overline{X}])$;
      MINCUT-COGRAPH-EDITING$(R[\overline{Y}])$;
  END IF
END ALGORITHM

Complexity: A MinCut of a given graph of n vertices and m edges can be found in time $O(nm + n^2 \log n)$ using the Stoer-Wagner algorithm [21]. In the MinCut-Cograph-Editing algorithm, MinCut is applied to both R and $\overline{R}$. As at least one of these two graphs has $\Omega(n^2)$ edges, the required time for MinCut is therefore

$O(n^3)$. This step is repeated at most n times, hence the overall time complexity of MinCut-Cograph-Editing is $O(n^4)$.

The remaining of this section is dedicated to proving Theorem 2, which states that MinCut-Cograph-Editing is an n-approximation algorithm. We denote by σ_R the minimum weight of a P_4-free edge-editing of R. If $X \subseteq V_R$, we denote $\sigma_{R[X]}$ by σ_X.

Lemma 4. *Let C be a minimum cut of R, and let $\overline{C}$ be a minimum cut of $\overline{R}$. Then $\sigma_R \geq \min\{w(C), w(\overline{C})\}$.*

Proof. Let E_R^* be a P_4-free edge-editing of R. By Lemma 3, either $R(E_R^*)$ or its complement is disconnected, implying that E_R^* must apply some cut on either R or $\overline{R}$. This cut is at best a minimum cut. $\square$

Lemma 5. *Let $\{X, Y\}$ be a partition of V. Then, $\sigma_R \geq \sigma_X + \sigma_Y$.*

Proof. Let E_R^* be a P_4-free edge-editing of weight σ_R, and let $R' = R(E_R^*)$. Assume that E_R^* has a weight stricly smaller than $\sigma_X + \sigma_Y$. Then, since $R'[X]$ and $R'[Y]$ are P_4-free, there must either be an edge-editing of $R[X]$ of weight smaller than σ_X, or an edge-editing of $R[Y]$ of weight smaller than σ_Y, contradicting the definition of σ_X and σ_Y. $\square$

Theorem 2. MinCut-Cograph-Editing *is an n factor approximation algorithm for* MinWES.

Proof. Denote by $\beta(R)$ the weight of the edge-editing found by the algorithm on R. We proceed by induction on $n = |V_R|$ to show that $\beta(R) \leq n\sigma_R$. The statement is trivial for $n \leq 3$ (as there is nothing to correct), so assume that the algorithm finds a solution of weight $\beta(R) \leq k\sigma_R$ for any graph of size at most $k < n$. The algorithm applies a minimum cut $C = \{X, Y\}$ on R or $\overline{R}$, and proceeds recursively on X and Y, with $|X|, |Y| \leq n - 1$. By the induction hypothesis, we have

$$\beta(R) \leq |X|\sigma_X + |Y|\sigma_Y + w(C) \leq (n-1)(\sigma_X + \sigma_Y) + w(C)$$
$$\leq (n-1)\sigma_R + \sigma_R = n\sigma_R$$

where the last inequality holds due to Lemmas 4 and 5. $\square$

It is possible to show that the approximation factor of MinCut-Cograph-Editing is tight.

By modifying MinCut-Cograph-Editing, it is possible to design an n factor approximation algorithm for MinWEC. The main difference with respect to MinCut-Cograph-Editing, is that the algorithm considers a minimum cut on a subset of R and a cut on a subset of $\overline{R}$ induced by the species tree S. The detailed algorithm, along with the proof of the following Theorem, are given in the Appendix. It also requires time $O(n^4)$.

Theorem 3. MinCut-Cograph-Editing-Cons *is an n factor approximation algorithm for* MinWEC.

5   Polynomial Time Approximation Schemes for the Maximization Variant of Graph Correction

Here, we consider the complementary maximization problem, which consists in maximizing conservation between the original and corrected graphs. Although sharing the same objectives, the minimization and maximization variants are not equivalent from an approximation point of view.

Below is a formal statement of the corresponding maximization version of MinWES (see Sect. 2) for unweighted graphs. Remember that edges represent orthologies, while non-edges are paralogies. Maximizing conservation therefore requires accounting for both edges and non-edges.

Maximum Editing for Satisfiability (MaxES):
Input: A relation graph $R = (V_R, E_R)$;
Output: A satisfiable relation graph $R' = (V_R, E_{R'})$ obtained from R by an edge-editing, such that its *value* $|E_R \cap E_{R'}| + |(\overline{E_R} \cap \overline{E_{R'}})|$ is maximized.

Given a relation graph R, the value of a solution R' for MaxES over instance R is called the *agreement* value of R'.

Lemma 6. *Given a relation graph R, an optimal solution of MaxES over instance R has an agreement value of at least $\frac{n^2}{8}$.*

Proof sketch: Consider the two 'extreme' solutions: either make all genes from two distinct species orthologs, or all genes paralogs. In R, either at least half the genes are orthologs, or at least half the genes are paralogs. Thus one extreme solution preserves at least half the total number of relations, which is $\binom{n}{2}/2 > \frac{n^2}{8}$. The detailed proof is in the Appendix.   □

Note that Lemma 6 gives, almost trivially, a factor $1/2$ approximation (i.e. preserving at least half as many relations as the optimal). Using Lemma 6 and results from [1], one can devise a PTAS for MaxES in the case that every gene belongs to a distinct species. Let $OPT(R)$ be the value of an optimal solution on R, and let c be such that $OPT(R) = cn^2$. The additive εn^2 approximation algorithm for cograph editing [1] yields a solution of value $(c - \varepsilon)n^2$. As $c \geq 1/8$ by Lemma 6, ε can be adjusted so that, for any $0 < \varepsilon' < 1$, $(c - \varepsilon)n^2 \geq (1 - \varepsilon')cn^2$, hence yielding a PTAS. In the more general case, this algorithm does not ensure that genes from the same species remain paralogs. However, the authors of [1] claim that their approximation algorithm applies to any hereditary graph property (i.e. preserved after vertex-deletion), which holds for satisfiability.

Finally, we end this paper with few insights on the maximization version of graph correction for consistency, that we call MaxEC. Notice that the lower bound $\frac{n^2}{8}$ of Lemma 6 also holds for an optimal solution of MaxEC. However, the PTAS for MaxES does not guarantee that the returned relation graph R' is S-consistent with the given species tree S. We can show however that a PTAS for MaxEC can be obtained, based on smooth-polynomial integer programming [2], a technique that has been applied to problems like Maximum Quartet Consistency [12].

Proofs are quite involved, and require several technical arguments, that will be included in a journal version of this extended abstract.

6   Conclusion

This paper explores a new direction in the field of orthology and paralogy prediction. Taking advantage of the many existing prediction tools, a set of relations is better represented as a weighted relation graph, where the weight of a relation represents its degree of confidence. In case of non-satisfiability or unconsistency, the goal is to minimally correct the corresponding relation graph. While the problem has been largely explored in the case of unweighted graphs, the weighted version of the problem remains largely unexplored. Here, we provide complexity results and polynomial approximation algorithms for this problem.

For real application to biological datasets, the challenge remains to assign appropriate weights to relations. This can be done by weighting relations according to sequence similarity scores, or in a more sophisticated way by incorporating various information from different prediction tools, depending on the degree of confidence given to each of them. A full bioinformatics study on simulated and real datasets remains to be undertaken for this purpose.

Appendix

A   Proof of Lemma 1

We first bound the number of edges of weight 1 in R.

Claim. Let $H = (V_H, E_H, X)$ be an instance of Minimum Multi-Cut and let $R = (V_R, E_R, w)$ be the corresponding instance of MinWRS. Then, R contains at most $2\left(\binom{|V|}{2} - |E_H|\right)$ edges of weight 1.

Proof. Consider the edges connecting vertices $v_{i,R}$ and $v_{j,R}$; $v_{i,R}$ and $v_{j,R}$ are connected by an edge of weight 1 if and only if $\{v_i, v_j\} \notin E_H$ and $\{v_i, v_j\} \in X$.

Consider the edges connecting vertices $v_{i,R}$ and $v'_{j,R}$, $v'_{i,R}$ and $v_{j,R}$. $v_{i,R}$, $v'_{j,R}$ (and $v'_{i,R}, v_{j,R}$) are connected by an edge of weight 1 if $\{v_i, v_j\} \notin E_H$ and $\{v_i, v_j\} \notin X$.

Any other edge has weight greater than 1, hence the lemma follows.   □

We are now ready to prove Lemma 1.

Proof. Given a set E' that defines a multicut in H, let $V_{H,1}, \ldots, V_{H,p}$ be the sets of vertices of the connected components in the graph $V'_H = (V'_H, E_H \setminus E'_H)$.

We define a solution of MinWRS over instance R as follows. We construct the partition $V_{R,1}, \ldots, V_{R,p}$ of the vertices of R such that $v_{j,R}$ and $v'_{j,R}$ belong to set $V_{R,i}$ if and only if $v_j \in V_{H,i}$. All edges having their endpoints in two distinct $V_{R,i}, V_{R,j}$ are removed.

We claim that the computed graph R' induced by the partition is P_4-free. By construction, for each $v_{j,R}$, $v'_{j,R}$, $v_{h,R}$, $v'_{h,R}$ that belong to $V_{R,i}$, the edges $\{v_{j,R}, v'_{h,R}\}$ and $\{v'_{j,R}, v_{h,R}\}$ belong to E_R (because $\{v_j, v_h\} \notin X$). Moreover, there is no edge between $v_{j,R}$ and $v_{h,R}$, nor between $v'_{j,R}$ and $v'_{h,R}$. Thus any path on four vertices in the graph on vertex set $V_{i,R}$ must be either of the form $v_{j,R} v'_{h,R} v_{k,R} v'_{\ell,R}$, or of the form $v'_{j,R} v_{h,R} v'_{k,R} v_{\ell,R}$. In both cases, the endpoints of the path share an edge, and thus cannot induce a P_4.

Now, consider the edges $\{v_i, v_j\} \in E'_H$. If $\{v_i, v_j\} \in X$, the corresponding solution of MinWRS removes an edge of weight q, namely $\{v_{i,R}, v_{j,R}\}$. If $\{v_i, v_j\} \notin X$, the corresponding solution of MinWRS removes two edges of weight $q/2$, namely $\{v_{i,R}, v'_{j,R}\}$ and $\{v'_{i,R}, v_{j,R}\}$. Hence those edges have a total weight $q|E'_H|$. Since at most $2\left(\binom{|V_H|}{2} - |E_H|\right)$ edges of weight 1 are removed (see Claim A), we can conclude that the lemma holds. $\qquad\square$

B   Proof of Lemma 2

Proof. Consider a solution $R' = (V_R, E'_R, w)$ of MinWRS over instance $R = (V_R, E_R, w)$ of weight at most $qW + 2\left(\binom{|V_H|}{2} - |E_H|\right)$, with $W \leq |E_H|$. First, notice that no edge $\{v_{i,R}, v'_{i,R}\}$, with $1 \leq i \leq |V|$, is removed to obtain R', since the weight of such an edge is greater than $qW + 2\left(\binom{|V_H|}{2} - |E_H|\right)$.

Consider now two vertices $v'_{i,R}$, $v'_{j,R}$, such that, given the corresponding vertices v_i, v_j in H, we have $\{v_i, v_j\} \in X$. By construction there is a P_4 in R, namely $v'_{i,R}, v_{i,R}, v_{R,j}, v'_{j,R}$. It follows that the edge $\{v_{i,R}, v_{j,R}\}$ must be removed in R'. Moreover, we claim that in R', the vertices $v'_{i,R}$, $v'_{j,R}$ must be disconnected. Assume by contradiction that this does not hold, and that $v'_{i,R}$, $v'_{j,R}$ belong to the same connected component of R'. Consider the shortest path P that connects vertices $v_{i,R}$ and $v_{j,R}$ in R'. Then P has length at least 2. Note that as P is a shortest path, it has no chord, i.e. non-consecutive vertices of P cannot share an edge.

Suppose that P does not include the vertex $v'_{i,R}$. Then we can assume that $v_{i,R}$ is adjacent in P to a vertex $v'_{t,R}$, since if it is adjacent to a vertex $v_{q,R}$, then the vertices $v_{i,R}$, $v'_{i,R}$, $v_{q,R}$, and $v'_{q,R}$ would induce a P_4. Now, if $v'_{t,R}$ is adjacent to $v_{j,R}$, then $v'_{i,R}$, $v_{i,R}$, $v'_{t,R}$ and $v_{j,R}$ induce a P_4. If there is no such $v'_{t,R}$, then P has length at least 3 and it must therefore contain an induced P_4.

So suppose instead that P includes the vertex $v'_{i,R}$. Since by construction $v'_{i,R}$ is not adjacent to $v_{j,R}$ and it is not adjacent to any $v'_{t,R}$, with $t \neq i$, while it is adjacent to $v_{i,R}$, P has length at least 3, and again must have an induced P_4.

We can conclude that when $\{v_i, v_j\} \in X$, the corresponding vertices $v'_{i,R}$, $v'_{j,R}$ belong to disconnected connected components of R'. Hence we can compute a multi-cut of H as follows:

$$E'_H = \{\{v_i, v_j\} : \{v_{i,R}, v_{j,R}\}, \text{ of weight } q, \text{ or } \{v_{i,R}, v'_{j,R}\}, \{v'_{i,R}, v_{j,R}\},$$
$$\text{of weight } \frac{q}{2}, \text{ are removed in } R'.\}$$

E'_H is a multi-cut, since each $\{v_i, v_j\} \in X$ is disconnected. Now, recall that R' is obtained by removing edges of overall weight at most $qW + 2\left(\binom{|V_H|}{2} - |E_H|\right)$. Since edge edge in E'_H corresponds to edges of overall weight q in R (an edge $\{v_{i,R}, v_{j,R}\}$ of weight q if $\{v_i, v_j\} \in X$, or two edges of weight $q/2$, namely $\{v_{i,R}, v'_{j,R}\}$ and $\{v'_{i,R}, v_{j,R}\}$ if $\{v_i, v_j\} \notin X$), we must have $|E'_H| \leq W$. ☐

C   Proof of Theorem 1

Proof. Given a graph H instance of Minimum Multi-Cut and the corresponding instance R of MinWRS, denote by OPT_M (AP_M, respectively) the value of an optimal solution (of an approximation solution, respectively) of Minimum Multi-Cut on instance H, and denote by OPT_C (AP_C, respectively) the value of an optimal solution (of an approximation solution, respectively) of MinWRS on instance R. Define $z = 2\left(\binom{|V_H|}{2} - |E_H|\right)$. By Lemma 1, we assume that $AP_C(R) \leq AP_M(H)/q$, as there exists an algorithm that always outputs at most such a value, and thus any approximation algorithm can be adapted to output at most this value. Also, by Lemma 2, we have $OPT_C(R) \leq OPT_M(H)q + z$. We have that

$$\frac{AP_C(R)}{OPT_C(R)} \geq \frac{AP_M(H)q}{OPT_M(H)q + z} = \frac{AP_M(H)q + AP_M(H)z - AP_M(H)z}{OPT_M(H)q + z} =$$

$$= \frac{AP_M(H)q + AP_M(H)z}{OPT_M(H)q + z} - \frac{AP_M(H)z}{OPT_M(H)q + z}$$

$$\geq \frac{AP_M(H)q + AP_M(H)z}{OPT_M(H)q + OPT_M(H)z} - \frac{AP_M(H)z}{OPT_M(H)q + z}$$

$$= \frac{AP_M(H)(q + z)}{OPT_M(H)(q + z)} - \frac{AP_M(H)z}{OPT_M(H)q + z}$$

$$= \frac{AP_M(H)}{OPT_M(H)} - \frac{AP_M(H)z}{OPT_M(H)q + z}$$

where we assume $OPT_M(H) \geq 1$ for the second inequality (the case $OPT_M(H) = 0$ can be checked in polynomial time). Since Minimum Multi-Cut is not approximable within a constant factor assuming the Unique Games Conjecture [4], even on unweighted graphs, it follows that

$$\frac{AP_M(H)}{OPT_M(H)} \geq \alpha$$

on an infinity of instances of H for any constant $\alpha \geq 1$. As a consequence, for any constant $\alpha \geq 1$, an infinity of instances of R yield:

$$\frac{AP_C(R)}{OPT_C(R)} \geq \alpha - \frac{AP_M(H)z}{OPT_M(H)q + z}$$

Since $q = n^5 + 1$, $AP_M(H) \leq n^2$ and $z \leq n^2$, it follows that $\frac{AP_M(H)z}{OPT_M(H)q+z} \leq 1/n$. Combining the last two inequalities, we have that

$$\frac{AP_C(R)}{OPT_C(R)} \geq \alpha - 1/n \geq \beta$$

for any constant $\beta \geq 1$, which concludes the proof.   □

D   Proof of Corollary 2

Proof. The result follows by a gap-preserving reduction similar to that for MinWRS and MinWES. Define a species tree S on V_R such that the root of S is connected to two subtrees, one with leafset $\{v_{i,R} : v_i \in V_H\}$, one with leafset $\{v'_{i,R} : v_i \in V_H\}$.

Consider the partition $V_{R,1}, \ldots, V_{R,p}$ of the vertices of a solution R' of MinWRS and MinWES. Each connected component $V_{R,t}$ that contains vertices $v_{i,R}$, $v'_{i,R}$, $v_{j,R}$, $v'_{j,R}$, contains only edges $\{v_{i,R}, v'_{i,R}\}$, $\{v_{j,R}, v'_{j,R}\}$, $\{v_{i,R}, v'_{j,R}\}$, $\{v_{j,R}, v'_{i,R}\}$.

For each set $V_{R,i}$, we construct a tree $G_{R,i}$ by defining two subtrees $G^1_{R,i}$ and $G^2_{R,i}$ such that $G^1_{R,i}$ has leafset $\{v_{j,R} : v_{j,R} \in V_{R,i}\}$ and $G^2_{R,i}$ has leafset $\{v'_{j,R} : v'_{j,R} \in V_{R,i}\}$. Each node of $G^1_{R,i}$ and $G^2_{R,i}$ is associated with a duplication. $G_{R,i}$ is obtained by joining $G^1_{R,i}$ and $G^2_{R,i}$ in a root, associated with a speciation. Finally, the subtrees $G_{R,1}, \ldots, G_{R,p}$ are joined in a gene tree G by duplication nodes (with any topology). By construction, G is S-consistent, thus the hardness result can be extended to MinWEC.   □

E   Proof of Theorem 3

We first provide the detailed MinCut-Cograph-Editing-Cons algorithm, and show that it also is a n-factor approximation.

Given a species tree S and a set $Z \subseteq V_R$, let $\Sigma(Z) = \{s(x) : x \in Z\}$. Let $S|\Sigma(Z)$ be the subtree of S restricted to $\Sigma(Z)$ and let X_S, Y_S be the clades of the left and right child, respectively, of the root of $S|\Sigma(Z)$. Consider the sets $X = \{x : s(x) \in X_S\}$ and $Y = \{y : s(y) \in Y_S\}$, the cut $C_S(Z)$ on $\overline{R}[Z]$ is defined as $C_S(Z) = \{X_R, Y_R\}$. Observe that $C_S(Z)$ is the only possible cut on $\overline{R}$ that maintains S-consistency, as this cut corresponds to a speciation in a DS-tree, and speciations must separate genes according to S. Therefore, it suffices to modify MinCut-Cograph-Editing by forcing the cut $\overline{C}$ to be $C_S(Z)$. Call this modified algorithm MinCut-Cograph-Editing-Cons.

```
ALGORITHM MINCUT-COGRAPH-EDITING-CONS(R):
    IF R has at most 2 vertices THEN RETURN;
    Find a MinCut C = {X, Y} for R;
    Let C_S(V_R) = {X̄, Ȳ};
    IF w(C) < w(C_S(V_R)) THEN
        Remove all edges between X and Y in R;
        MINCUT-COGRAPH-EDITING-CONS(R[X]);
        MINCUT-COGRAPH-EDITING-CONS(R[Y]);
    ELSE
        Add all possible edges between X̄ and Ȳ in R;
        MINCUT-COGRAPH-EDITING-CONS(R[X̄]);
        MINCUT-COGRAPH-EDITING-CONS(R[Ȳ]);
    END IF
END ALGORITHM
```

Proof. Denote by $\beta(R)$ the weight of the edge-editing found by the algorithm on R. We proceed by induction on $n = |V_R|$ to show that $\beta(R) \leq n\sigma_R$. The statement is trivial for $n \leq 2$ (as there is nothing to correct), so assume that the algorithm finds a solution of weight $\beta(R) \leq k\sigma_R$ for any graph of size at most $k < n$.

The algorithm applies a cut $C = \{X, Y\}$ which is either a minimum cut on R or it is the cut $C_S(V_R)$, and proceeds recursively on X and Y, with $|X|, |Y| \leq n - 1$. By the induction hypothesis, we have

$$\beta(R) \leq |X|\sigma_X + |Y|\sigma_Y + w(C) \leq (n-1)(\sigma_X + \sigma_Y) + w(C)$$

Now, similarly to Lemma 4, we have that $w(C) \leq \sigma_R$. First, let G' be the gene tree associated with a solution of MinWEC over instance R. If C is a minimum cut on R, it holds due to the proof Lemma 4. If C is $C_S(V_R)$, then notice that, in order to guarantee the consistency with S, the root of G' must be exactly $C_S(V_R)$.

Lemma 5 holds also for MinWEC, hence

$$\beta(R) \leq |X|\sigma_X + |Y|\sigma_Y + w(C) \leq (n-1)(\sigma_X + \sigma_Y) + w(C)$$
$$\leq (n-1)\sigma_R + \sigma_R = n\sigma_R$$

hence the theorem holds.   □

F   Proof of Lemma 6

Given a relation graph R, the value of a solution R' for MaxES over instance R is called the *agreement* value of R' and it is denoted by $A(R', R)$. Moreover, given a gene tree G, we denote by $A(G, R)$ the agreement between the relation graph associated with G and R.

Proof. Let $X = \{\{u, v\} : u, v \in V_R$ and $s(u) = s(v)\}$ be the set of 'must-be' paralogs. Consider the relation graphs $R' = (V_R, \emptyset)$ and $R'' = (V_R, \binom{V_R}{2} \setminus X)$, where $\binom{V_R}{2}$ is the set of all unordered pairs of V_R. It is not hard to see that R' and R'' are both feasible solutions of MaxES and of MaxEC. For each $\{u, v\} \in \binom{V_R}{2} \setminus X$, the u, v relation in R agrees with exactly one of R' or R'', and for each $\{u, v\} \in X$, the u, v relation agrees with both R' and R''. It follows that

$$A(R, R') + A(R, R'') \geq \binom{n}{2}$$

But then, for this inequality to hold, at least one of R', R'' must have an agreement value of at least $\frac{1}{2}\binom{n}{2}$, hence an optimal solution of MaxES and MaxEC has an agreement value of at least $\frac{1}{2}\binom{n}{2} \geq \frac{n^2}{8}$. $\qquad\square$

References

1. Alon, N., Stav, U.: Hardness of edge-modification problems. Theor. Comput. Sci. **410**(47–49), 4920–4927 (2009)
2. Arora, S., Frieze, A.M., Kaplan, H.: A new rounding procedure for the assignment problem with applications to dense graph arrangement problems. Math. Program. **92**(1), 1–36 (2002)
3. Berglund, A., Sjolund, E., Ostlund, G., Sonnhammer, E.: InParanoid 6: eukaryotic ortholog clusters with inparalogs. Nucl. Acids Res. **36**, D263–D266 (2008)
4. Chawla, S., Krauthgamer, R., Kumar, R., Rabani, Y., Sivakumar, D.: On the hardness of approximating multicut and sparsest-cut. Comput. Complex. **15**(2), 94–114 (2006)
5. The Gene Ontology Consortium: Gene ontology: tool for the unification of biology. Nat. Genet. **25**(1), 25–29 (2000)
6. Corneil, D.G., Perl, Y., Stewart, L.K.: A linear recognition algorithm for cographs. SIAM J. Comput. **14**(4), 926934 (1985)
7. Fitch, W.M.: Homology: a personal view on some of the problems. TIG **16**(5), 227–231 (2000)
8. Goodman, M., Czelusniak, J., Moore, G., Romero-Herrera, A., Matsuda, G.: Fitting the gene lineage into its species lineage, a parsimony strategy illustrated by cladograms constructed from globin sequences. Syst. Zool. **28**, 132–163 (1979)
9. Hellmuth, M., Hernandez-Rosales, M., Huber, K., Moulton, V., Stadler, P., Wieseke, N.: Orthology relations, symbolic ultrametrics, and cographs. J. Math. Biol. **66**(1–2), 399–420 (2013)
10. Hellmuth, M., Wieseke, N., Lechner, M., Lenhof, H.-P., Middendorf, M., Stadler, P.F.: Phylogenomics with paralogs. In: PNAS (2014)
11. Hernandez-Rosales, M., Hellmuth, M., Wieseke, N., Huber, K.T., Moulton, V., Stadler, P.F.: From event-labeled gene trees to species trees. BMC Bioinform. **13**(Suppl 19), S6 (2012)
12. Jiang, T., Kearney, P.E., Li, M.: A polynomial time approximation scheme for inferring evolutionary trees from quartet topologies and its application. SIAM J. Comput. **30**(6), 1942–1961 (2000). doi:10.1137/S0097539799361683
13. Lafond, M., Dondi, R., El-Mabrouk, N.: The link between orthology relations and gene trees: a correction perspective. Algorithms Mol. Biol. **11**(1), 1 (2016)

14. Lafond, M., El-Mabrouk, N.: Orthology and paralogy constraints: satisfiability and consistency. BMC Genom. **15**(Suppl. 6), S12 (2014)
15. Lafond, M., Semeria, M., Swenson, K., Tannier, E., El-Mabrouk, N.: Gene tree correction guided by orthology. BMC Bioinform. **14**(suppl. 15), S5 (2013)
16. Lafond, M., Swenson, K., El-Mabrouk, N.: Error detection and correction of gene trees. In: Chauve, C., El-Mabrouk, N., Tannier, E. (eds.) Models and Algorithms for Genome Evolution. Springer, London (2013)
17. Lechner, M., Findeiß, S., Steiner, L., Marz, M., Stadler, P.F., Prohaska, S.J.: Proteinortho: detection of co-orthologs in large-scale analysis. BMC Bioinform. **12**(1), 1 (2011)
18. Li, L., Stoeckert, C.J., Roos, D.: OrthoMCL: identification of ortholog groups for eukaryotic genomes. Genome Res. **13**, 2178–2189 (2003)
19. Liu, Y., Wang, J., Guo, J., Chen, J.: Complexity and parameterized algorithms for cograph editing. Theor. Comput. Sci. **461**, 45–54 (2012)
20. Natanzon, A., Shamir, R., Sharan, R.: Complexity classification of some edge modification problems. Discrete Appl. Math. **113**(1), 109–128 (2001)
21. Stoer, M., Wagner, F.: A simple min-cut algorithm. J. ACM **44**(4), 585–591 (1997)
22. Tatusov, R., Galperin, M., Natale, D., Koonin, E.: The COG database: a tool for genome-scale analysis of protein functions. Nucleic Acids Res. **28**, 33–36 (2000)

Copy-Number Evolution Problems: Complexity and Algorithms

Mohammed El-Kebir[1], Benjamin J. Raphael[1(✉)], Ron Shamir[2(✉)],
Roded Sharan[2], Simone Zaccaria[1,3], Meirav Zehavi[2], and Ron Zeira[2]

[1] Department of Computer Science, Center for Computational Molecular Biology,
Brown University, Providence, RI, USA
{melkebir,braphael,szaccari}@cs.brown.edu
[2] School of Computer Science, Tel Aviv University, 69978 Tel Aviv, Israel
{rshamir,roded,meizeh,ronzeira}@post.tau.ac.il
[3] Dipartimento di Informatica Sistemistica E Comunicazione (DISCo),
Univ. Degli Studi di Milano-Bicocca, Milan, Italy

Abstract. Cancer is an evolutionary process characterized by the accumulation of somatic mutations in a population of cells that form a tumor. One frequent type of mutations are copy number aberrations, which alter the number of copies of genomic regions. The number of copies of each position along a chromosome constitutes the chromosome's copy-number profile. Understanding how such profiles evolve in cancer can assist in both diagnosis and prognosis. We model the evolution of a tumor by segmental deletions and amplifications, and gauge distance from profile $\mathbf{a}$ to $\mathbf{b}$ by the minimum number of events needed to transform $\mathbf{a}$ into $\mathbf{b}$. Given two profiles, our first problem aims to find a parental profile that minimizes the sum of distances to its children. Given k profiles, the second, more general problem, seeks a phylogenetic tree, whose k leaves are labeled by the k given profiles and whose internal vertices are labeled by ancestral profiles such that the sum of edge distances is minimum. For the former problem we give a pseudo-polynomial dynamic programming algorithm that is linear in the profile length, and an integer linear program formulation. For the latter problem we show it is NP-hard and give an integer linear program formulation. We assess the efficiency and quality of our algorithms on simulated instances.

1 Introduction

The clonal theory of cancer posits that cancer results from an evolutionary process where somatic mutations that arise during the lifetime of an individual accumulate in a population of cells that form a tumor [9]. Consequently, a tumor consists of *clones*, which are subpopulations of cells sharing a unique combination of somatic mutations. The *evolutionary history* of the clones can be described by a phylogenetic tree whose leaves correspond to extant clones and whose edges are labeled by mutations. Computational inference of phylogenetic trees is a fundamental problem in species evolution [4], and has recently been

© Springer International Publishing Switzerland 2016
M. Frith and C.N.S. Pedersen (Eds.): WABI 2016, LNBI 9838, pp. 137–149, 2016.
DOI: 10.1007/978-3-319-43681-4_11

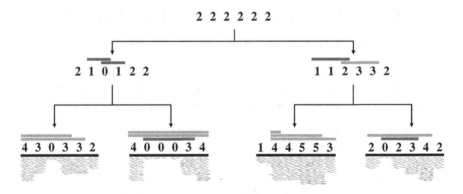

Fig. 1. Copy-Number Tree Problem. As input we are given the copy-number profiles of four leaves, each profile is an integer vector that is inferred from data; e.g. the coverage of mapped reads (blue segments). The tree topology and profiles at internal vertices are found to minimize the total number of amplifications (green bars) and deletions (red bars). The displayed scenario has 14 total events. (Color figure online)

studied extensively for tumor evolution in the case where mutations are single-nucleotide variants [3,7,8,10,15]. Here, we study the problem of constructing a phylogenetic tree of a tumor in the case where mutations are copy number aberrations.

Copy number aberrations include segmental deletions and amplifications that affect large genomic regions, and are common in many cancer types [2]. As a result of these events, the number of copies of genomic regions (*positions*) along a chromosome can deviate from the diploid, two-copy state of each position in a normal chromosome. Understanding these events and the underlying evolutionary tree that relates them is important in predicting disease progression and the outcome of medical interventions [5].

Several methods have been introduced to infer trees from copy number aberrations in cancer. In [1,16] the authors use fluorescent in situ hybridisation data to analyze gain and loss of whole chromosomes and single genes. However, due to technical limitations, this technology does not scale to a large number of positions. In addition, common deletions and amplifications that affect only a subset of the positions of a chromosome are not supported by the model. In another work, Schwartz *et al.* [12] introduced MEDICC, an algorithm that analyzes amplifications and deletions of contiguous segments. The input to MEDICC is a set of *copy-number profiles*, vectors of integers defining the copy-number state of each position. These profiles are measured for multiple samples from a tumor using DNA microarrays or DNA sequencing. The edit distance from profile **a** to **b** was defined as the minimum number of amplifications and deletions of segments required to transform **a** into **b**. Note that this distance is not symmetric. Using this distance measure, the authors applied heuristics to reconstruct phylogenetic trees. However, the complexity of their methods was not analyzed. Recently, Shamir *et al.* [13] analyzed some combinatorial aspects of this amplification/deletion distance model and proved that the distance from one profile to another can be computed in linear time.

In this work, we consider two problems in the evolutionary analysis of copy-number profiles: the Copy-Number Triplet (CN3) and Copy-Number Tree (CNT) problems. Given two profiles, the CN3 problem aims to find a parental profile that minimizes the sum of distances to its children. The CNT problem asks to construct a phylogenetic tree whose k leaves are labeled by the k given profiles, and to assign profiles to the internal vertices so that the sum of distances over all edges is minimum; such a tree describes the evolutionary history under a maximum parsimony assumption (Fig. 1). For the CN3 problem we give a pseudo-polynomial time algorithm that is linear in n, the number of positions in the profiles, along with an integer linear program (ILP) formulation whose number of variables and constraints is linear in n. We show that the CNT problem is NP-hard and present an ILP formulation that scales to practical problem instance sizes. Finally, we use simulations to test our algorithms. Due to space constraints, some details are omitted.

2   Preliminaries

Profiles and Events. We represent a reference chromosome as a sequence of intervals that we call *positions*, numbered from 1 to n in left to right order. We consider mutations that amplify or delete contiguous positions. The *copy-number profile*, or *profile* for short, of a clone specifies the number of copies of each of the n positions. Formally, a profile $\mathbf{y}_i = [y_{i,s}]$ is a vector of length n. An entry $y_{i,s} \in \mathbb{N}$ indicates the number of copies of position s in clone i. For simplicity, we consider a single chromosome only. The results can be easily extended to the case of multiple chromosomes.

An operation, or *event*, acting on profile $\mathbf{y}_i$ increases or decreases copy-numbers in a contiguous segment of $\mathbf{y}_i$. Formally, an event is a triple (s, t, b) where $s \leq t$ and $b \in \mathbb{Z}$. If b is positive then profile-valued positions $s, \ldots, t$ are incremented by b, whereas for negative b the positions $s, \ldots, t$ are decremented by at most $|b|$. That is, applying event (s, t, b) to $\mathbf{y}_i$ results in a new profile $\mathbf{y}_i'$ such that

$$y_{i,\ell}' = \begin{cases} \max\{y_{i,\ell} + b, 0\}, & \text{if } s \leq \ell \leq t \text{ and } y_{i,\ell} \neq 0, \\ y_{i,\ell}, & \text{otherwise.} \end{cases} \tag{1}$$

As indicated by the condition above, once a position ℓ has been lost, i.e. $y_{i,\ell} = 0$, it can never be regained (or deleted). Therefore, for a pair of profiles, there might not be any sequence of events that transform one into the other.

The Copy-Number Tree Problem. We describe the evolutionary process that led to the tumor clones by a *copy-number tree* T, which is a rooted full binary tree. As such, each vertex of T has either zero or two children. We denote the vertex set of T by $V(T)$, root vertex by $r(T)$, leaf set by $L(T)$ and edge set by $E(T)$. The vertices of T correspond to clones. Thus, each vertex $v_i \in V(T)$ is labeled by a profile $\mathbf{y}_i$. The root vertex $r(T)$ corresponds to the *normal clone*, which we assume to be diploid. As such, we have for the corresponding profile that $y_{r,s} = 2$ for all positions s. Note that we do not require vertices to be labeled by a unique profile.

Each edge $(v_i, v_j) \in E(T)$ relates a parent clone v_i to its child v_j, and is labeled by a sequence $\sigma(i,j) = (s_1, t_1, b_1), \ldots, (s_q, t_q, b_q)$ (where $q = |\sigma(i,j)|$) of events that yielded $\mathbf{y}_i$ from $\mathbf{y}_j$. These events are applied in order from 1 to q. Since events in $\sigma(i,j)$ may overlap, i.e. affect the same position, the order as specified by $\sigma(i,j)$ matters. The *cost* of an event (s,t,b) is the number of changes and is thus equal to $|b|$. Therefore, the *cost* $\delta_\sigma(i,j)$ of an edge (v_i, v_j) is the total cost of the events in $\sigma(i,j)$, i.e.

$$\delta_\sigma(i,j) = \sum_{(s,t,b) \in \sigma(i,j)} |b|. \tag{2}$$

Note that the cost is not symmetric. The cost $\Delta(T)$ of the tree T is the sum of the costs of all edges.

Our observations correspond to the profiles $\mathbf{c}_1, \ldots, \mathbf{c}_k$ of k extant clones. Under the assumption of parsimony, the goal is to find a copy-number tree T^* of minimum cost whose leaves correspond to the extant clones. Furthermore, we assume that the maximum copy-number in the phylogeny is bounded by $e \in \mathbb{N}$. We thus have the following problem.

Problem 1 (Copy-Number Tree (CNT)). Given profiles $\mathbf{c}_1, \ldots, \mathbf{c}_k$ on n positions and an integer $e \in \mathbb{N}$, find a copy-number tree T^*, vertex labeling $\mathbf{y}_i$ and edge labeling $\sigma(i,j)$ such that (1) T^* has k leaves labeled $1, \ldots, k$ and $\mathbf{y}_i = \mathbf{c}_i$ for all $i \in \{1, \ldots, k\}$, (2) $y_{i,s} \leq e$ for all $v_i \in V(T^*)$ and $s \in \{1, \ldots, n\}$, and (3) $\Delta(T^*)$ is minimum.

Note that by definition the profile of the root vertex $r(T)$ of any copy-number tree T is the vector whose entries are all 2's. As such, this must hold as well for the minimum-cost tree T^* which always exists. Additionally, the requirement of T being a binary tree can be made without loss of generality by splitting high degree vertices. Furthermore, the assumption that T is a *full* binary tree (i.e. each vertex has out-degree either 0 or 2) can also be made without loss of generality by collapsing degree-2 internal non-root vertices. To account for the case where $r(T)$ has out-degree 1, given an instance $(\mathbf{c}_1, \ldots, \mathbf{c}_k, e)$ we solve a second instance $(\mathbf{c}_1, \ldots, \mathbf{c}_k, \mathbf{c}_{k+1}, e)$ with an additional profile $\mathbf{c}_{k+1}$ consisting of 2's. The result is the minimum-cost tree among the two instances.

The Copy-Number Triplet Problem. The special case where $k = 2$ is the Copy-Number Triplet (CN3) problem. When discussing CN3, due to the fact that we consider only two input profiles, it is not necessary to explicitly refer to trees. Thus, we formulate CN3 as follows:

Problem 2 (Copy-Number Triplet (CN3)). Given profiles $\mathbf{u}$ and $\mathbf{v}$ on n positions, find a profile $\mathbf{m}$ on n positions and sequences of events, $\sigma(\mathbf{m}, \mathbf{u})$ an $\sigma(\mathbf{m}, \mathbf{v})$, such that (1) $\sigma(\mathbf{m}, \mathbf{u})$ yields $\mathbf{u}$ from $\mathbf{m}$ and $\sigma(\mathbf{m}, \mathbf{v})$ yields $\mathbf{v}$ from $\mathbf{m}$, and (2) $\delta_\sigma(\mathbf{m}, \mathbf{u}) + \delta_\sigma(\mathbf{m}, \mathbf{v})$ is minimum.

Instances to both CNT and CN3 always have a solution as the diploid profile is an ancestor to any other profile. Next, we present definitions that will allow us to describe results specific to CN3 in a compact manner. We denote the minimum value $\delta_\sigma(\mathbf{m}, \mathbf{u}) + \delta_\sigma(\mathbf{m}, \mathbf{v})$ associated with a solution $(\mathbf{m}, \sigma(\mathbf{m}, \mathbf{u}), \sigma(\mathbf{m}, \mathbf{v}))$ by $\Delta(\mathbf{u}, \mathbf{v})$. We say that a triple $(\mathbf{m}, \sigma(\mathbf{m}, \mathbf{u}), \sigma(\mathbf{m}, \mathbf{v}))$ is *optimal* if it realizes $\Delta(\mathbf{u}, \mathbf{v})$. Note that $\Delta(\mathbf{u}, \mathbf{v})$ is symmetric and finite. Moreover, if $\delta_\sigma(\mathbf{u}, \mathbf{v})$ (resp. $\delta_\sigma(\mathbf{v}, \mathbf{u})$) is finite then $\mathbf{m} \leftarrow \mathbf{u}$ (resp. $\mathbf{m} \leftarrow \mathbf{v}$) gives a trivial solution to CN3. Let $N = \max\{\max_{i=1}^n\{u_i\}, \max_{i=1}^n\{v_i\}\}$ denote the maximum copy-number in the input. Finally, given $\alpha \in \{\sigma(\mathbf{m}, \mathbf{u}), \sigma(\mathbf{m}, \mathbf{v})\}$ and $w \in \{-, +\}$, we denote the cost of deletions/amplifications affecting position i by

$$co(\alpha, w, i) = \sum_{(s,t,b) \in \alpha \, : \, s \le i \le t, \text{sign}(b) = w} |b|.$$

Previous Results. We now turn to present three results incorporated in the design of our dynamic programming and ILP algorithms for CN3 and CNT. The first one relies on the observation that if $u_i = v_i = 0$, then $\Delta(\mathbf{u}, \mathbf{v}) = \Delta((u_1, \ldots, u_{i-1}, u_{i+1}, \ldots, u_n), (v_1, \ldots, v_{i-1}, v_{i+1}, \ldots, v_n))$, i.e. it is safe to fix $m_i = 0$. Therefore, we have the following straightforward, yet useful result.

Lemma 1. *Without loss of generality, it can be assumed that for all $1 \le i \le n$, at least one value among u_i and v_i is positive.*

This lemma also implies that we can assume that the profile $\mathbf{m}$ of any optimal triple $(\mathbf{m}, \sigma(\mathbf{m}, \mathbf{u}), \sigma(\mathbf{m}, \mathbf{v}))$ consists only of positive values (since for a position i such that $m_i = 0$, it holds that $v_i = u_i = 0$).

We say that a sequence of events where all of the deletions precede all of the amplifications is *sorted*. Formally, let $\sigma(\mathbf{p}, \mathbf{q})$ be a sequence of events that yields $\mathbf{q}$ from $\mathbf{p}$. Then, if there exist a sequence α^- of deletion events and a sequence α^+ of amplification events such that $\sigma(\mathbf{p}, \mathbf{q}) = \alpha^- \alpha^+$, we say that $\sigma(\mathbf{p}, \mathbf{q})$ is sorted. The following lemma states that we can focus on sorted sequences of events:

Lemma 2. [13] *Given a sequence of events $\sigma(\mathbf{p}, \mathbf{q})$ that yields $\mathbf{q}$ from $\mathbf{p}$, there exists a sorted sequence of cost at most $\delta_\sigma(\mathbf{p}, \mathbf{q})$ that yields $\mathbf{q}$ from $\mathbf{p}$.*

Shamir et al. [13] also showed that the minimum cost of a sequence yielding $\mathbf{q}$ from $\mathbf{p}$ is computable by the recursive formula given below. Here, we let $G[i, d, a]$ be the minimum cost of a sequence of events σ that from the prefix $\mathbf{p}^i = (p_1, \ldots, p_i)$ of $\mathbf{p}$ yields the prefix $\mathbf{q}^i = (q_1, \ldots, q_i)$ of $\mathbf{q}$ and which satisfies $co(\sigma, -, i) = d$ and $co(\sigma, +, i) = a$. In case such a sequence does not exist, we let $G[i, d, a] = \infty$.

Lemma 3. [13] *Let* $\mathbf{p}$ *and* $\mathbf{q}$ *be two profiles, and let* $0 \leq d, a \leq N$. *Then,*

1. *If* $q_i > 0$ *and either* $d \geq p_i$ *or* $q_i \neq p_i - d + a$: $G[i, d, a] = \infty$.
2. *Else if* $q_i = 0$ *and* $d < p_i$: $G[i, d, a] = \infty$.
3. *Else if* $i = 1$: $G[i, d, a] = d + a$.
4. *Else:* $G[i, d, a] = \min\limits_{0 \leq d', a' \leq N} \{G[i - 1, d', a'] + \max\{d - d', 0\} + \max\{a - a', 0\}\}$.

The minimum cost of a sequence yielding $\mathbf{q}$ *from* $\mathbf{p}$ *is* $\min_{0 \leq d, a \leq N} G[n, d, a]$.

3 Complexity

In this section we show that CNT is NP-hard by reduction from the Maximum Parsimony Phylogeny (MPP) problem [6]. In MPP, we seek to find a *binary phylogeny* T, which is a full binary tree whose vertices are labeled by binary vectors of size n. The cost of a binary phylogeny T is defined as the sum of the Hamming distances of the two binary vectors associated with each edge. We are only given the leaves of an unknown binary phylogeny in the form of k binary vectors $\mathbf{b}_1, \ldots, \mathbf{b}_k$ of size n, and the task is to find a minimum-cost binary phylogeny T with k leaves such that each leaf $v_i \in L(T)$ is labeled by $\mathbf{b}_i$ and the root is labeled by a vector of all 0s. We consider the decision version where we are asked whether there exists a binary phylogeny T with cost at most h. This problem is NP-complete [6].

We start by defining the transformation (Fig. 2). Let $\mathbf{b}_1, \ldots, \mathbf{b}_k$ be an instance of MPP such that $|\mathbf{b}_i| = n$. The corresponding CNT-instance has parameter $e = 2$ and profiles $\mathbf{c}_1, \ldots, \mathbf{c}_{k+1}$ of length $n + (n - 1)nk$. Each input profile $\mathbf{c}_i$, where $i \in \{1, \ldots, k\}$, is defined as

$$\mathbf{c}_i = \phi(\mathbf{b}_i) = \big(\phi(b_{i,1}) \; \Omega \; \phi(b_{i,2}) \; \Omega \cdots \Omega \; \phi(b_{i,k})\big) \tag{3}$$

where

$$\phi(b_{i,s}) = \begin{cases} 1, & \text{if } b_{i,s} = 1, \\ 2, & \text{otherwise} \end{cases} \tag{4}$$

and Ω, called a *wall*, is a vector of size nk such that for each $j \in \{1, \ldots, nk\}$

$$\Omega_j = \begin{cases} 2, & \text{if } j \text{ is odd,} \\ 1, & \text{otherwise.} \end{cases} \tag{5}$$

Informally, $\mathbf{c}_i$ is defined as a vector consisting of *true positions* (which correspond to the original values) that are separated by *walls* (which are vectors Ω of alternating $2, 1$ values of length nk). The purpose of wall positions Ω is to prevent an event from spanning more than one true position. Profile $\mathbf{c}_{k+1}$ consists of only 2's, and plays a role in initializing the wall elements Ω immediately from the all 2's root. This transformation can be computed in polynomial time, and it is used in the hardness proof (omitted).

Theorem 1. *The CNT problem is NP-hard.*

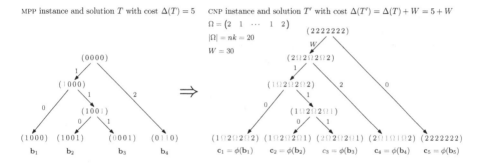

Fig. 2. Transformation of an MPP instance and solution T (left) to a CNT instance and solution T' (right). Edges are labeled by the cost of the associated events and their affected positions are colored in blue. (Color figure online)

4 Algorithms

4.1 Copy-Number Triplet Problem: DP

In this section we develop a DP algorithm, called **DP-Alg1**, that solves the CN3 problem in time $O(nN^{10})$ and space $O(nN^5)$. We will assume w.l.o.g. that sequences of events consist only of events of the form (s, t, b) where $b \in \{-1, 1\}$. Events with $|b| > 1$ can be replaced by $|b|$ events of that form, having the same total cost. DP-Alg1 is based on Lemma 3 and the following claim.

Lemma 4. *Let* $\mathbf{u}$ *and* $\mathbf{v}$ *be two profiles. Then, there exists an optimal triple* $(\mathbf{m}, \sigma(\mathbf{m}, \mathbf{u}), \sigma(\mathbf{m}, \mathbf{v}))$ *where both* $\sigma(\mathbf{m}, \mathbf{u})$ *and* $\sigma(\mathbf{m}, \mathbf{v})$ *are sorted sequences of events, and such that each position* i *of* $\mathbf{m}$ *has at most* N *copies and the cost of amplifications/deletions affecting* i *(in both* $\sigma(\mathbf{m}, \mathbf{u})$ *and* $\sigma(\mathbf{m}, \mathbf{v})$*) is at most* N.

Let $\mathbf{u}^i = (u_1, \dots, u_i)$ and $\mathbf{v}^i = (v_1, \dots, v_i)$ be the prefixes consisting of the first i positions of $\mathbf{u}$ and $\mathbf{v}$, respectively. We will store costs corresponding to partial solutions in a table L (see Fig. 3). This table has an entry $L[i, m, d^\mathbf{u}, a^\mathbf{u}, d^\mathbf{v}, a^\mathbf{v}]$ for all $1 \leq i \leq n$, $0 \leq m \leq N$ and $0 \leq d^\mathbf{u}, a^\mathbf{u}, d^\mathbf{v}, a^\mathbf{v} \leq N$. At such an entry, we will store the the minimum total cost, $\delta_\sigma(\mathbf{m}, \mathbf{u}^i) + \delta_\sigma(\mathbf{m}, \mathbf{v}^i)$

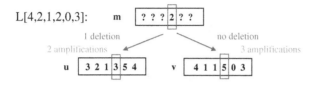

Fig. 3. Illustration of an item in the DP table: Given that the 4th position of $\mathbf{m}$ is 2, one of the combinations considered is 1 deletion and 2 amplifications on the path to $\mathbf{u}$, and 3 amplifications on the path to $\mathbf{v}$. The best cost of that combination is computed by DP-Alg1 based on the L entries for position 3.

of a triple $(\mathbf{m}, \sigma(\mathbf{m}, \mathbf{u}^i), \sigma(\mathbf{m}, \mathbf{v}^i))$ in the set $S(i, m, d^{\mathbf{u}}, a^{\mathbf{u}}, d^{\mathbf{v}}, a^{\mathbf{v}})$, which is defined as follows. This set contains all triples $(\mathbf{m}, \sigma(\mathbf{m}, \mathbf{u}^i), \sigma(\mathbf{m}, \mathbf{v}^i))$ such the numbers of deletions/amplifications affecting i are given by $d^{\mathbf{u}}, a^{\mathbf{u}}, d^{\mathbf{v}}, a^{\mathbf{v}}$, where the notation d/a and $\mathbf{v}/\mathbf{u}$ indicate whether we consider amplifications or deletions as well as $\sigma(\mathbf{m}, \mathbf{u}^i)$ or $\sigma(\mathbf{m}, \mathbf{v}^i)$, $m_i = m$ and for all $j \in \{1, \ldots, n\}$, $m_j \leq N$.

By Lemma 4, $\Delta(\mathbf{u}, \mathbf{v})$ is the minimum cost stored in an entry where $i = n$. Thus, it remains to show how to correctly compute the entries of L efficiently. We use the following base cases, whose correctness follows from Lemma 3:

1. If $u_i > 0$, and $d^{\mathbf{u}} \geq m_i$ or $u_i \neq m_i - d^{\mathbf{u}} + a^{\mathbf{u}}$: $L[i, m, d^{\mathbf{u}}, a^{\mathbf{u}}, d^{\mathbf{v}}, a^{\mathbf{v}}] = \infty$.
2. Else if $v_i > 0$, and $d^{\mathbf{v}} \geq m_i$ or $v_i \neq m_i - d^{\mathbf{v}} + a^{\mathbf{v}}$: $L[i, m, d^{\mathbf{u}}, a^{\mathbf{u}}, d^{\mathbf{v}}, a^{\mathbf{v}}] = \infty$.
3. Else if $u_i = 0$ and $d^{\mathbf{u}} < m_i$: $L[i, m, d^{\mathbf{u}}, a^{\mathbf{u}}, d^{\mathbf{v}}, a^{\mathbf{v}}] = \infty$.
4. Else if $v_i = 0$ and $d^{\mathbf{v}} < m_i$: $L[i, m, d^{\mathbf{u}}, a^{\mathbf{u}}, d^{\mathbf{v}}, a^{\mathbf{v}}] = \infty$.
5. Else if $i = 1$: $L[i, m, d^{\mathbf{u}}, a^{\mathbf{u}}, d^{\mathbf{v}}, a^{\mathbf{v}}] = d^{\mathbf{u}} + a^{\mathbf{u}} + d^{\mathbf{v}} + a^{\mathbf{v}}$.

Now, consider entries $L[i, m, d^{\mathbf{u}}, a^{\mathbf{u}}, d^{\mathbf{v}}, a^{\mathbf{v}}]$ that are not filled by the base cases. We compute them using the following formula:

$$L[i, m, d^{\mathbf{u}}, a^{\mathbf{u}}, d^{\mathbf{v}}, a^{\mathbf{v}}] = \min_{\substack{0 \leq m' \leq N \\ 0 \leq d^{\mathbf{u}'}, a^{\mathbf{u}'}, d^{\mathbf{v}'}, a^{\mathbf{v}'} \leq N}} \left\{ L[i-1, m', d^{\mathbf{u}'}, a^{\mathbf{u}'}, d^{\mathbf{v}'}, a^{\mathbf{v}'}] \right.$$
$$+ \max\{d^{\mathbf{u}} - d^{\mathbf{u}'}, 0\} + \max\{a^{\mathbf{u}} - a^{\mathbf{u}'}, 0\}$$
$$\left. + \max\{d^{\mathbf{v}} - d^{\mathbf{v}'}, 0\} + \max\{a^{\mathbf{v}} - a^{\mathbf{v}'}, 0\} \right\}.$$

The correctness of this formula follows from Lemma 3 and since in light of Lemma 4, it exhaustively searches for the best choice for the previous value of $\mathbf{m}$. By computing the entries of L in an ascending order according to their first argument i, we have that the computation of each entry relies only on entries that are computed before it. The table L consists of $O(nN^5)$ entries, and each of them can be computed in time $O(nN^5)$. Thus, we obtain the following lemma.

Lemma 5. DP-Alg1 *solves CN3 in time* $O(nN^{10})$ *and space* $O(nN^5)$.

We can show that DP-Alg1 can be modified to obtain a DP algorithm, called DP-Alg2, for which we prove the following result.

Theorem 2. DP-Alg2 *solves CN3 in time* $O(nN^7)$ *and space* $O(nN^4)$.

We also devised an ILP formulation for CN3 using only $O(n)$ variables. Details are omitted.

4.2   Copy-Number Tree Problem: ILP

In this section we describe an ILP for CNT consisting of $O(k^2n + kn\log e)$ variables and $O(k^2n + kn\log e)$ constraints. Let $(\mathbf{c}_1, \ldots, \mathbf{c}_k, e)$ be an instance of CNT. Recall that we seek to find a full binary tree with k leaves. We define a directed graph G that contains any full binary tree with k leaves as a spanning tree. As such, $|V(G)| = 2k - 1$. The vertex set $V(G)$ consists of a subset $L(G)$ of leaves such that $|L(G)| = k$. We denote by $r(T) \in V(G) \backslash L(G)$ the vertex that corresponds to the root vertex. Throughout the following, we consider an order $v_1, \ldots, v_k, \ldots, v_{2k-1}$ of the vertices in $V(G)$ such that $v_1 = r(T)$ and $\{v_k, \ldots, v_{2k-1}\} = L(G)$. The edge set $E(G)$ has edges $\{(v_i, v_j) \mid 1 \le i < k, 1 \le i < j \le 2k - 1\}$. We denote by $N^-(j)$ the set of vertices incident to an outgoing edge to j. Conversely, $N^+(i)$ denotes the set of vertices incident to an incoming edge from i. We make the following two observations.

Observation 1. *G is a directed acyclic graph.*

Observation 2. *Any copy-number tree T is a spanning tree of G.*

We now proceed to define the set of feasible solutions (X, Y) to a CNT instance $(\mathbf{c}_1, \ldots, \mathbf{c}_k, e)$ by introducing constraints and variables modeling the tree topology, and vertex labeling and edge costs. More specifically, variables $X = [x_{i,j}]$ encode a spanning tree T of G and variables $Y = [y_{i,s}]$ encode the profiles of each vertex such that X and Y combined induce edge costs. In the following we provide more details.

Tree Topology. The goal is to enforce that we select a spanning tree T of G that is a full binary tree. To do so, we introduce a binary variable $x_{i,j} \in \{0, 1\}$ for each edge $(v_i, v_j) \in E(G)$ indicating whether the corresponding edge (v_i, v_j) is in T. Note that by construction $i < j$. We require that each vertex $v \in V(G) \backslash \{v_1\}$ has exactly one incoming edge in T.

$$\sum_{i \in N^-(j)} x_{i,j} = 1 \qquad\qquad 1 < j \le 2k - 1 \qquad\qquad (6)$$

We require that each vertex $v \in V(G) \backslash L(G)$ has two outgoing edges in T.

$$\sum_{j \in N^+(i)} x_{i,j} = 2 \qquad\qquad 1 \le i < k \qquad\qquad (7)$$

Vertex Labeling and Edge Costs. We introduce variables $y_{i,s} \in \{0, \ldots, e\}$ that encode the copy-number state of position s of vertex v_i. Since the profiles of each leaf as well as the root vertex are given, we have the following constraints.

$$y_{1,s} = 2 \qquad\qquad\qquad 1 \le s \le n \qquad\qquad (8)$$
$$y_{i,s} = c_{i-k+1,s} \qquad\qquad k \le i \le 2k - 1, 1 \le s \le n \qquad\qquad (9)$$

Next, we encode a set $\sigma(v_i, v_j)$ of events that transform the profile $\mathbf{y}_i$ of v_i into profile $\mathbf{y}_j$ of v_j. Recall that an event is a triple (s, t, b) and corresponds to an amplification if $b > 0$ and a deletion otherwise. We model the cost of the amplifications and the cost of the deletions covering any position s with two separate variables. Variables $a_{i,j,s} \in \{0, \ldots, e\}$ correspond to the cost of the amplifications in $\sigma(v_i, v_j)$ covering position s. Variables $d_{i,j,s} \in \{0, \ldots, e\}$ correspond to the cost of the deletions in $\sigma(v_i, v_j)$ covering position s.

Now, we consider the effect of amplifications and deletions on a position s. By Lemma 2, we have that there exists an optimal solution such that for each edge (v_i, v_j) there are two sets of events $\sigma^-(v_i, v_j)$ and $\sigma^+(v_i, v_j)$ that yield $y_{j,s}$ from $y_{i,s}$ by first applying $\sigma^-(v_i, v_j)$ followed by $\sigma^+(v_i, v_j)$. If a subset of the events in $\sigma^-(v_i, v_j)$ results in position s reaching value 0, the remaining amplifications and deletions will not change the value of that position. We distinguish the following four different cases.

(a) $y_{i,s} = 0$ and $y_{j,s} = 0$: Since both positions have value 0, the number of amplifications $a_{i,j,s}$ and deletions $d_{i,j,s}$ are between 0 and e.
(b) $y_{i,s} \neq 0$ and $y_{j,s} \neq 0$: Since $y_{j,s} > 0$, the number of deletions $d_{i,j,s}$ must be strictly smaller than $y_{i,s}$. Moreover, it must hold that $y_{j,s} + d_{i,j,s} = y_{i,s} + a_{i,j,s}$.
(c) $y_{i,s} \neq 0$ and $y_{j,s} = 0$: Recall that by Lemma 2 deletions precede amplifications. As such, the number of deletions $d_{i,j,s}$ must be at least $y_{i,s}$.
(d) $y_{i,s} = 0$ and $y_{j,s} \neq 0$: Once a position s has been lost it cannot be regained. As such, this case is infeasible.

The full description of the constraints and variables that model these cases and the objective function are omitted.

5 Experimental Evaluation

Copy-Number Triplet (CN3) Problem. We compared the running times of our DP and ILP algorithms for the CN3 problem as a function of n and N. Our results on simulations (omitted) show that while the running time of the DP algorithm highly depends on the copy-number range N, the ILP time is almost independent of N. With the exception of the case of $N = 2$, the ILP is faster.

Copy-Number Tree (CNT) Problem. To assess the performance of the ILP for CNT, we simulated instances by randomly generating a full binary tree T with k leaves. We randomly labeled edges by events according to a specified maximum number m of events per edge with amplifications/deletions ratio ρ. Specifically, we label an edge by d events where d is drawn uniformly from the set $\{1, \ldots, m\}$. For each event (s, t, b) we uniformly at random draw an interval $s \leq t$ and decide with probability ρ whether $b = 1$ (amplification) or $b = -1$ (deletion). The resulting instance of CNT is composed of the profiles $\mathbf{c}_1, \ldots, \mathbf{c}_k$ of the k leaves of T and e is set to the maximum value of the input profiles.

We considered varying numbers of leaves $k \in \{4, 6, 8\}$ and of segments $n \in \{5, 10, 15, 20, 30, 40\}$. In addition, we varied the number of events $m \in \{1, 2, 3\}$ and varied the ratio $\rho \in \{0.2, 0.4\}$. We generated three instances for each combination of k, n, m and ρ, resulting in a total of 324 instances.

We implemented the ILP in C++ using CPLEX v12.6 (www.cplex.com). The implementation is available upon request. We ran the simulated instances on a compute cluster with 2.6 GHz processors (16 cores) and 32 GB of RAM each. We solved 302 instances (93.2 %) to optimality within the specified time limit of 5 hours. Computations exceeding this limit were aborted and the best identified solution was considered. The instances that were not solved to optimality are a subset of the larger instances with $k = 8$ and $n \in \{20, 30, 40\}$.

For 323 out of 324 instances (99.7 %) the tree inferred by the ILP has a cost that was at most the simulated tree cost. The only exception is an instance with $k = 8$ leaves and $n = 40$ positions that was not solved to optimality, and where the inferred cost was 15 vs. a simulated cost of 14. These results empirically validate the correctness of our ILP implementation.

We observe that the running time increases with the number of leaves and to a lesser extent with the number of positions (Fig. 4a). In addition, we assessed the distance between topologies of the inferred and simulated trees using the Robinson-Foulds (RF) metric [11]. To allow for a comparison across varying number of leaves, we normalized by the total number of splits to the range [0,1] such that a value of 0 corresponds to the same topology of both trees. For 264 instances (81.4 %) the normalized RF was at most 0.35. For $k = 4$ leaves the median RF value was 0, which indicates that for at least 50 % of these instances the simulated tree topology was recovered. Figure 4b shows the distribution of normalized RF values with varying numbers of leaves and positions. Given a fixed number of leaves, the normalized RF value decreases with increasing number of positions. This indicates that the maximum parsimony assumption becomes more appropriate with larger number of positions, which is not surprising since amplifications and deletions are less likely to overlap. In addition, we observed (data not shown) that running time and RF values are not affected by varying values of m and ρ. In summary, we have shown that our ILP scales to practical problem instance sizes with $k = 6$ and up to $n = 40$ positions, which is a reasonable size for applications to real data [12,14].

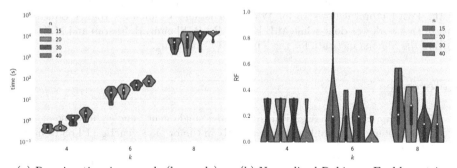

(a) Running time in seconds (log scale)   (b) Normalized Robinson-Foulds metric

Fig. 4. Violin plots of running time (a) and tree distance (b) for varying number k of leaves and number n of positions. Median values are indicated by a white dot in each plot. (Color figure online)

6   Discussion

In this paper we studied two problems in the evolution of copy-number profiles. For the CN3 problem, we gave a pseudo-polynomial DP algorithm and an ILP formulation, and compared their efficiency on simulated data. Determining the computational complexity of CN3 remains an open problem. We showed that the general CNT problem is NP-hard and gave an ILP solution. Finally, we assessed the performance of our tree reconstruction on simulated data. While all formulations describe copy-number profiles on a single chromosome, our results readily generalize to multiple chromosomes. In addition, while our formulations presently lack the phasing step performed in [12], both the DP algorithm and the ILP formulations can be extended to support phasing.

We note that experiments on real cancer sample data are required to establish the relevance of our formulations. To this end, several extensions to our models might be required. These include handling fractional copy-number values that are a result of most experiments and handling missing data for some positions. Moreover, since tumor samples are often impure, each sample may actually represent a mixture of several clones. In such situations, different objectives might try to decompose the clone mixture in order to reconstruct the evolutionary tree as has been investigated for single-nucleotide variants [3,7,8,10,15].

Acknowledgments. B.J.R. is supported by a National Science Foundation CAREER Award CCF-1053753, NIH RO1HG005690 a Career Award at the Scientific Interface from the Burroughs Wellcome Fund, and an Alfred P Sloan Research Fellowship. R. Shamir is supported by the Israeli Science Foundation (grant 317/13) and the Dotan Hemato-Oncology Research Center at Tel Aviv University. R.Z. is supported by fellowships from the Edmond J. Safra Center for Bioinformatics at Tel Aviv University and from the Israeli Center of Research Excellence (I-CORE) Gene Regulation in Complex Human Disease (Center No. 41/11). M.Z. is supported by a fellowship from the I-CORE in Algorithms and the Simons Institute for the Theory of Computing in Berkeley and by the Postdoctoral Fellowship for Women of Israel's Council for Higher Education. Part of this work was done while M.E-K., B.J.R., R. Shamir, R. Sharan and M.Z. were visiting the Simons Institute for the Theory of Computing.

References

1. Chowdhury, S., et al.: Algorithms to model single gene, single chromosome, and whole genome copy number changes jointly in tumor phylogenetics. PLoS Comput. Biol. **10**(7), 1–19 (2014)
2. Ciriello, G., et al.: Emerging landscape of oncogenic signatures across human cancers. Nat. Genet. **45**, 1127–1133 (2013)
3. El-Kebir, M., et al.: Reconstruction of clonal trees and tumor composition from multi-sample sequencing data. Bioinformatics **31**(12), i62–i70 (2015)
4. Felsenstein, J.: Inferring Phylogenies. Sinauer Associates, Sunderland (2004)
5. Fisher, R., et al.: Cancer heterogeneity: implications for targeted therapeutics. Br. J. Cancer **108**(3), 479–485 (2013)

6. Foulds, L.R., Graham, R.L.: The Steiner problem in phylogeny is NP-complete. Adv. Appl. Math. **3**, 43–49 (1982)
7. Jiao, W., et al.: Inferring clonal evolution of tumors from single nucleotide somatic mutations. BMC Bioinform. **15**(1), 1–16 (2014)
8. Malikic, S., et al.: Clonality inference in multiple tumor samples using phylogeny. Bioinformatics **31**(9), 1349–1356 (2015)
9. Nowell, P.C.: The clonal evolution of tumor cell populations. Science **194**, 23–28 (1976)
10. Popic, V., et al.: Fast and scalable inference of multi-sample cancer lineages. Genome Biol. **16**, 91 (2015)
11. Robinson, D.F., Foulds, L.R.: Comparison of phylogenetic trees. Math. Biosci. **53**, 131–147 (1981)
12. Schwarz, R., et al.: Phylogenetic quantification of intra-tumour heterogeneity. PLoS Comput. Biol. **10**(4), 1–11 (2014)
13. Shamir, R., et al.: A linear-time algorithm for the copy number transformation problem. In: Grossi, R., Lewenstein, M., et al. (eds.) 27th Annual Symposium on Combinatorial Pattern Matching (CPM 2016), LIPIcs, vol. 54, pp. 16:1–16:13. Dagstuhl, Germany (2016)
14. Sottoriva, A., et al.: A Big Bang model of human colorectal tumor growth. Nat. Genet. **47**(3), 209–216 (2015)
15. Yuan, K., et al.: BitPhylogeny: a probabilistic framework for reconstructing intra-tumor phylogenies. Genome Biol. **16**(1), 1–16 (2015)
16. Zhou, J., Lin, Y., Rajan, V., Hoskins, W., Tang, J.: Maximum parsimony analysis of gene copy number changes. In: Pop, M., Touzet, H. (eds.) WABI 2015. LNCS, vol. 9289, pp. 108–120. Springer, Heidelberg (2015)

Gerbil: A Fast and Memory-Efficient k-mer Counter with GPU-Support

Marius Erbert, Steffen Rechner$^{(\boxtimes)}$, and Matthias Müller-Hannemann

Institute of Computer Science, Martin Luther University Halle-Wittenberg,
Halle (Saale), Germany
{erbert,rechner,muellerh}@informatik.uni-halle.de

Abstract. A basic task in bioinformatics is the counting of k-mers in genome strings. The k-*mer counting problem* is to build a histogram of all substrings of length k in a given genome sequence. We present the open source k-mer counting software *Gerbil* that has been designed for the efficient counting of k-mers for $k \geq 32$. Given the technology trend towards long reads of next-generation sequencers, support for large k becomes increasingly important. While existing k-mer counting tools suffer from excessive memory resource consumption or degrading performance for large k, *Gerbil* is able to efficiently support large k without much loss of performance. Our software implements a two-disk approach. In the first step, DNA reads are loaded from disk and distributed to temporary files that are stored at a working disk. In a second step, the temporary files are read again, split into k-mers and counted via a hash table approach. In addition, *Gerbil* can optionally use GPUs to accelerate the counting step. For large k, we outperform state-of-the-art open source k-mer counting tools by up to a factor of 4 for large genome data sets.

1 Introduction

The counting of k-mers in large amounts of reads is a common task in bioinformatics. The problem is to count the occurrences of all k-long substrings in a large amount of sequencing reads. Its most prominent application is de novo assembly of genome sequences. Although building a histogram of k-mers seems to be quite a simple task from an algorithmic point of view, it has attracted a considerably amount of attention in recent years. In fact, the counting of k-mers becomes a challenging problem for large instances, if it is to be both resource- and time-efficient and therefore makes it an interesting object of study for algorithm engineering. Existing tools for k-mer counting are often optimized for $k < 32$ and lack good performance for larger k. Recent advances in technology towards larger read lengths are leading to the quest to cope with values of k exceeding 32. Studies elaborating on the optimal choice for the value of k recommend for various applications relatively high values [1,13]. In particular, working with long sequencing reads helps to improve accuracy and contig assembly (with k values in the hundreds) [11]. In this paper, we develop a tool with a high performance for such large values of k.

© Springer International Publishing Switzerland 2016
M. Frith and C.N.S. Pedersen (Eds.): WABI 2016, LNBI 9838, pp. 150–161, 2016.
DOI: 10.1007/978-3-319-43681-4_12

Related Work. Among the first software tools that succeeded in counting the *k*-mers of large genome data sets was Jellyfish [5], which uses a lock-free hash table that allows parallel insertion. In the following years, several tools were published, successively reducing running time and required memory. BFCounter [6] uses Bloom filters for *k*-mer counting to filter out rarely occurring *k*-mers stemming from sequencing errors. Other tools like DSK [7] and KMC [2] exploit a two-disk architecture and aim at reducing expensive IO operations. Turtle [10] replaces a standard Bloom filter by a cache-efficient counterpart. MSPKmer-Counter [4] introduces the concept of minimizers to the *k*-mer counting, thus further optimizing the disk-based approach. The minimizer approach was later on refined to *signatures* within KMC2 [3]. Up to now, the two most efficient open source software tools have been KMC2 and DSK. KMC2 uses a sorting based counting approach that has been optimized for $k < 32$. However, its performance drops when k grows larger. Instead, DSK uses a single large hash table and is therefore efficient for large k (but does not support $k > 127$). However, for small k, it is clearly slower than KMC2. To the best of our knowledge, the only existing approach that uses GPUs for counting *k*-mers is the work by Suzuki et al. [12].

Contribution. In this article we present the open source *k*-mer counting tool *Gerbil*. Our software is the result of an extensive process of algorithm engineering that tried to bring together the best ideas from the literature. The result is a *k*-mer counting tool that is both time efficient and memory frugal.[1] In addition, *Gerbil* can optionally use GPUs to accelerate the counting step. It outperforms its strongest competitors both in efficiency and resource consumption significantly. For large values of k, it reduces the runtime by up to a factor of four. The software is written in C++ and CUDA and is freely available at https://github.com/uni-halle/gerbil.

In the next section we describe the general algorithmic work flow of *Gerbil*. Thereafter, in Sect. 3, we focus on algorithm engineering aspects that proved essential for high performance and describe details, like the integration of a GPU into the counting process. In Sect. 4, we evaluate *Gerbil*'s performance in a set of experiments and compare it with those of KMC2 and DSK. We conclude this article by a short summary and a glance on future work.

2   Methods

Gerbil is divided into two phases: (1) Distribution and (2) Counting. In this section, we give a high-level description of *Gerbil*'s work flow.

Distribution. Whole genome data sets typically do not fit into the main memory. Hence, it is necessary to split the input data into a couple of smaller

---

[1] The tool is named *Gerbil* because of its modest resource requirements, which it has in common with the name-giving mammal.

temporary files. *Gerbil* uses a two-disk approach that is similar to those of most contemporary k-mer counting tools [3,4,7]. The first disk contains the input read data and is used to store the counted k-mer values. We call this disk input/output-disk. The second disk, which we call working disk, is used to store temporary files. The key idea is to assure that the temporary files partition the input reads in such a way, that all occurrences of a certain k-mer are stored in the same temporary file. This way, one can simply count the k-mers of the temporary files independently of each other, with small main memory requirements. To split the genome data into temporary files, we make use of the *minimizer* approach that has been proposed by [9] and later on refined by [3]. A genome sequence can be decomposed into a number of overlapping *super-mers*. Each super-mer is a substring of maximal length such that all k-mers on that substring share the same minimizer. Hereby, a minimizer of a k-mer is defined as its lexicographically smallest substring of a fixed length $m < k$ with respect to some total ordering on strings of length m. See Fig. 1 for an example. It suffices to partition the set of super-mers into different temporary files to achieve a partitioning of all different k-mers [3]. In our experiments, we found that choosing minimizer length $m = 7$ is most efficient.

```
CAAGAACAGTG
CAAG                    1. CAAGA
AAGA
   AGAA                 2. AGAA
    GAAC                3. GAACA
     AACA
      ACAG              4. ACAG
       CAGT             5. CAGTG
        AGTG
```

Fig. 1. Minimizers and super-mers of the DNA string CAAGAACAGTG. Here, $k = 4$ and $m = 3$. For each k-mer, the bold part is its minimizer. The example uses the lexicographic ordering on 3-mers based on $A < C < G < T$. The sequence is divided into the five super-mers CAAGA, AGAA, GAACA, ACAG, and CAGTG that would be stored in temporary files.

Counting. The counting of k-mers is typically done by one of two approaches: Sorting and Compressing [3] or using a hash table with k-mers as keys and counters as values [5,7]. The efficiency of the sorting approach typically relies on the sorting algorithm Radix Sort, whose running time increases with the length of k-mers. Since we aim at high efficiency for large k, we decided to implement the hash table approach. Therefore, we use a specialized hash table with k-mers as keys and counters as values.

2.1 Work Flow

Although the following description of the main process is sequential, all of the steps are interleaved and therefore executed in parallel. This is done by a classical pipeline architecture. Each output of a step makes the input of the next.

Fig. 2. Work flow of phase one.

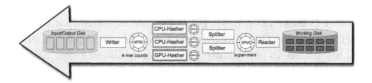

Fig. 3. Work flow of phase two.

We use ring buffers to connect the steps of the pipeline. Such buffers are specialized for all combinations of single (S)/multiple (M) producers (P) and single (S)/multiple (M) consumers (C). The actual number of parallel threads depend on the system and is determined by the software at runtime to achieve optimal memory throughput.

Phase One: Distribution. The goal of the first phase is to split the input data into a number of temporary files. Figure 2 visualizes the first phase.

1. A group of reader threads read the genome reads from the input disk into the main memory. For compressed input, these threads also decompress it.
2. A second group of parser threads convert the read data from the input format into an internal read bundle format.
3. A group of splitter threads compute the minimizers of the reads. All subsequent substrings of a read that share the same minimizer are stored as a super-mer into an output buffer.
4. A single writer thread stores the output buffers to a variable number of temporary files at the working disk.

Phase Two: Counting. After the first phase has been completed, the temporary files are sequentially re-read from working disk and processed in the following manner (see Fig. 3).

1. A single reader thread reads the super-mers of a temporary file and stores them in main memory.
2. A group of threads split the super-mers into *k*-mers. Each *k*-mer is distributed to one of multiple hasher threads by using a hash function on each *k*-mer. This ensures that multiple occurrences of the same *k*-mer are assigned to the

same hasher thread and allows the distribution of separated hash tables to different memory spaces.

3. A group of hasher threads insert the k-mers into their thread-own hash tables. After a temporary file has been completely processed, each hasher thread sends the content of its hash table to an output buffer.
4. A single writer thread writes from the output buffer to the output disk.

2.2 DNA Sequence Handling

Undetermined Bases. DNA reads typically contain bases that could not been identified correctly during the sequencing process. Usually, such bases are marked N in FASTQ input files. In accordance with the established k-mer counting tools, we ignore all k-mers that contain an undetermined base.

Reverse-Complement. Since DNA is organized in double helix form, each k-mer $x \in \{A, C, G, T\}^k$ corresponds to its *reverse-complement* that is defined by reversing x and replacing $A \Leftrightarrow T$ and $C \Leftrightarrow G$. Thus, the k-mer $ACCG$ corresponds to $CGGT$. Many applications do not distinguish between a k-mer and its reverse-complement. Thus, each occurrence of $ACCG$ and $CGGT$ is counted as occurrences of their unique *canonical* representation. *Gerbil* uses the lexicographically smaller k-mer as canonical representation. The use of reverse complement normalization can be turned off by command flag.

3 Implementation Details

We now want to point out several details on the algorithm engineering process that were essential to gain high performance.

3.1 Total Ordering on Minimizers

The choice of a total ordering has large effects on the size of temporary files and thus, also on the performance. To find a good total ordering, we have to balance various aspects. On the one hand, the total number of resulting super-mers are to be minimized to reduce the total size of disk memory that is needed by temporary files. On the other hand, the maximal number of distinct k-mers that share the same minimizer should not be too large since we want an approximately uniform distribution of k-mers to the temporary files. An "ideal" total ordering would have both a large total number of super-mers and a small maximal number of distinct k-mers per minimizer. Since these requirements contradict each other, we experimentally evaluated the pros and cons of various ordering strategies.

CGAT. The lexicographic ordering of minimizers based on $C < G < A < T$.

Roberts et al. [8]. They propose the lexicographic ordering of minimizers with respect to $C < A < T < G$. Furthermore, within the minimizer computation all bases at even positions are to be replaced by their reverse complement. Thus, rare minimizers like $CGCGCG$ are preferred.

KMC2. The ordering that is proposed by [3] is a lexicographic ordering with $A < C < G < T$ and some built-in exceptions to eliminate the large number of minimizers that start with AAA or ACA.

Random. A random order of all string of fixed length m is unlikely to have both a small number of super-mers and a highly imbalanced distribution of distinct k-mers. It is simple to establish, since we do not need frequency samples or further assumptions about the distribution of minimizers.

Distance from Pivot (dfp(p)). To explain this strategy, consider the following observations: Ascendingly sorting the minimizers by their frequency favors rare minimizers. As a consequence, the maximal number of distinct k-mers per minimizer is small. However, the total number of super-mers can be very large. Similarly, an descendingly sorted ordering results in quite the opposite effect. To find a compromise between both extremes, we initially sort the set of minimizers by their frequency. Since the frequencies depend on the data set, we approximate them by taking samples during runtime. We fix a pivot factor $0 \leq p \leq 1$ and re-sort the minimizers by the absolute difference of their initial position to the pivot position $4^m p$. The result is an ordering that does neither prefer very rare nor very common minimizers and therefore makes a good compromise.

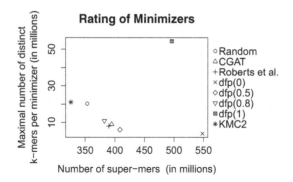

Fig. 4. Evaluation of various total ordering strategies for minimizers (F *Vesca*, $m = 6$, $k = 28$). Strategy dfp(p) has been tested with $p \in \{0, 0.5, 0.8, 1\}$.

Evaluation. See Fig. 4 for a rating of each strategy. The value on the x-axis corresponds to the expected temporary disk memory, whereas the value on the y-axis is correlated with the maximal main memory consumption of our program. A perfect strategy would be located at the bottom left corner. Several strategies seem to be reasonable choices. We evaluated each strategy and found that a small number of super-mers is more important than a small maximal number of k-mer per minimizer for most data sets. As a result, we confirm that the total ordering that is already been used by KMC2 is a good choice for most data sets. Therefore, *Gerbil* uses the strategy from KMC2 for its ranking of minimizers.

3.2   GPU Integration

To integrate one or more GPUs into the process of k-mer counting, several problems have to be dealt with. Typically, a GPU performs well only if it deals with data in a parallel manner. In addition, memory bound tasks (i.e. tasks that do not require a lot of arithmetic operations) like the counting of k-mers require a carefully chosen memory access pattern to minimize the number of the accesses to the GPU's global memory. We decided to transfer the hash table based counting approach to the GPU.

GPU Hash Tables. When compiled and executed with GPU support, *Gerbil* automatically detects CUDA capable GPUs. For each GPU, *Gerbil* replaces a CPU hasher thread by a GPU hasher thread which maintains its own hash table in GPU memory. Each GPU hash table is similar in function to a traditional hash table. However, unlike the traditional approach, we add a large number of k-mers in parallel. Therefore, the insertion procedure is slightly changed.

First, a bundle of several thousand k-mers is copied to the GPU global memory space. Afterwards, we launch a large number of CUDA blocks, each consisting of 32 threads. Each block sequentially inserts a few k-mers into the GPU hash table. Since with increasing running time, it becomes more and more probable to find a mismatch when probing a hash table position, we additionally scan adjacent table positions in a range of 128 bytes when probing a hash table entry (see Fig. 5). Due to the architecture of a GPU, this can be done within the same global memory access. Thus, we scan up to 16 table entries in parallel, thereby reducing the number of accesses to a GPU's global memory. To eliminate race conditions between CUDA blocks, we synchronize the probing of the hash table by using atomic operations to lock and unlock hash table entries. Since such operations are efficiently implemented in hardware, a large number of CUDA blocks can be executed in parallel.

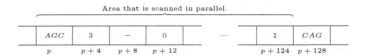

Fig. 5. GPU memory access pattern. The figure shows the memory area that is being scanned while probing a hash table entry that is stored at memory address p. In this example, $k = 3$ and each table entry needs four bytes for the key and four bytes for the counter. Therefore, 16 entries can be loaded from global memory within one step and are scanned in parallel.

Load Balancing. We dynamically balance the amount of k-mers that are assigned to the various CPU and GPU hasher threads. Therefore, we constantly measure the throughput of each hasher thread, i.e. the CPU-time needed to insert a certain number of *k*-mers. Whenever a new temporary file is loaded from disk, we rebalance the number of *k*-mers that are assigned to each hasher thread, considering the throughput and capacity of each hash table. By that, we automatically determine a good division of labour between CPU and GPU hasher threads without the need of careful hand-tuning.

3.3 Hash Table Details

Estimating Table Sizes. We aim at estimating the expected size of each hash table as closely as possible to save main memory. We do so since reduced memory consumption leaves more memory to the operating system that can be used as cache when writing temporary files. Therefore, we approximate the number of expected distinct *k*-mers in each temporary file. We use a simple approximation mechanism that predicts the number of distinct *k*-mers in a file by multiplying the number of *k*-mers in each file with a constant that has been determined experimentally (see Fig. 6). Since this ratio depends on properties of the data set, we dynamically adjust the ratio during runtime.

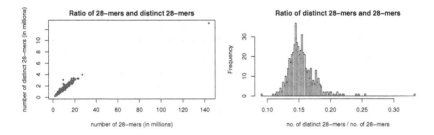

Fig. 6. Left: Number of 28-mers and number of distinct 28-mers in the 512 temporary files that have been created while processing the *F Vesca* data set. Each point corresponds to a temporary file. Here, the KMC2 minimizer ordering did not succeed in creating uniformly sized temporary files since a single file contains far more 28-mers than the other 511 files. Right: Dividing the number of 28-mers in each file by number of its distinct 28-mers leads to a ratio that is used to determine the size of the hash tables. A ratio between 0.15 and 0.2 is a proper choice for the *F Vesca* data set.

Probing Strategy. As a general strategy we use double hashing. We stop the probing of the hash table after a constant number of trials. Therefore, it is possible that *k*-mers could not be inserted into a hash table. For that reason, *Gerbil* has a built-in emergency mechanism that handles such *k*-mers to prevent them from getting lost. Hereby, CPU and GPU hasher threads have different strategies. CPU hasher threads store such *k*-mers in an additional temporary file, which is processed after the work with the current temporary file has been completed. In contrast, GPU hasher threads use part of free GPU memory to sequentially store those k-mers that could not be inserted. After all k-mers of a temporary file have been processed, the k-mers in this area are counted via a sorting and compression approach. However, it is still possible to exceed the available GPU memory. In such a case, we copy the whole amount of k-mers in that area back to main memory and store them in a temporary file, similar to the CPU emergency handling. Such an operation is very costly. However, we have never observed a single GPU error handling and only few executions of CPU error handling when processing real world data sets.

4   Results

We tested our implementation in a set of experiments, using the same instances as Deorowicz et al. [3] (see Table 1). For each data set we counted all k-mers for $k = 28, 40, 56$, and 65 and compared *Gerbil*'s running time with those of KMC2 in version 2.3.0 and DSK in version 2.0.7. In addition, we used a synthesized test set *GRCh38*, created from Genome Reference Consortium Human Reference 38 (GCA_000001405.2), from which we uniformly sampled k-mers of size 1000. The purpose of this data set is to have longer reads allowing to test the performance for larger values of k. To judge performance on various types of hardware, we executed the experiments on two different desktop computers. See Table 2 for details about the hardware configuration of the test systems.

Table 3 and Fig. 7 show the results of the performance evaluation. We want to point out several interesting observations.

Table 1. Data sets.

Data set	Format	Size (GB)	Read length	28-mers	Distinct 28-mers
F Vesca	FASTQ	10.2	353	4 134 078 256	632 436 468
M Balbisiana	FASTQ	98.6	100	20 531 572 597	965 691 662
G Gallus	FASTQ	115.9	101	25 337 974 831	2 727 529 829
H Sapiens	FASTQ	223.3	100	62 739 461 708	6 336 805 684
H Sapiens 2	FASTQ	339.5	101	98 892 620 173	6 634 382 141
GRCh38	FASTA	100.0	1000	97 300 000 000	1 802 953 276

Table 2. Test systems.

	System one	System two
CPU	Intel Core-i5 2550k (4 cores)	Intel Xeon(R) E3-1231 v3 (8 cores)
RAM	16 GB DDR3	32 GB DDR3
GPU	GeForce GTX 970	GeForce GTX TITAN X
		GeForce GTX 970
Working-disk	256 GB Crucial M550	2x Samsung 850 EVO 500 GB (RAID-0)
Free disk space	128 GB	1000 GB
OS	Ubuntu 14.04 LTS	
In/Out-disk	Transcend StoreJet 35T3 USB 3.0 (External HDD)	

Table 3. Running times in the format mm:ss (the best performing in bold). Each entry is the average over three runs. Missing running times for DSK are due to insufficient disk space. The label 'gGerbil' stands for *Gerbil* with activated GPU mode. Instead, standard 'Gerbil' does not use any GPU.

Data set	k	System one				System two			
		Gerbil	gGerbil	KMC2	DSK	Gerbil	gGerbil	KMC2	DSK
F Vesca	28	02:08	**01:40**	02:01	03:00	01:36	**01:18**	01:32	02:05
	40	02:34	**01:53**	03:03	04:14	02:01	**01:38**	02:12	02:52
	56	02:58	**01:53**	03:19	03:55	02:25	**01:39**	02:30	02:50
	65	03:05	**01:59**	04:34	05:23	02:16	**01:42**	03:35	03:37
M Balbisiana	28	13:37	**11:42**	12:54	14:49	11:17	**10:07**	10:50	11:06
	40	13:48	**12:24**	16:15	16:12	11:46	**10:59**	13:46	12:26
	56	12:46	**11:36**	16:06	14:56	10:50	**10:18**	13:36	11:44
	65	12:32	**11:28**	18:33	15:52	10:46	**10:16**	15:47	12:34
G Gallus	28	18:41	**14:25**	15:39	26:54	15:47	**12:31**	13:10	21:00
	40	19:55	**16:00**	19:44	29:42	16:29	**14:10**	16:49	23:48
	56	18:12	**14:48**	19:48	24:11	15:38	**13:12**	16:48	19:59
	65	18:27	**15:22**	22:49	26:50	15:41	**13:08**	19:25	21:33
H Sapiens	28	41:10	**30:04**	32:18	-	33:26	**25:16**	26:44	50:15
	40	45:02	**35:52**	43:19	-	35:20	**29:00**	35:59	54:21
	56	39:47	**33:21**	42:53	-	32:21	**26:46**	35:25	45:32
	65	38:09	**35:32**	51:23	-	32:09	**26:27**	42:19	47:50
H Sapiens 2	28	65:33	49:41	**49:17**	-	53:40	**39:24**	41:47	76:50
	40	72:06	**66:04**	70:33	-	57:03	**46:00**	57:02	83:59
	56	64:00	**60:27**	69:58	-	51:34	**42:15**	56:28	72:35
	65	**61:05**	64:44	87:24	-	51:16	**41:30**	68:10	78:13

- *Gerbil* with GPU support (*gGerbil*) is the most efficient tool in almost all cases. Exceptions occur for small $k = 28$, where the sorting based approach KMC2 is sometimes slightly more efficient.
- When k grows, KMC2 becomes more and more inefficient, while *Gerbil* stays efficient. When counting the 200-mers in the *GRCh38* data set, KMC2 did not finish within 20 h, whereas *Gerbil* required only 98 min (Fig. 7). The running time of DSK grows similarly fast as that of KMC2. Recall that DSK does not support values of $k > 127$.
- For small k, the use of a GPU decreases the running time by a significant amount of time. However, with growing k, the data structure that stores k-mers grows larger. Therefore, the number of table entries that can be scanned in parallel decreases. Experimentally, we found that the GPU induced speedup vanishes when k exceeds 150.

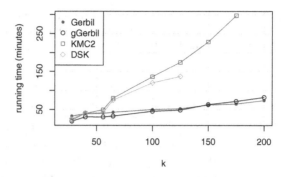

Fig. 7. Running times for $28 \leq k \leq 200$ (*GRCh38*, test system two).

We gain some additional interesting insights when we take a closer look into Table 4 that shows detailed information on running time and memory usage. The use of a GPU accelerates *Gerbil*'s second phase by up to a factor of two, whereas the additional speedup given by a second GPU is only moderate. All tools were called with an option that sets the maximal memory size to 14 GB on Test System One and 30 GB on Test System Two. However, *Gerbil* typically uses much less memory due to its dynamic prediction of the hash table size. In contrast, both KMC2 and DSK use more main memory. *Gerbil*'s disk usage is comparable to KMC2's disk usage, whereas the disk usage of DSK is much larger. *Gerbil*'s frugal use of disk- *and* main memory is a main reason for its high performance. The use of little main memory gives the operating system opportunity to use the remaining main memory for buffering disk operations. A small disk space consumption is essential since disk operations are far more expensive than the actual counting.

Table 4. Detailed running times (in format mm:ss) and maximal main memory and disk space consumption (in GB) for the *G Gallus* instance. Each entry is the average of three runs.

		System one				System two			
	k	Gerbil	gGerbil	KMC	DSK	Gerbil	gGerbil	KMC	DSK
Phase 1	28	10:08	10:06	10:51	10:22	09:46	09:43	09:52	09:30
Phase 2	28	08:32	04:19	04:46	16:00	06:01	02:47	03:16	11:01
Main memory	28	2.36	1.77	14.28	15.28	2.20	2.01	26.99	16.69
Disk space	28	23.66	23.66	24.86	37.30	23.66	23.66	24.86	37.30
Phase 1	56	10:07	10:06	10:40	10:26	09:47	09:43	09:47	09:30
Phase 2	56	08:05	04:42	09:08	13:13	05:50	03:28	06:59	10:00
Main memory	56	4.24	3.20	14.29	15.00	4.00	3.40	26.98	14.78
Disk space	56	16.25	16.25	17.02	57.20	16.25	16.25	17.02	57.20

5 Conclusion

We introduced the *k*-mer counting software *Gerbil* that uses a hash table based approach for the counting of *k*-mers. For large *k*, a use case that becomes important for long reads, we are able to clearly outperform the state-of-the-art open source *k*-mer counting tools, while using significantly less resources. We showed that *Gerbil*'s running time can be accelerated by the use of GPUs. However, since this only affects the second phase, the overall additional speedup is only very moderate. As future work, we plan to evaluate strategies to use GPUs to accelerate also the first phase. Another option for further speed-up would be to give up exactness by using Bloom filters.

References

1. Chikhi, R., Medvedev, P.: Informed and automated *k*-mer size selection for genome assembly. Bioinformatics **30**(1), 31–37 (2014)
2. Deorowicz, S., Debudaj-Grabysz, A., Grabowski, S.: Disk-based *k*-mer counting on a PC. BMC Bioinform. **14**(1), 1–12 (2013)
3. Deorowicz, S., Kokot, M., Grabowski, S., Debudaj-Grabysz, A.: KMC 2: fast and resource-frugal *k*-mer counting. Bioinformatics **31**(10), 1569–1576 (2015)
4. Li, Y., et al.: MSPKmerCounter: a fast and memory efficient approach for *k*-mer counting (2015). arXiv preprint arXiv:1505.06550
5. Marçais, G., Kingsford, C.: A fast, lock-free approach for efficient parallel counting of occurrences of *k*-mers. Bioinformatics **27**(6), 764–770 (2011)
6. Melsted, P., Pritchard, J.K.: Efficient counting of *k*-mers in DNA sequences using a bloom filter. BMC Bioinform. **12**(1), 1–7 (2011)
7. Rizk, G., Lavenier, D., Chikhi, R.: DSK: *k*-mer counting with very low memory usage. Bioinformatics **29**(5), 652–653 (2013)
8. Roberts, M., Hayes, W., Hunt, B.R., Mount, S.M., Yorke, J.A.: Reducing storage requirements for biological sequence comparison. Bioinformatics **20**(18), 3363–3369 (2004)
9. Roberts, M., Hunt, B.R., Yorke, J.A., Bolanos, R.A., Delcher, A.L.: A preprocessor for shotgun assembly of large genomes. J. Comput. Biol. **11**(4), 734–752 (2004)
10. Roy, R.S., Bhattacharya, D., Schliep, A.: Turtle: identifying frequent *k*-mers with cache-efficient algorithms. Bioinformatics **30**(14), 1950–1957 (2014)
11. Sameith, K., Roscito, J.G., Hiller, M.: Iterative error correction of long sequencing reads maximizes accuracy and improves contig assembly. Briefings in Bioinformatics (2016). http://dx.org/10.1093/bib/bbw003
12. Suzuki, S., Kakuta, M., Ishida, T., Akiyama, Y.: Accelerating identification of frequent *k*-mers in DNA sequences with GPU. In: GTC 2014 (2014)
13. Xavier, B.B., Sabirova, J., Pieter, M., Hernalsteens, J.P., de Greve, H., Goossens, H., Malhotra-Kumar, S.: Employing whole genome mapping for optimal de novo assembly of bacterial genomes. BMC Res. Notes **7**(1), 1–4 (2014)

Genome Rearrangements on Both Gene Order and Intergenic Regions

Guillaume Fertin[1], Géraldine Jean[1(✉)], and Eric Tannier[2,3]

[1] LINA UMR CNRS 6241, Université de Nantes, Nantes, France
{guillaume.fertin,geraldine.jean}@univ-nantes.fr
[2] INRIA Grenoble Rhône-Alpes, Montbonnot, France
[3] LBBE UMR CNRS 5558, Université Lyon 1, Villeurbanne, France
eric.tannier@inria.fr

Abstract. All combinatorial works on genome rearrangements have so far ignored the influence of intergene sizes, i.e. the number of nucleotides between consecutive genes, although it was recently shown decisive for the accuracy of the inference methods [3,4]. In this line, we define a new genome rearrangement model called wDCJ, a generalization of the well-known Double Cut and Join (or DCJ) model that allows for modifying both the gene order *and* the intergene size distribution of a genome. We first provide a generic formula for the wDCJ distance between two genomes, and show that computing this distance is strongly NP-complete. We then propose an approximation algorithm of ratio 3/2, and two exact ones: a fixed parameterized (FPT) algorithm and an ILP formulation. We finally provide theoretical and empirical bounds on the expected growth of the parameter at the center of our FPT and ILP algorithms, assuming a probabilistic model of evolution under wDCJ, which shows that both these algorithms should run reasonably fast in practice.

1 Introduction

General Context. Mathematical models for genome evolution by rearrangements have defined a genome as a linear or circular ordering of genes[1] [5]. These orderings have first been seen as (possibly signed) permutations, or strings if duplicate genes are present, or disjoint paths and cycles in graphs in order to allow multiple chromosomes. However, the organization of a genome is not entirely subsumed in gene orders. In particular, consecutive genes are separated by an intergenic region, and intergenic regions have diverse sizes [7]. Besides, it was recently shown that integrating intergene sizes in the models radically changes the distance estimations between genomes, as usual rearrangement distance estimators

---

Supported by GRIOTE project, funded by Région Pays de la Loire, and the ANCE-STROME project, Investissement d'avenir ANR-10-BINF-01-01.

[1] The word *gene* is as usual in genome rearrangement studies taken in a liberal meaning, as any segment of DNA, computed from homologous genes or synteny blocks, which is not touched by a rearrangement in the considered history.

© Springer International Publishing Switzerland 2016
M. Frith and C.N.S. Pedersen (Eds.): WABI 2016, LNBI 9838, pp. 162–173, 2016.
DOI: 10.1007/978-3-319-43681-4_13

ignoring intergene sizes do not estimate well on realistic data [3,4]. We thus propose to re-examine the standard models and algorithms in this light. A first step is to define and compute standard distances, such as Double Cut and Join (or DCJ) [9], taking into account intergene sizes. In this setting, two genomes are considered, which are composed of gene orders *and* intergene sizes. One is transformed into the other by wDCJ operations, which consist in the usual DCJ operations, which additionally change the sizes of the intergenes it affects.

Genomes and Rearrangements. Given a set V of vertices such that $|V| = 2n$, we define a *genome* g as a set of n disjoint edges, i.e. a perfect matching on V. A genome is *weighted* if each edge e of g is assigned an integer weight $w(e) \geq 0$, and we define $W(g)$ as the sum of all weights of the edges of g. The union of two genomes g_1 and g_2 on the same set V thus forms a set of disjoint even size cycles called the *breakpoint graph* $BG(g_1, g_2)$ of g_1 and g_2, in which each cycle is *alternating*, i.e. is composed of edges alternately belonging to g_1 and g_2. Note that in the rest of the paper, we will be only interested in evenly weighted genomes, i.e. genomes g_1 and g_2 such that $W(g_1) = W(g_2)$.

A *Double cut-and-join* (DCJ) [9] is an operation on an unweighted genome g, which transforms it into another genome g' by deleting two edges ab and cd and by adding either (i) edges ac and bd, or (ii) edges ad and bc. If g is weighted, the operation we introduce in this paper is called wDCJ: wDCJ is a DCJ that additionally modifies the weights of the resulting genome in the following way: if we are in case (i), (1) any edge but ac and bd is assigned the same weight as in g, and (2) $w(ac)$ and $w(bd)$ are assigned arbitrary non negative integer weights, with the constraint that $w(ac) + w(bd) = w(ab) + w(cd)$. If we are in case (ii), a similar rule applies by replacing ac by ad and bd by bc. Note that wDCJ clearly generalizes the usual DCJ, since any unweighted genome g can be seen as a weighted one in which $w(e) = 0$ for any edge e in g.

Motivation for These Definitions. This representation of a genome supposes that each vertex is a *gene extremity* (a gene being a segment, it has two extremities, which explains the even number of vertices), and an edge means that two gene extremities are contiguous on a chromosome. This representation generalizes signed permutations, and allows for an arbitrary number of circular and linear chromosomes. The fact that there should be n edges in a genome means that chromosomes are circular, or that extremities of chromosomes are not in the vertex set. It is possible to suppose so when the genomes we compare are *co-tailed*, i.e. the same gene extremities are extremities of chromosomes in both genomes. In this way, a wDCJ on a circular (resp. co-tailed) genome always yields a circular (resp. co-tailed) genome, which, in our terminology, just means that a weighted perfect matching stays a weighted perfect matching through wDCJ. So all along this paper we suppose that we are in the particular case of classical genomic studies where genomes are co-tailed or circular. Each edge represents an intergenic region. Weights on edges are intergene sizes, that is, the number of nucleotides separating two genes. The way weights are distributed after a wDCJ models a breakage inside an intergene between two nucleotides.

Statement of the Problem. Given two evenly weighted genomes g_1 and g_2 expressed on the same set V of $2n$ vertices, a sequence of wDCJ that transforms g_1 into g_2 is called a *wDCJ sorting scenario*. Note that any sequence transforming g_1 into g_2 can be easily transformed into a sequence of same length transforming g_2 into g_1, as the problem is fully symmetric. Thus, in the following, we will always suppose that g_2 is fixed and that the wDCJ are applied on g_1. The wDCJ *distance* between g_1 and g_2, denoted $wDCJ(g_1, g_2)$ is defined as the number of wDCJ of a smallest wDCJ sorting scenario. Note that when genomes are unweighted, computing the usual DCJ distance is tractable, as $DCJ(g_1, g_2) = n - c$, where c is the number of cycles of $BG(g_1, g_2)$ [9]. The problem we consider in this paper, that we denote by wDCJ-DIST, is the following: given two evenly weighted genomes g_1 and g_2 defined on the same set V of $2n$ vertices, determine $wDCJ(g_1, g_2)$. We need further notations. The *imbalance* of a cycle C in $BG(g_1, g_2)$ is denoted $I(C)$, and is defined as follows: $I(C) = w_1(C) - w_2(C)$, where $w_1(C)$ (resp. $w_2(C)$) is the sum of the weights of the edges of C which belong to g_1 (resp. g_2). A cycle C of the breakpoint graph is said to be *balanced* if $I(C) = 0$, and *unbalanced* otherwise. We will denote by $\mathcal{C}_u$ the set of unbalanced cycles in $BG(g_1, g_2)$, and by $n_u = |\mathcal{C}_u|$ its cardinality. Similarly, n_b denotes the number of balanced cycles in $BG(g_1, g_2)$, and $c = n_u + n_b$ denotes the (total) number of cycles in $BG(g_1, g_2)$.

Related Works. In the recent past, generalizations of standard models integrate more realistic features in order to be closer to real genome evolution. It concerns, among others, models where inversions are considered, that are weighted by their length or symmetry around a replication origin [2], by the proximity of their extremities in the cell [8], or by their use of hot regions for rearrangement breakages [1].

Our Results. We explore the algorithmic properties of wDCJ-DIST. We first provide the main properties of (optimal) wDCJ sorting scenarios in Sect. 2. We then show in Sect. 3 that the wDCJ-DIST problem is strongly NP-complete, 3/2 approximable, and we provide two exact (FPT and ILP) algorithms. By simulations and analytic studies on a probabilistic model of genome evolution, in Sect. 4 we bound the parameter at the center of both our FPT and ILP algorithms, and conclude that they should run reasonably fast in practice. Note that due to space constraints, some proofs are omitted from this paper.

2   Main Properties of Sorting by wDCJ

The present section is devoted to providing properties of any (optimal) wDCJ sorting scenario. These properties mainly concern the way the breakpoint graph evolves, whenever one or several wDCJ is/are applied. These will lead to a close formula for the wDCJ distance (Theorem 1). Moreover, they will also be essential in the algorithmic study of the wDCJ-DIST problem that will be developed in Sect. 3. We first show the following lemma.

Lemma 1. *Let C be a balanced cycle of some breakpoint graph $BG(g_1, g_2)$. Then there exist three consecutive edges e, f, g in C such that (i) e and g belong to g_1 and (ii) $w(e) + w(g) \geq w(f)$.*

Proof. Suppose, aiming at a contradiction, that for any three consecutive edges e, f, g in C with $e, g \in E(g_1)$, we have $w(e) + w(g) < w(f)$. Summing this inequality over all such triplets of consecutive edges of C, we obtain the following inequality: $2 \cdot w_1(C) < w_2(C)$. Since C is balanced, by definition we have $w_1(C) - w_2(C) = 0$. Hence we obtain $w_1(C) < 0$, a contradiction since all edge weights are non negative by definition. □

Note that any wDCJ can act on the number of cycles of the breakpoint graph in only three possible ways: either this number is increased by one (cycle *split*), decreased by one (cycle *merge*), or remains the same (cycle *freeze*). We now show that if a breakpoint graph only contains balanced cycles, then any optimal wDCJ sorting scenario only uses cycle splits.

Proposition 1. *Let $BG(g_1, g_2)$ be a breakpoint graph that contains balanced cycles only – in which case $c = n_b$. Then $wDCJ(g_1, g_2) = n - n_b$.*

In the following, we are interested in the sequences of two wDCJ formed by a cycle split s *directly followed* by a cycle merge m, to the exception of *df-sequences* (for *double-freeze*), which is the special case where s is applied on a cycle C (forming cycles C_a and C_b) and m merges back C_a and C_b to give a new cycle C' built on the same set of vertices than C. The name derives from the fact that a *df-sequence* acts as a freeze, except that it can involve up to 4 edges in the cycle, as opposed to only 2 edges for a freeze.

Proposition 2. *In a wDCJ sorting scenario, if there is a sequence of two operations formed by a cycle split s directly followed by a cycle merge m that is not a df-sequence, then there exists a wDCJ sorting scenario of same length where s and m are replaced by a cycle merge m' followed by a cycle split s'.*

Proof. Let s and m be two consecutive wDCJ in a sorting scenario that do not form a df-sequence, where s is a split, m is a merge, and s is applied before m. Let also G (resp. G') be the breakpoint graph before s (resp. after m) is applied. We will show that there always exist two wDCJ m' and s', such that (i) m' is a cycle merge, (ii) s' is a cycle split and (iii) starting from G, applying m' then s' gives G'. First, if none of the two cycles produced by s is used by m, then the two wDCJ are independent, and it suffices to set $m' = m$ and $s' = s$ to conclude.

Now suppose one of the two cycles produced by s is involved in m. Let C_1 denote the cycle on which s is applied, and let us assume s cuts ab and cd, of respective weights w_1 and w_2, and joins ac and bd, of respective weights w_1' and w_2' – thus $w_1 + w_2 = w_1' + w_2'$ (a). We will denote by C_a (resp. C_b) the two cycles obtained by s from C_1; see Fig. 1 for an illustration. Now let us consider m. Wlog, let us suppose that m acts on C_b and another cycle $C_2 \neq C_a$ (since df-sequences are excluded), in order to produce cycle C_3. It is easy to see that if m cuts an edge different from bd in C_b, then s and m are two independent wDCJ,

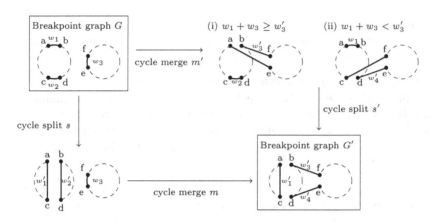

Fig. 1. Two different scenarios that lead to G' starting from G: (downward) a split s followed by a merge m; (rightward) a merge m' followed by a split s'.

and thus can be safely swapped. Thus we now assume that m cuts bd. Suppose the edge that is cut in C_2 is ef, of weight w_3, and that the joins are edges bf and de, of respective weights w_3' and w_4'. We thus have $w_3' + w_4' = w_2' + w_3$ (b). Moreover, adding (a) and (b) gives $w_1 + w_2 + w_3 = w_1' + w_3' + w_4'$ (c). Now let us show that there exists a scenario that allows to obtain C_a and C_3 from C_1 and C_2, which begins by a merge followed by a split. For this, we consider two cases:

- $w_1 + w_3 \geq w_3'$ (see Fig. 1(i)): m' consists in cutting ab from C_1 and ef from C_2, then forming ae and bf, so as to obtain a unique cycle C. Note that C now contains edges cd (of weight w_2), bf (of weight w_3') and ae (of weight $w_1 + w_3 - w_3'$, which is non negative by hypothesis). Then, s' is defined as follows: cut ae and cd, form edges ac, de. Finally, note that assigning w_1' to ac and w_4' to de is possible, since ae is of weight $w_1 + w_3 - w_3'$, cd is of weight w_2, and since $w_1 + w_3 - w_3' + w_2 = w_1' + w_4'$ by (c).
- $w_1 + w_3 < w_3'$ (see Fig. 1(ii)). Consider the following merge m': cut edges cd and ef, and form the edges de of weight w_4', and cf of weight $w = w_2 + w_3 - w_4'$. This merge is feasible because $w \geq 0$: indeed, by hypothesis $w_1 + w_3 < w_3'$, i.e. $w_1 + w_2 + w_3 < w_2 + w_3'$, which by (c) implies $w_1' + w_4' < w_2$. Thus $w_4' < w_2$, and consequently $w > w_3 \geq 0$. Now let s' be as follows: cut ab (of weight w_1) and cf (of weight $w = w_2 + w_3 - w_4'$) to form edges ac and bf of respective weights w_1' and w_3'. Note that s' is always feasible since $w_1 + w = w_1 + w_2 + w_3 - w_4' = w_1' + w_3'$ by (c).

In all cases, it is always possible to obtain G', starting from G, using a merge m' followed by a split s', rather than s followed by m, and the result is proved.   □

Proposition 3. *In an optimal wDCJ sorting scenario, no cycle freeze or df-sequence occurs.*

Proof. Suppose a wDCJ sorting scenario contains at least one cycle freeze or df-sequence, and let us consider the last such event f that appears in it. We will show that there also exists a sorting scenario that does not contain f, and whose length is decreased by at least one. For this, note that the sequence of wDCJ that follow f, say $\mathcal{S}$, is only composed of cycle splits and merges which do not form df-sequences. By Proposition 2, in $\mathcal{S}$ any split that precedes a merge can be replaced by a merge that precedes a split, in such a way that the new scenario is a sorting one, and of same length. By iterating this process, we end up with a sequence $\mathcal{S}'$ in which, after f, we operate a series M of merges, followed by a series S of splits. Let G_M be the breakpoint graph obtained after all M merges are applied. If a cycle was unbalanced in G_M, any split would leave at least one unbalanced cycle, and it would be impossible to finish the sorting by applying the splits in S. Thus G_M must contain only balanced cycles. Recall that f acts inside a given cycle C, while maintaining its imbalance $I(C)$ unchanged. C may be iteratively merged with other cycles during M, but we know that, in G_M, the cycle C' that finally "contains" C is balanced. Thus, if we remove f from the scenario, the breakpoint graph G'_M we obtain only differs from G_M by the fact that C' is now replaced by another cycle C'', which contains the same vertices and is balanced. However, by Proposition 1, we know that G'_M can be optimally sorted using the same number of splits than G_M, which allows us to conclude that there exists a shorter sorting scenario that does not use f.   □

Proposition 4. *Any wDCJ sorting scenario can be transformed into another wDCJ sorting scenario of same or shorter length, and in which any cycle merge occurs before any cycle split.*

Proposition 5. *In an optimal wDCJ sorting scenario, no balanced cycle is ever merged.*

Based on the above results, we are now able to derive a formula for the wDCJ distance, which is somewhat similar to the "classical" DCJ distance formula [9].

Theorem 1. *Let $BG(g_1, g_2)$ be the breakpoint graph of two genomes g_1 and g_2, and let c be the number of cycles in $BG(g_1, g_2)$. Then $wDCJ(g_1, g_2) = n - c + 2m$, where m is the minimum number of cycle merges needed to obtain a set of balanced cycles from the unbalanced cycles of $BG(g_1, g_2)$.*

Proof. By the previous study, we know that there exists an optimal wDCJ scenario without cycle freezes or df-sequences, and in which merges occur before splits (Propositions 3 and 4). We also know that before the splits start, the graph G_M we obtain is a collection of balanced cycles, and that the split sequence that follows is optimal and only creates balanced cycles (Proposition 1). Thus the optimal distance is obtained when the merges are as few as possible. By Proposition 5, we know that no balanced cycle is ever used in a cycle merge in an optimal scenario. Hence an optimal sequence of merges consists in creating balanced cycles from the unbalanced cycles of $BG(g_1, g_2)$ only, using a minimum number m of merges. Altogether, we have (i) m merges that lead to $c - m$ cycles, then (ii) $n - (c - m)$ splits by Proposition 1. Hence the result.   □

3 Algorithmic Aspects of wDCJ-Dist

Based on the properties of a(n optimal) wDCJ sorting scenario given in Sect. 2, we are now able to provide algorithmic results concerning the wDCJ-DIST problem. We begin by assessing its computational complexity.

Theorem 2. *The* wDCJ-DIST *problem is strongly* NP-complete.

Proof. The proof is by reduction from the strongly NP-complete 3-PARTITION problem [6], whose instance is a multiset $A = \{a_1, a_2 \ldots a_{3n}\}$ of $3n$ positive integers such that (i) $\sum_{i=1}^{3n} a_i = B \cdot n$ and (ii) $\frac{B}{4} < a_i < \frac{B}{2}$ for any $1 \le i \le 3n$, and where the question is whether one can partition A into n multisets $A_1 \ldots A_n$, such that for each $1 \le i \le n$, $\sum_{a_j \in A_i} a_j = B$. Given any instance A of 3-PARTITION, we construct two genomes g_1 and g_2 as follows: g_1 and g_2 are built on a vertex set V of cardinality $8n$, and consist of the same perfect matching. Thus $BG(g_1, g_2)$ is composed of $4n$ trivial cycles, that is cycles of length 2, say $C_1, C_2 \ldots C_{4n}$. The only difference between g_1 and g_2 thus lies on the weights of their edges. For any $1 \le i \le 4n$, let e_i^1 (resp. e_i^2) be the edge from C_i that belongs to g_1 (resp. g_2). The weight we give to each edge is the following: for any $1 \le i \le 3n$, $w(e_i^1) = a_i$ and $w(e_i^2) = 0$; for any $3n + 1 \le i \le 4n$, $w(e_i^1) = 0$ and $w(e_i^2) = B$. As a consequence, the imbalance of each cycle is $I(C_i) = a_i$ for any $1 \le i \le 3n$, and $I(C_i) = -B$ for any $3n + 1 \le i \le 4n$. Now we will prove the following equivalence: 3-PARTITION is satisfied iff $wDCJ(g_1, g_2) \le 6n$.

($\Rightarrow$) Suppose there exists a partition $A_1 \ldots A_n$ of A such that for each $1 \le i \le n$, $\sum_{a_j \in A_i} a_j = B$. For any $1 \le i \le n$, let $A_i = \{a_{i_1}, a_{i_2}, a_{i_3}\}$. Then, for any $1 \le i \le n$, we merge cycles C_{i_1}, C_{i_2} and C_{i_3}, then apply a third merge with C_{3n+i}. For each $1 \le i \le n$, these three merges lead to a balanced cycle, since after the two first merges, the obtained weight is $a_{i_1} + a_{i_2} + a_{i_3} = B$. After these $3n$ merges (in total) have been applied, we obtain n balanced cycles, from which $4n - n = 3n$ splits suffice to end the sorting, as stated by Proposition 1. Thus, altogether we have used $6n$ wDCJ, and consequently $wDCJ(g_1, g_2) \le 6n$.

($\Leftarrow$) Suppose that $wDCJ(g_1, g_2) \le 6n$. Recall that in the breakpoint graph $BG(g_1, g_2)$, we have $c = 4n$ cycles and $8n$ vertices. Thus, by Theorem 1, we know that $wDCJ(g_1, g_2) = 4n - 4n + 2m = 2m$, where m is the smallest number of merges that are necessary to obtain a set of balanced cycles from $BG(g_1, g_2)$. Since we suppose $wDCJ(g_1, g_2) \le 6n$, we conclude that $m \le 3n$. Otherwise stated, the number of balanced cycles we obtain after the merges cannot be less than n, because we start with $4n$ cycles and apply at most $3n$ merges. However, at least 4 cycles from $C_1, C_2 \ldots C_{4n}$ must be merged in order to obtain a single balanced cycle: at least 3 from $C_1, C_2 \ldots C_{3n}$ (since any a_i satisfies $\frac{B}{4} < a_i < \frac{B}{2}$ by definition), and at least one from $C_{3n+1}, C_{3n+2} \ldots C_{4n}$ (in order to end up with an imbalance equal to zero). Thus any balanced cycle is obtained using exactly 4 cycles (and thus 3 merges), which in turn implies that there exists a way to partition the multiset A into $A_1 \ldots A_n$ in such a way that for any $1 \le i \le n$, $(\sum_{a_j \in A_i}) - B = 0$, which positively answers the 3-PARTITION problem. □

Since wDCJ-DIST is NP-complete, we now seek for algorithms that compute, either approximately or exactly, the wDCJ distance.

Theorem 3. *The* wDCJ-DIST *problem is $\frac{3}{2}$-approximable.*

Proof. Given two weighted genomes g_1 and g_2, let $\mathcal{C}_u = \{C_1, C_2 \ldots C_{n_u}\}$ be the set of unbalanced cycles in $BG(g_1, g_2)$. First, compute a maximum cardinality set S_2 of independent pairs $\{I(C_i), I(C_j)\}$ (with $i \neq j$) such that $I(C_i) + I(C_j) = 0$, where by independent we mean that any cycle C_i is used at most once in S_2; let $n_2 = |S_2|$. Intuitively, each pair in S_2 represents two unbalanced cycles that become balanced when merged. Note that S_2 can be easily computed by iteratively searching for a number and its opposite among the imbalances in $\mathcal{C}_u$. Now, our approximation algorithm does the following: merge the cycles of S_2 by pairs, then merge the remaining unbalanced cycles into a unique (balanced) cycle. At this stage we have performed $m_A = \frac{n_2}{2} + (n_u - n_2 - 1)$ cycle merges, and we obtain $n_b + \frac{n_2}{2} + 1$ cycles which are all balanced. Then we perform $s_A = n - n_b - \frac{n_2}{2} - 1$ splits in order to finish the sorting (see Proposition 1). Our algorithm thus uses $dcj_A(g_1, g_2) = m_A + s_A = n - n_b + n_u - n_2 - 2$ wDCJ.

Now let us observe an optimal sorting scenario of length $wDCJ(g_1, g_2)$, which, as we know by the results in Sect. 2, can be assumed to contain m_{opt} merges followed by s_{opt} splits. Note that we can also safely assume that this scenario merges the pairs of cycles in S_2 – if not, a scenario that does it, and is of same length, exists. Concerning the remaining $n_u - n_2$ unbalanced cycles, the best case scenario is when we are able to obtain balanced cycles by merging them three by three; thus at least $\frac{2(n_u - n_2)}{3}$ extra cycles merges are necessary, leading to $m_{opt} \geq \frac{n_2}{2} + \frac{2(n_u - n_2)}{3}$. In any case, we end up with at most $n_b + \frac{n_2}{2} + \frac{(n_u - n_2)}{3}$ balanced cycles, and thus $s_{opt} \geq n - n_b - \frac{n_2}{2} - \frac{(n_u - n_2)}{3}$. Altogether, we have that $wDCJ(g_1, g_2) \geq n - n_b + \frac{(n_u - n_2)}{3}$.

Our goal is now to show that $dcj_A(g_1, g_2) \leq \frac{3}{2} \cdot wDCJ(g_1, g_2)$. First, since $dcj_A(g_1, g_2) = n - n_b + n_u - n_2 - 2$, we have $dcj_A(g_1, g_2) \leq (n - n_b + \frac{(n_u - n_2)}{3}) + \frac{2(n_u - n_2)}{3}$, that is $dcj_A(g_1, g_2) \leq wDCJ(g_1, g_2) + \frac{2(n_u - n_2)}{3}$. Hence, it suffices to show that $\frac{2(n_u - n_2)}{3} \leq \frac{wDCJ(g_1, g_2)}{2}$ to conclude. For this, we note that we always have $n \geq n_b + n_u$, since n is the maximum number of possible cycles in $BG(g_1, g_2)$. In other words, $n - n_b \geq n_u$, which we can write $n - n_b + \frac{(n_u - n_2)}{3} \geq \frac{4(n_u - n_2)}{3}$. Since we have $wDCJ(g_1, g_2) \geq n - n_b + \frac{(n_u - n_2)}{3}$, we conclude that $\frac{wDCJ(g_1, g_2)}{2} \geq \frac{2(n_u - n_2)}{3}$, and we are done. $\square$

We now turn to exact algorithms for computing wDCJ-DIST.

Theorem 4. *The* wDCJ-DIST *problem is FPT when parameterized by the number n_u of unbalanced cycles in $BG(g_1, g_2)$.*

Proof. By Theorem 1, we know that, given g_1 and g_2, $wDCJ(g_1, g_2)$ can be computed from the three parameters n, c and m. Clearly, n and c are computed from $BG(g_1, g_2)$ in polynomial time; since wDCJ-DIST is NP-complete, the "hard" part consists in computing m. We recall that m is the minimum number of cycle merges that are necessary to transform the set $\mathcal{C}_u$ of unbalanced cycles of $BG(g_1, g_2)$ into balanced ones. Equivalently, we seek for a maximum number of balanced cycles we can obtain from $\mathcal{C}_u$ using merges only, i.e. we want to partition $\mathcal{C}_u$ into $\{\mathcal{C}_1, \mathcal{C}_2 \ldots \mathcal{C}_p\}$ such that (1) for each $1 \leq i \leq p$, merging all cycles in $\mathcal{C}_i$ leads to a balanced cycle and (2) p is maximized. Thus, $m = n_u - p$, and consequently $wDCJ(g_1, g_2) = n - c + 2(n_u - p)$. The algorithm thus works as follows: exhaustively generate all the partitions of $\mathcal{C}_u$, and output the solution S that satisfies (1) and (2) above (such a solution exists, since taking $p = 1$ satisfies Condition (1)). From S, the wDCJ sorting scenario follows, and it is optimal of length $n - c + 2(n_u - p)$ as argued above. The exponential part of the algorithm is clearly the generation of all partitions of $\mathcal{C}_u$, which depends only on $n_u = |\mathcal{C}_u|$, hence the result. $\qquad\square$

An Integer Linear Programming for Solving wDCJ-DIST. The ILP we propose here actually consists in computing the number p described in proof of Theorem 4 above, i.e. the maximum number of sets of a partition $\{\mathcal{C}_1, \mathcal{C}_2 \ldots \mathcal{C}_p\}$ of $\mathcal{C}_u$ for which for any $1 \leq i \leq p$, the sum of the imbalances of the cycles in $\mathcal{C}_i$ is equal to zero. As argued in proof of Theorem 4, once this number p is computed, one can compute $wDCJ(g_1, g_2)$ in polynomial time, as $wDCJ(g_1, g_2) = n - c + 2(n_u - p)$, where c (resp. n_u) is the number of cycles (resp. unbalanced cycles) in $BG(g_1, g_2)$. We note that $p \leq \frac{n_u}{2}$, since it takes at least two unbalanced cycles to create a balanced one.

$$\text{maximize} \quad \sum_{1 \leq i \leq n_u/2} p_i \tag{1}$$

$$\text{subject to} \quad \sum_{1 \leq j \leq n_u/2} x_{i,j} = 1 \ \forall 1 \leq i \leq n_u \tag{2}$$

$$\sum_{1 \leq i \leq n_u} I(C_i) \cdot x_{i,j} = 0 \ \forall 1 \leq j \leq n_u/2 \tag{3}$$

$$\sum_{1 \leq i \leq n_u} x_{i,j} \geq p_j \ \forall 1 \leq j \leq n_u/2 \tag{4}$$

$$x_{i,j} \in \{0, 1\} \ \forall 1 \leq i \leq n_u \text{ and } \forall 1 \leq j \leq n_u/2 \tag{5}$$

$$p_i \in \{0, 1\} \ \forall 1 \leq i \leq n_u/2 \tag{6}$$

Fig. 2. ILP description for the computation of parameter p.

Let us now describe our ILP (see also Fig. 2): we first define binary variables $x_{i,j}$, for $1 \leq i \leq n_u$ and $1 \leq j \leq \frac{n_u}{2}$, that will be set to 1 if the unbalanced cycle $C_i \in \mathcal{C}_u$ belongs to subset $\mathcal{C}_j$, and 0 otherwise. The binary variables p_i, $1 \leq i \leq \frac{n_u}{2}$, will simply indicate whether C_i is "used" in the solution, i.e. $p_i = 1$ if $\mathcal{C}_i \neq \emptyset$, and 0 otherwise. In our ILP, (2) ensures that each unbalanced cycle is assigned to exactly one subset $\mathcal{C}_i$; (3) requires that the sum of the imbalances of the cycles from $\mathcal{C}_i$ is equal to zero. Finally, (4) ensures that a subset $\mathcal{C}_i$ is marked as unused if no unbalanced cycle has been assigned to it. Moreover, since the objective is to maximize the number of non-empty subsets, p_i will necessary be set to 1 whenever $C_i \neq \emptyset$. Note that the size of the above ILP depends only on n_u, as it contains $\Theta(n_u^2)$ variables and $\Theta(n_u)$ constraints.

4   A Probabilistic Model of Evolution by wDCJ

In this section, we define a model of evolution by wDCJ, in order to derive theoretical and empirical bounds for the parameter n_u on which both the FPT and ILP algorithms depend. The model is a Markov chain on all weighted genomes (that is, all weighted perfect matchings) on $2n$ vertices. Transitions are wDCJ, such that from one state, two distinct edges ab and cd are chosen uniformly at random, and replaced by either ac and bd or by ad and cb (with probability 0.5 each). Weights of the new edges are computed by drawing two numbers x and y uniformly at random in respectively $[0, w(ab)]$ and $[0, w(cd)]$, and assigning $x+y$ to one edge, and $w(ab) + w(cd) - x - y$ to the other (with probability 0.5 each).

Proposition 6. *The equilibrium distribution of this Markov chain is such that a genome has a probability proportional to the product of the weights on its edges.*

As a consequence, the weight distributions follow a symmetric Dirichlet law with parameter $\alpha = 2$. It is possible to draw a genome at random in the equilibrium distribution by drawing a perfect matching uniformly at random and distributing its weights with a Gamma law of parameters 1 and 2.

We first prove a theoretical bound on the number of expected unbalanced cycles, and then show by simulations that this number probably stays far under this theoretical bound on evolutionary experiments.

Theorem 5. *Given a weighted genome g_1 with n edges, if k random wDCJ are applied to g_1 to give a weighted genome g_2, then the expected number of unbalanced cycles in $BG(g_1, g_2)$ satisfies $\mathbb{E}(n_u) = O(k/\sqrt{n})$.*

Proof. In this proof, for simplicity, let us redefine the *size of a cycle* as half the number of its edges. Let n_u^+ (respectively n_u^-) be the number of unbalanced cycles of size greater than or equal to (respectively less than) $\sqrt{n}$. We thus have $n_u = n_u^+ + n_u^-$. We will prove that (i) $n_u^+ \leq k/\sqrt{n}$ and (ii) $n_u^- = O(k/\sqrt{n})$.

First, if the breakpoint graph contains u unbalanced cycles of size at least s, then the number k of wDCJ is at least us. Indeed from Theorem 1, the DCJ distance is at least $n - c + u$, and as $n \geq us + (c - u)$, we have $k \geq us + (c - u) - c + u = us$. As a consequence, $k \geq n_u^+ \cdot \sqrt{n}$, and (i) is proved.

Second, any unbalanced cycle of size less than s is the product of a cycle split. Given a cycle C of size $r > s$ with $r \neq 2s$, there are r possible wDCJ which can split C and produce one cycle of size s. If $r = 2s$, there are $r/2$ possible splits which result in 2 cycles of size s. So there are $O(sr)$ ways of splitting C and obtaining an unbalanced cycle of size less than s. If we sum over all cycles, this makes $O(sn)$ ways because the sum of the sizes of all cycles is bounded by n. As there are $O(n^2)$ possible wDCJ in total, the probability to split a cycle of size r and obtain an unbalanced cycle of size less than s at a certain point of a scenario is $O(s/n)$. If we sum over all the scenarios of k wDCJ, this makes an expected number of unbalanced cycles in $O(ks/n)$, which implies (ii) since $s < \sqrt{n}$.   □

We simulated a genome evolution with $n = 1\,000$, and the weights on a genome drawn from the above discussed equilibrium distribution. Then we applied $k = 10\,000$ wDCJ, and we measured the value of n_u on the way. As shown in Fig. 3, n_u was always below 13, and stopped growing after $k = 2\,000$. This tends to show that the theoretical bound given in Theorem 5 is far from being reached in reality, and that the parameter n_u is very low is this model. We actually conjecture that the expected number $\mathbb{E}(n_u) = o(n)$ and in particular does not depend on k. Nevertheless, this shows that, in practice, both the FPT and ILP algorithms from the previous section should run in reasonable time on this type of instances. As an illustration, we ran the ILP algorithm on a set of $10\,000$ instances, generated as described above. For each of these instances, the execution time for computing parameter p (discussed in the description of the ILP in Sect. 3) on a standard computer never exceeded $8\,\mathrm{ms}$.

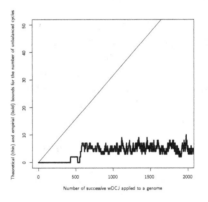

Fig. 3. Number of unbalanced cycles, in a simulation on genomes with $n = 1\,000$ edges where wDCJ operations are applied successively. The number of unbalanced cycles is computed (i) according to the theoretical bound $k/\sqrt{n}$ (in thin), and (ii) from the simulated genomes (in bold).

As a side remark, we note that the model presented here is different from the one used in Biller et al. [4], in which rearrangements are drawn with a probability proportional to the product of the weights of the involved edges. We checked that the behavior concerning n_u was the same in both models; however, we were unable to adapt proof of Theorem 5 to that case.

5   Conclusion and Perspectives

We made the first steps in the combinatorial study of rearrangement operations which depend on and affect intergene sizes. We leave open many problems and extensions based on this study. First, is wDCJ-DIST APX-hard? Is it FPT in the sought distance? Second, the applicability of our model to biological data lacks additional flexibility, thus we suggest two (non exclusive) possible extensions: (a) give a weight to every wDCJ, e.g. a function of the weights of the involved edges; (b) instead of assuming that the total intergene size is conservative (which is not the case in biological data), consider a model in which intergene size may be altered by deletions, insertions and duplications. Third, generalizing the model to non co-tailed genomes (in our terminology, matchings that are not perfect) remains an open problem. It is clearly NP-complete, as it generalizes our model, but other algorithmic questions, such as approximability and fixed-parameter tractability, remain to be answered.

References

1. Alexeev, N., Alekseyev, M.A.: Estimation of the true evolutionary distance under the fragile breakage model. In: Proceedings of the 5th IEEE International Conference on Computational Advances in Bio and Medical Sciences (ICCABS) (2015)
2. Baudet, C., Dias, U., Dias, Z.: Length and symmetry on the sorting by weighted inversions problem. In: Campos, S. (ed.) BSB 2014. LNCS, vol. 8826, pp. 99–106. Springer, Heidelberg (2014)
3. Biller, P., Knibbe, C., Beslon, G., Tannier, E.: Comparative genomics on artificial life. In: Proceedings of Computability in Europe (2016)
4. Biller, P., Guéguen, L., Knibbe, C., Tannier, E.: Breaking good: accounting for the diversity of fragile regions for estimating rearrangement distances. Genome Biol. Evol. **8**, 1427–1439 (2016)
5. Fertin, G., Labarre, A., Rusu, I., Tannier, E., Vialette, S.: Combinatorics of Genome Rearrangements. Computational Molecular Biology. MIT Press, Cambridge (2009)
6. Garey, M.R., Johnson, D.S.: Computers and Intractability: A Guide to the Theory of NP-Completeness. W.H. Freeman & Co., New York (1990)
7. Lynch, M.: The Origin of Genome Architecture. Sinauer, Sunderland (2007)
8. Swenson, K.M., Blanchette, M.: Models and algorithms for genome rearrangement with positional constraints. In: Pop, M., Touzet, H. (eds.) WABI 2015. LNCS, vol. 9289, pp. 243–256. Springer, Heidelberg (2015)
9. Yancopoulos, S., Attie, O., Friedberg, R.: Efficient sorting of genomic permutations by translocation, inversion and block interchange. Bioinformatics **21**(16), 3340–3346 (2005)

Better Identification of Repeats
in Metagenomic Scaffolding

Jay Ghurye and Mihai Pop$^{(\boxtimes)}$

Department of Computer Science and Center for Bioinformatics and Computational
Biology, University of Maryland, College Park, USA
jayg@cs.umd.edu, mpop@umiacs.umd.edu

Abstract. Genomic repeats are the most important challenge in
genomic assembly. While for single genomes the effect of repeats is largely
addressed by modern long-read sequencing technologies, in metagenomic
data intra-genome and, more importantly, inter-genome repeats continue to be a significant impediment to effective genome reconstruction.
Detecting repeats in metagenomic samples is complicated by characteristic features of these data, primarily uneven depths of coverage and the
presence of genomic polymorphisms. The scaffolder Bambus 2 introduced
a new strategy for repeat detection based on the betweenness centrality
measure – a concept originally used in social network analysis. The exact
computation of the betweenness centrality measure is, however, computationally intensive and impractical in large metagenomic datasets. Here
we explore the effectiveness of approximate algorithms for network centrality to accurately detect genomic repeats within metagenomic samples. We show that an approximate measure of centrality achieves much
higher computational efficiencies with a minimal loss in the accuracy of
detecting repeats in metagenomic data. We also show that the combination of multiple features of the scaffold graph provides a more effective
strategy for identifying metagenomic repeats, significantly outperforming
all other commonly used approaches.

Keywords: Metagenomics · Random forest · Betweenness centrality ·
Scaffolding · Algorithms · Graph

1 Introduction

Genomic repeats are the most important challenge in genomic assembly even
for isolate genomes. When reads are shorter than the repeats (a common situation until the recent development of long read sequencing technologies) it can
be shown that the number of genome reconstructions consistent with the read
data grows exponentially with the number of repeats [10]. The use of additional
information to constrain the one genome reconstruction representing the actual
genome being assembled leads to computationally intractable problems. In other
words, when reads are shorter than repeats the correct and complete reconstruction of a genome is impossible. In the case of isolate genomes, long read technologies have largely addressed this challenge, at least for bacteria where the

© Springer International Publishing Switzerland 2016
M. Frith and C.N.S. Pedersen (Eds.): WABI 2016, LNBI 9838, pp. 174–184, 2016.
DOI: 10.1007/978-3-319-43681-4_14

majority of genomic repeats fall within the range of achievable read lengths [11]. In metagenomics, however, the problem is compounded by the fact that microbial mixtures often include multiple closely-related genomes differing in just a few locations. The genomic segments shared by closely related organisms – inter-genomic repeats – are substantially larger than intra-genomic repeats and cannot be fully resolved even if long read data were available. Instead, the best hope is to identify and flag these repeats in order to avoid mis-assemblies that incorrectly span across genomes.

To date, most approaches for repeat detection have been based on the basic observation that repetitive segments have unusual coverage depth, fact which is usually ascertained through simple statistical tests. These approaches, however, fail in the context of metagenomic data as well as in other settings (e.g., single cell genomics) that violate the assumption of uniform depth of coverage within the genome, assumption that is critical for the correctness of statistical tests. Furthermore, the challenges posed by repeats to assembly algorithms are not directly related to the depth of sequencing coverage within contigs, rather they result from the fact that repeats "tangle" the assembly graph. More specifically, the correct genomic sequence (whether of a single genome or mixture of genomes) can be represented as one or more linear sub-paths of the graph. Repeats induce links within the graph that are inconsistent with this linear structure, making it difficult for algorithms to reconstruct the true genomic structure. We, therefore

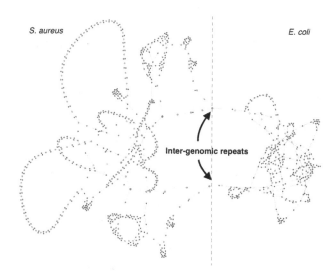

Fig. 1. Assembly graph of a simulated community consisting of 200 Kbp subsets of *Escherichia coli* str. K-12 MG1655 and *Staphylococcus aureus*. Nodes are colored and sized based on their relative betweenness centrality with larger, green nodes indicating a higher centrality. The highlighted nodes are inter-genomic repeats whose deletion would separate the graph. Note that the betweenness centrality measure correctly identifies these nodes.

propose an operational definition of genomic repeats as those nodes in the graph that induce inconsistencies. This definition is orthogonal to depth of coverage considerations - high coverage contigs that do not "tangle" the graph do not impact assembly algorithms, while contigs that confuse the assembly need to be removed whether or not they can be conclusively labeled as "high coverage".

We have previously proposed an operational definition of repeats in terms of betweenness centrality. This approach was implemented in the Bambus 2 [12] scaffolder and is a key component of the MetAMOS metagenomic assembly pipeline [24]. An example of the effectiveness of this approach in a simple community composed of two genomes is shown in Fig. 1. The full implementation of betweenness centrality, however, requires an all-pairs shortest path computation which is computationally too intensive for typical metagenomic datasets. In Bambus 2, for example, repeat finding in a typical stool sample requires days of computation. To overcome this limitation, we demonstrate here that substantial speed-ups can be obtained through the use of approximate betweenness centrality algorithms without sacrificing accuracy. We further extend this operational definition of repeats by integrating a larger set of graph properties to construct an efficient and accurate repeat detection strategy.

2   Related Work

Repeat Detection in Scaffolding

Scaffolding involves using the connectivity information from mate pairs to orient and order pre-assembled contigs obtained from an assembler to reconstruct a genome. This problem of orienting and ordering contigs was shown to be NP-Hard [9]. Various scaffolding methods have been designed based on different heuristics to obtain approximate solutions to the problem. However, all of these methods face difficulties when dealing with contigs originating from repetitive regions in the genome. A common strategy for handling repeats is to identify and remove them from the graph prior to the scaffolding process, then re-introduce them after the contigs have been properly ordered and oriented. Most of the existing scaffolders use depth of coverage information to classify a contig as a repeat. For example, Opera [4] and SOPRA [2] filter out as repetitive contigs with coverage 1.5 and 2.5 times more than average coverage, respectively. The MIP scaffolder [22] uses high coverage (greater than 2.5 times average) as well as high degree (≥ 50) of nodes within scaffold graph to determine repeats. Bambus 2 [12] – a scaffolder specifically designed for metagenomic data – uses a notion of betweenness centrality [1] along with global coverage information to find out repeats.

Betweenness Centrality

In network analysis, metrics of centrality are used to identify the most important nodes within a graph. Several metrics to measure centrality have been proposed, but in this work, we use betweenness centrality. The betweenness centrality of a particular node is equal to the number of shortest paths from all nodes to all others that pass through that node. Intuitively, a node that is frequently found on paths connecting other nodes is a potential repeat, as along a simple path all nodes should have roughly the same centrality value. The algorithm for computing exact centrality [1] takes $\Theta(mn)$ time on a graph with m nodes and n edges. Several solutions were proposed to overcome this computational cost of computing network centrality, including and exact massively parallel implementation [16], and an approximate solution based on sampling a subset of the nodes [6]. Recently, a better parallel approximation algorithm was proposed by Riondato and Kornaropoulos [21] which uses a strategy for sampling from among the shortest paths in the graph to compute betweenness centrality. The size of chosen sample of paths can provide provable bounds on the accuracy of the centrality value given by the algorithm. The sample size is determined as a function of an approximation factor ϵ and the diameter of the graph.

3   Methods

Construction of Scaffold Graph

A scaffold graph is defined as a graph $G(V, E)$, where V is set of all the contigs. The edges represent links between the contigs inferred from read pairing information – if the opposite ends of a read pair map to different contigs we can infer the possible adjacency of these contigs within the genome. Since most genome assemblers do not report the location of reads within contigs, we infer this information by mapping using bowtie2 [13]. Experimental library size estimates are often incorrect, and we re-estimate here the distance between the paired reads from pairs of reads mapped to a same contig. We record the average insert size l and standard deviation $\sigma(l)$ within a library. For each pair of contigs we retain the maximal set of links that are consistent in terms of the implied distance between the contigs for each implied relative placement of the contigs. Since contigs can be oriented in forward or reverse direction depending on the orientation implied by mapped mate pairs, there exist 4 possible orientations of adjacent contigs (forward-forward, forward-reverse, reverse-forward and reverse-reverse). For each of the possible relative orientation, we need to find a maximal set of consistent links implying that orientation. This set can be identified in $O(nlogn)$ time using an algorithm to find maximal clique in an interval graph [20]. The distance between the contigs implied by the resulting "bundle" of links has mean $l(e) = \frac{\sum \frac{l}{\sigma(l)}}{\sum \frac{1}{\sigma(l)^2}}$ and standard deviation $\sigma(l) = \frac{1}{\frac{1}{\sigma(l)^2}}$, as suggested by Huson et al. [7].

Orienting the Bidirected Scaffold Graph

The scaffold graph derived from the process outlined above is birected [17]. It can be converted into a directed graph by assigning an orientation to each node,

reflecting the strand of the DNA molecule that is represented by the corresponding contig. In computational terms, we need to embed a bipartite graph (the two
sets corresponding to the two strands of DNA being reconstructed) within the
scaffold graph. In the general case, such an embedding is not possible without
removing edges in order to break all odd-length cycles in the original graph. Finding such a minimum set of edges is NP-Hard [5]. We use here a greedy heuristic
proposed by Kececioglu and Myers [9] which achieves a 2−approximation and
runs in $O(V + E)$ time.

Repeat Detection Through Betweenness Centrality

We start by calculating centrality values for all the nodes in the graph using
either an exact or approximate centrality algorithm as outlined in the introduction. Let μ be the mean and σ be the standard deviation of the resulting
centrality values. A contig is marked as repeat if its centrality value is greater
than $\mu + 3 * \sigma$. This cutoff criterion is the same as the one used in Bambus 2. We
have also experimented with other definitions of outliers (such as interquartile
range), however the original definition used in Bambus 2 performed better than
the interquartile range cutoff (data not shown).

Repeat Detection with an Expanded Feature Set

Centrality is just one of the possible signatures that a node in the graph "tangles"
the graph structure, making it harder to identify a correct genomic reconstruction. At a high level, one can view centrality to relate to difficulties in ordering
genomic contigs along a chromosome. The orientation procedure outlined above
provides potential insights into contigs that may prevent the correct orientation
of contigs – contigs adjacent to a large number of edges invalidated by the orientation procedure are possible repeats. Other potential signatures we consider
include the degree of graph nodes (highly connected nodes are potential repeats)
as well as abrupt changes in coverage between adjacent nodes. The latter information is defined as follows. For each contig we capture the distribution of read
coverage values. We then use a Kolmogorov-Smirnov test [15] to identify pairs of
contigs that have statistically different distributions of coverage values. We flag
all edges that exceed a pre-defined p-value cutoff (in the results presented here
we simply use 0.05). We combine these different measures (contig length, centrality, node degree, fraction of number of edges invalidated by the orientation
routine that are adjacent to a node, fraction of number of edges with abrupt
changes in coverage, and ratio of contig coverage to average coverage) within a
Random Forest classifier [14].

To generate training information for the classifier we aligned the contigs to
an appropriate set of reference genomes using MUMmer [3] dependent on the
data being assembled, and flagged as repetitive all contigs that had more than
one match with greater than 95 % identity over 90 % of the length within the
reference collection.

4   Results

Dataset and Assembly

To test our methods, we used a synthetic metagenomic community dataset (S1) by Shakya et al. [23] that was derived from a mixture of cells from 83 organisms with known genomes. Reads in the datasets were cleaned and trimmed using Sickle [8]. Assembly was performed using IDBA-UD [19] with default parameters. The assembly of S1 yielded 47,767 contigs.

Extended Feature Set Improves Repeat Detection

We trained a Random Forest classifier that takes into account the various measures outlined above as follows. We simulated a low coverage (10x) dataset using a read simulator provided with the IDBA assembler from the set of 40 genomes downloaded from NCBI[1]. We constructed contigs from the simulated reads and mapped them to reference sequences to identify which contigs are repetitive (have ambiguous placement in the reference set). We used this information to train the classifier, then used the resulting classifier to predict repeats within the synthetic community S1 described above. As can be seen in Fig. 2 the accuracy of the classifier based on multiple graph properties is higher than that of approaches that rely on just coverage as a criterion to classify a contig as a repeat. Classification of repeats using approximate centrality provides higher specificity compared to the coverage approach at the cost of slightly lower sensitivity. The Random Forest approach leverages the advantage of high sensitivity from the coverage approach and high specificity from the centrality approach along with some additional features to provide better overall classification.

Important Parameters in Determining Repeats

We further explored the features of the data that contribute to the better performance of the classifier. In Fig. 3 we show the contribution of each feature to the classifier. The length of contigs, factor not usually taken into account when detecting repeats, appears to have the largest influence. This is perhaps unsurprising as repeats confuse the assembly process as well, fragmenting the assembly. In other words, longer contigs are less likely to represent repetitive sequences. The second most important features is the fraction of edges adjacent to a contig that indicate an abrupt change in coverage. Contigs with unusual coverage in comparison to their neighbors can also be reasonably assumed to be repetitive. Centrality was the third most important factor, as expected. Perhaps surprising, overall depth of coverage or node degree are not as important as features despite these measures being among the most widely used signatures of "repetitiveness" by existing tools.

---

[1] ftp://ftp.ncbi.nlm.nih.gov/genomes/bacteria/all.fna.tar.gz.

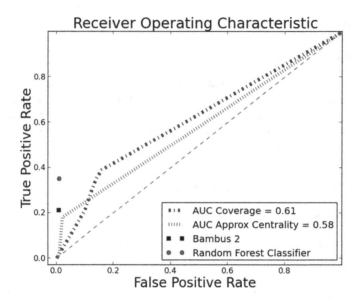

Fig. 2. Plot for comparison of Random Forest classifier with the coverage and centrality approach. The red circle in the plot indicates the sensitivity and specificity obtained by using the Random Forest approach. The black square in the plot indicates the sensitivity and specificity obtained by using Bambus 2.

Comparison of Incorrectly Oriented Pair of Contigs

Beyond testing the simple classification power of different approaches, we also evaluated the different methods in terms of whether the removal of nodes marked as repeats makes the scaffolding process more accurate. Specifically, we explored how different repeat removal strategies affect the contig orientation process. The scaffold graph for the S1 dataset had 21,950 nodes and 31,059 edges. We removed the repeats reported by the different methods from this graph and oriented the resulting graph. We then tracked the accuracy of the results in terms of the number of edges that imply a different relative orientation of the adjacent nodes than the correct one, inferred by mapping the contigs to the reference genomes. Here the relative orientation can either be same if both the contigs on the edge have same orientation (forward-forward and reverse-reverse) and different if the contigs on the edge different orientations (forward-reverse and reverse-forward). The results are shown in Table 1. The centrality based methods and the Random Forest classifier based methods resulted in lower error rates and retained a higher percentage of the edges in the original graph than coverage based methods.

Comparison of Runtime with Bambus 2

The results above show that Bambus 2 has, unsurprisingly, a similar level of accuracy with the approximate centrality approach. We have already mentioned,

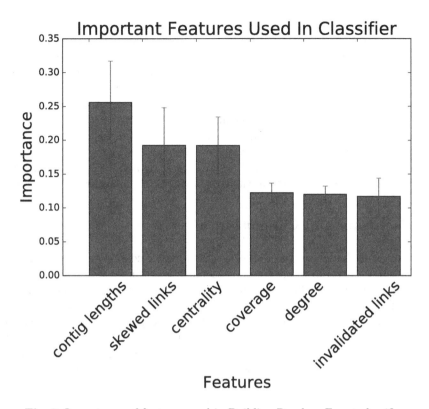

Fig. 3. Importance of features used in Building Random Forest classifier

Table 1. Number of correctly and incorrectly oriented links in scaffold graph using various repeat removal strategies. The % correct column represents the percentage of correctly oriented links as a function of the total number of edges in the original scaffold graph. % wrong column represents the percentage of incorrectly oriented links in the graph obtained by removing repeats.

Method	Correct	Wrong	% correct	% wrong
Bambus 2	12042	867	38.77 %	4.11 %
Approximate betweenness centrality	12336	917	39.71 %	3.94 %
Coverage (MIP, SOPRA)	3840	315	17.49 %	4.72 %
Coverage (Opera)	2007	165	6.46 %	5.62 %
Random forest	12255	807	39.45 %	3.52 %

however, that Bambus 2 is inefficient on large datasets. To explore the efficiency of the approximate centrality approach, we used a real metagenomic dataset (SRX024329 from NCBI) representing sequencing data from the tongue dorsum of a female patient. Assembly of these reads was performed using IDBA yielding 106,380 contigs in total. The scaffold graph constructed from these contigs had 112,502 edges. The 'MarkRepeats' module of Bambus 2 took almost 2 h to detect repeats, whereas the approximate betweenness centrality algorithm found repeats in approximately 5 min, a substantial improvement in speed without a loss of accuracy as shown above. To compare the runtime with training of Random Forest classifier, we trained the classifier on contigs in this dataset. Since we did not have reference sequences for this dataset, we randomly marked a subset of contigs as repeats and performed training. It took about 20 min to calculate features and fit a classifier which was still faster than time taken by Bambus 2.

5   Discussion and Conclusion

Our prior work had introduced the use of network centrality as an approach for detecting repeats in metagenomic assembly, a setting where coverage-based approaches are often ineffective. This approach, implemented in the scaffolder Bambus 2, was, however, inefficient for large datasets, fact that has limited its use. Here we extend our original approach by incorporating multiple features of the scaffold graph (including centrality) that may be signatures of repetitive sequences within a Random Forest classifier. We also show that an approximate calculation of network centrality based on the random sampling of paths obtains similar accuracy as the full centrality computation at a fraction of computational time.

Our results demonstrate that methods that directly capture the effect of repeats on the assembly graph are more effective at detecting repeats than indirect measures such as depth of coverage, particularly in the context of metagenomic assembly. Our new approach improves in both accuracy and efficiency over existing methods for repeat detection, and we plan to incorporate it within the MetAMOS metagenomic assembly pipeline as a replacement for the exisiting code within Bambus 2. We note that the classification accuracy was surprisingly high despite the fact that the classifier was trained on purely simulated data yet applied to real dataset. This underscores the robustness of the feature set we have identified. At the same time the graph features that we have identified as useful in detecting repeats are just a first step towards a better understanding of the features of the data that most influence the ability of assembly algorithms to accurately reconstruct metagenomic sequences. Also classifiers like Random Forest can be implemented in parallel [18] which can provide significant runtime speedups for large metagenomic datasets. We plan in future work to further explore both the feature set and the approaches used to build and train the classifier to increase accuracy and ultimately improve the quality of metagenomic reconstructions.

Acknowledgements. We thank Chris Hill for helping us with generating Fig. 1 and experiments. We also thank Todd Treangen for helping us to improve the manuscript and design experiments.

References

1. Brandes, U.: A faster algorithm for betweenness centrality*. J. Math. Sociol. **25**(2), 163–177 (2001)
2. Dayarian, A., Michael, T.P., Sengupta, A.M.: SOPRA: scaffolding algorithm for paired reads via statistical optimization. BMC Bioinform. **11**(1), 1 (2010)
3. Delcher, A.L., Salzberg, S.L., Phillippy, A.M.: Using MUMmer to identify similar regions in large sequence sets. Curr. Protocols Bioinform. 10.3.1–10.3.18 (2003). Chapter 10:Unit 10.3
4. Gao, S., Sung, W.-K., Nagarajan, N.: Opera: reconstructing optimal genomic scaffolds with high-throughput paired-end sequences. J. Comput. Biol. **18**(11), 1681–1691 (2011)
5. Garey, M., Johnson, D.: Computers and Intractability - A Guide to NP-Completeness. W.H. Freeman & Co., New York (1979)
6. Geisberger, R., Sanders, P., Schultes, D.: Better approximation of betweenness centrality. In: ALENEX, pp. 90–100. SIAM (2008)
7. Huson, D.H., Reinert, K., Myers, E.W.: The greedy path-merging algorithm for contig scaffolding. J. ACM (JACM) **49**(5), 603–615 (2002)
8. Fass, J.N., Joshi, N.A.: Sickle: a sliding-window, adaptive, quality-based trimming tool for FastQ files (version 1.33)
9. Kececioglu, J.D., Myers, E.W.: Combinatorial algorithms for DNA sequence assembly. Algorithmica **13**(1–2), 7–51 (1995)
10. Kingsford, C., Schatz, M.C., Pop, M.: Assembly complexity of prokaryotic genomes using short reads. BMC Bioinform. **11**(1), 21 (2010)
11. Koren, S., Phillippy, A.M.: One chromosome, one contig: complete microbial genomes from long-read sequencing and assembly. Curr. Opin. Microbiol. **23**, 110–120 (2015)
12. Koren, S., Treangen, T.J., Pop, M.: Bambus 2: scaffolding metagenomes. Bioinformatics **27**(21), 2964–2971 (2011)
13. Langmead, B., Salzberg, S.L.: Fast gapped-read alignment with Bowtie 2. Nat. Methods **9**(4), 357–359 (2012)
14. Liaw, A., Wiener, M.: Classification and regression by randomforest. R News **2**(3), 18–22 (2002)
15. Lilliefors, H.W.: On the Kolmogorov-Smirnov test for normality with mean and variance unknown. J. Am. Stat. Assoc. **62**(318), 399–402 (1967)
16. Madduri, K., Ediger, D., Jiang, K., Bader, D.A., Chavarria-Miranda, D.: A faster parallel algorithm and efficient multithreaded implementations for evaluating betweenness centrality on massive datasets. In: 2009 IEEE International Symposium on Parallel and Distributed Processing, IPDPS 2009, pp. 1–8. IEEE (2009)
17. Medvedev, P., Georgiou, K., Myers, G., Brudno, M.: Computability of models for sequence assembly. In: Giancarlo, R., Hannenhalli, S. (eds.) WABI 2007. LNCS (LNBI), vol. 4645, pp. 289–301. Springer, Heidelberg (2007)
18. Mitchell, L., Sloan, T.M., Mewissen, M., Ghazal, P., Forster, T., Piotrowski, M., Trew, A.S.: A parallel random forest classifier for R. In: Proceedings of the Second International Workshop on Emerging Computational Methods for the Life Sciences, pp. 1–6. ACM (2011)

19. Peng, Y., Leung, H.C., Yiu, S.-M., Chin, F.Y.: Meta-IDBA: a de novo assembler for metagenomic data. Bioinformatics **27**(13), i94–i101 (2011)
20. Pop, M., Kosack, D.S., Salzberg, S.L.: Hierarchical scaffolding with bambus. Genome Res. **14**(1), 149–159 (2004)
21. Riondato, M., Kornaropoulos, E.M.: Fast approximation of betweenness centrality through sampling. In: Proceedings of the 7th ACM International Conference on Web Search and Data Mining, pp. 413–422. ACM (2014)
22. Salmela, L., Mäkinen, V., Välimäki, N., Ylinen, J., Ukkonen, E.: Fast scaffolding with small independent mixed integer programs. Bioinformatics **27**(23), 3259–3265 (2011)
23. Shakya, M., Quince, C., Campbell, J.H., Yang, Z.K., Schadt, C.W., Podar, M.: Comparative metagenomic and RRNA microbial diversity characterization using archaeal and bacterial synthetic communities. Environ. Microbiol. **15**(6), 1882–1899 (2013)
24. Treangen, T.J., Koren, S., Sommer, D.D., Liu, B., Astrovskaya, I., Ondov, B., Darling, A.E., Phillippy, A.M., Pop, M.: MetAMOS: a modular and open source metagenomic assembly and analysis pipeline. Genome Biol. **14**(1), R2 (2013)

A Better Scoring Model for De Novo Peptide Sequencing: The Symmetric Difference Between Explained and Measured Masses

Ludovic Gillet[1], Simon Rösch[2], Thomas Tschager[2(✉)], and Peter Widmayer[2]

[1] Department of Biology, ETH Zürich, 8092 Zürich, Switzerland
`gillet@imsb.biol.ethz.ch`
[2] Department of Computer Science, ETH Zürich, 8092 Zürich, Switzerland
`roeschs@ethz.ch`, {`tschager,widmayer`}`@inf.ethz.ch`

Abstract. Given a peptide as a string of amino acids, the masses of all its prefixes and suffixes can be found by a trivial linear scan through the amino acid masses. The inverse problem is the *ideal de novo peptide sequencing problem*: Given all prefix and suffix masses, determine the string of amino acids. In biological reality, the given masses are measured in a lab experiment, and measurements by necessity are noisy. The (real, noisy) *de novo peptide sequencing problem* therefore has a noisy input: a few of the prefix and suffix masses of the peptide are missing and a few others are given in addition. For this setting we ask for an amino acid string that explains the given masses as accurately as possible. Past approaches interpreted accuracy by searching for a string that explains as many masses as possible. We feel, however, that it is not only bad to not explain a mass that appears, but also to explain a mass that does not appear. That is, we propose to minimize the symmetric difference between the set of given masses and the set of masses that the string explains. For this new optimization problem, we propose an efficient algorithm that computes both the best and the k best solutions. Experiments on measurements of 342 synthesized peptides show that our approach leads to better results compared to finding a string that explains as many given masses as possible.

1 Introduction

The determination of the amino acid string of a peptide based on mass spectrometric data is an important task in proteomics. A typical tandem mass spectrometry experiment consists of three steps [6,8]: First, the mass spectrometer measures the mass-to-charge ratio and the abundance of the analyzed peptide. Then, several techniques can be applied to fragment multiple copies of this peptide at random positions into charged prefix and suffix fragments. Finally, the mass spectrometer measures the mass-to-charge ratios and abundances of the resulting fragments. Standard data preprocessing converts mass-to-charge ratios to masses. There are several sources of errors in every step of this experiment.

© Springer International Publishing Switzerland 2016
M. Frith and C.N.S. Pedersen (Eds.): WABI 2016, LNBI 9838, pp. 185–196, 2016.
DOI: 10.1007/978-3-319-43681-4_15

Therefore, some masses of prefix and suffix fragments are missing, while other masses are given in addition.

In this noisy setting, *de novo sequencing* is the problem to compute as accurately as possible the amino acid string of the analyzed peptide given the mass M of the peptide measured in the first step of the experiment and the set X of prefix and suffix masses measured in the third step. Several approaches [1–3,6,10] attack this problem by computing an amino acid string S with mass M, such that the set TS(S) of all prefix and suffix masses of S contains as many masses as possible of the set X (often referred to as *shared peaks count*). Besides only considering the size of the intersection TS(S) $\cap$ X, several of these approaches [3,7,11] can also maximize a more elaborate score on the masses in TS(S) $\cap$ X.

However, considering only the intersection of TS(S) and X might lead to a bias towards the use of amino acids with small masses: For example, the amino acid Glutamine has the same mass as a Glycine and an Alanine. When maximizing $|\text{TS(S)} \cap X|$, one can always replace a Glutamine by a Glycine and an Alanine in the string S without decreasing the size of the intersection. In an ideal experiment, where all prefix and suffix masses and no other masses are given in X, there exists a string S with TS(S) $= X$. However, in a real experiment with missing masses, we want to explain masses that are in X, but not to explain masses that are not in X. Dančík et al. [3] noted this problem and proposed a probabilistic scoring model incorporating penalty scores for some specific fragment masses present in TS(S) but not in X. However, current algorithms do not focus on systematically accounting for exactly those masses in TS(S) $\setminus X$. We propose a different optimization goal and want to compute a string S that minimizes the size of the symmetric difference $|\text{TS(S)} \triangle X| = |\text{TS(S)} \setminus X| + |X \setminus \text{TS(S)}|$.

In this paper we first give a precise definition of our new optimization problem for de novo peptide sequencing (Sect. 2). In Sect. 3 we develop a dynamic programming algorithm to find the best string with respect to our objective function. It is of great interest to not only compute the best solution, but also the k best solutions, e.g. to detect ambiguities due to missing prefix and suffix masses. Our algorithm can compute both the best and the k best solutions. In mass spectrometry experiments, a peptide can fragment at different chemical bonds between two amino acids and molecular losses can happen during the fragmentation process. Both issues affect the mass of the resulting fragments. We do not consider these aspects in Sect. 3 and assume that the masses in X are masses of prefixes and suffixes of the analyzed peptide. In Sect. 4 we study a more general version of the problem that also considers mass offsets due to the mentioned aspects of real experiments. Finally, we compare the performance of our algorithm with Chen's seminal algorithm [1] that aims to explain as many masses in X as possible for experimental data from 342 synthesized peptides (SWATH Gold Standard dataset [12]) in Sect. 5.

Preliminary data cleaning: The molecular composition of a real peptide consists of a chain of amino acid molecules and, additionally, an oxygen and two hydrogen atoms. The mass of an uncharged peptide is the sum of its amino acid masses and the mass of the additional H_2O molecule (18 Dalton). For the sake of simplicity of exposition, we assume that the given mass M (measured in the first step of the experiment) corresponds only to the sum of the amino acid masses without this additional mass. We can always fulfill this assumption by subtracting the mass of the water molecule from the measured mass M after the preprocessing step. Moreover, we assume that the set X of masses measured in the third step of the experiment additionally contains both 0 and the peptide mass M.

2 Problem Definition

We consider a peptide as a string S of characters (amino acids) of an alphabet Σ. Each character $a \in \Sigma$ has its own mass $m(a) \in \mathbb{R}^+$. For a string $S = a_1, \ldots, a_n$, we denote a substring by $S_{i,j} = a_i, \ldots, a_j$ for $1 \leq i \leq j \leq n$. The mass of S is the sum of its characters' masses, i.e. $m(S) = \sum_{i=1}^{n} m(a_i)$. The set $\mathrm{Pre}(S)$ of prefixes of S contains every string $S_{1,i}$ for $1 \leq i \leq n$ and the set $\mathrm{Suf}(S)$ of suffixes of S every string $S_{j,n}$ with $1 \leq j \leq n$. Both $\mathrm{Pre}(S)$ and $\mathrm{Suf}(S)$ additionally contain the empty string whose mass is zero. A *fragment* of S is a prefix or a suffix of S. The *theoretical spectrum* of S is the union of all fragment masses $\mathrm{TS}(S) = \{m(T) \mid T \in (\mathrm{Pre}(S) \cup \mathrm{Suf}(S))\}$. A mass is *explained* by S if it is in $\mathrm{TS}(S)$.

The de novo sequencing problem. *Let Σ be a set of characters, with a mass $m(a) \in \mathbb{R}^+$ for each $a \in \Sigma$. Given the peptide mass $M \in \mathbb{R}^+$ (measured in the first step of the experiment) and a set $X = \{x_i \in \mathbb{R}^+ \mid i = 1, \ldots, k\}$ of fragment masses (measured in the third step), find a string S of characters in Σ with $m(S) = M$ that minimizes $|\mathrm{TS}(S) \triangle X|$.*

We will solve the equivalent problem of finding a string S that maximizes $|X \cap \mathrm{TS}(S)| - |\mathrm{TS}(S) \setminus X|$. The reason is that for a fixed X, a chosen S that maximizes the latter also minimizes the symmetric difference.

3 An Algorithm for the De Novo Sequencing Problem

In this section we present a new dynamic programming algorithm for the de novo sequencing problem. Our algorithm builds on Chen's algorithm [1], a seminal graph-based algorithm for de novo sequencing that computes a string that maximizes the number of explained masses that are in X. We will briefly present Chen's algorithm and then propose a new algorithm that also accounts for masses that are explained by the solution, but are not in the set of measured masses X.

Chen's algorithm [1] models the set X as a directed acyclic graph (*NC-spectrum graph*, Fig. 1), where the vertices are masses and a path represents a string. The problem of computing a string S that maximizes $|\mathrm{TS}(S) \cap X|$ is then reduced to the *longest path avoiding forbidden pairs problem*, that is the problem of finding a longest path between two vertices s and t (vertices 0 and M in this case) such that at most one vertex of every given forbidden pair of vertices is used. This problem is NP-hard in general [5] and Chen's algorithm [1] solves the problem for a special structure of forbidden pairs.

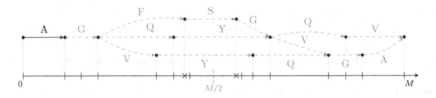

Fig. 1. Masses in X are denoted by vertical bars on the real line and masses in $\bar{X}_M \setminus X$ by crosses. Only a subgraph of the NC-spectrum graph is shown.

The *NC-spectrum graph* is a directed acyclic graph on the vertex set $\bar{X}_M = \{m, M - m \mid m \in X\}$. There is a directed edge from a vertex v to a vertex w if $w - v$ is equal to the mass of some string. Every edge is labeled with a string that has a mass equal to the mass difference of the terminal vertices. A path from v to w represents a string of mass $w - v$ and for every vertex a traversed by the path $a - v$ is a prefix mass of this string. If a path starting at vertex 0 traverses a vertex a, the string it represents explains both a (as a prefix mass) and $M - a$ (as the complementary suffix mass). The forbidden pairs are defined as the complementary masses $\{m, M - m\}$ for all $m \in X \setminus \{0, M\}$. Thus, a path avoiding forbidden pairs does not use both complementary masses, because it is sufficient to only use one of them to explain both. The longest path avoiding forbidden pairs then represents a string that maximizes $|TS(S) \cap X|$ for appropriately defined vertex weights: A vertex v has weight $|\{v, M - v\} \cap X|$ if $v \notin \{0, M\}$ and weight 1 otherwise.

Chen's algorithm [1] computes a path avoiding forbidden pairs by extending two subpaths: One starts at vertex 0 and represents a prefix and the other one ends in vertex M and represents a suffix of the solution. The algorithm repeatedly extends one of both paths by an edge until the masses of the prefix and the suffix sum up to M. The paths are extended such that the string represented by the extended path has a larger mass than the string of the other path. By this, the paths never use both vertices of a forbidden pair. The algorithm has found a solution if both paths can be merged using an edge.

In Fig. 1, one path represents the prefix A and the other path the empty suffix. In the next step, the algorithm either extends the prefix A by G or the empty suffix by V. The algorithm does not extend the suffix by A, because the corresponding mass has already been explained by the prefix. For simplicity, we only consider edges labeled with a single character. In this example, two strings that maximize the number of explained masses in X are AGFSGQV or AGQYQV (Fig. 2). While the first string explains masses that are not in X (crosses), all explained masses of the second string are in X. In our view, the second string is more likely to be the right answer and we are interested in a string that minimizes the symmetric difference between the explained masses and the measured masses. At first sight, one might think that Chen's algorithm can be easily modified to additionally consider how many explained masses are not in X. However, this is

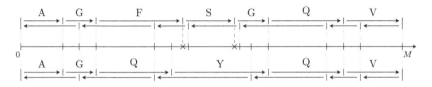

Fig. 2. Two strings that maximize the number of explained masses in X (Fig. 1). While the upper string has two prefixes (**AGF** and **AGFS**) with masses that are not in X (crosses), the lower string explains no mass that is not in X.

not obvious as the algorithm needs to check whether an explained mass that is not in X has already been explained in a different way in a previous step.

3.1   An Algorithm that Minimizes the Symmetric Difference

We propose the algorithm **DeNovo△** that solves the de novo peptide sequencing problem as defined in Sect. 2. Similar to Chen's algorithm, **DeNovo△** repeatedly extends two paths until the sum of the masses of the corresponding strings is equal to M. In contrast to Chen's algorithm, our algorithm considers a directed acyclic *multigraph* $G = (\bar{X}_M, E)$. For every pair of vertices v and w and for every string **S** with mass $w - v$ there is a directed edge labeled with **S** from v to w. Note that all edges are directed from the smaller to the larger mass. G is a multigraph, because there can exist multiple strings with equal mass, i.e. multiple edges can connect the same pair of vertices. We denote the label of an edge (v, w) by $l(v, w)$ and the concatenation of the edge labels of a path P by $l(P)$. A path in G from v to w represents a string with mass $w - v$.

To simplify the description, we consider two paths that both start from vertex 0, such that one path represents a prefix and the other path a reversed suffix of the solution. In contrast to Chen's algorithm, our algorithm always extends the path that represents the string of smaller mass. Thus, the following holds: Assume that v and b are the two last vertices of both paths with $v \leq b$ and let a be the second-to-last vertex of the path ending in b. Then, $a \leq v$, because otherwise the algorithm would not have extended this path by the edge (a, b) in a previous step, but the path ending in v. Based on this observation, we will argue that **DeNovo△** can update the number of explained masses that are in X and of explained masses that are not in X efficiently while extending the paths.

Consider a prefix string with mass v and a suffix string with mass b after some steps of our algorithm, such that $v \leq b$ and $v + b < M$. These strings correspond to paths $P = (0, \ldots, v)$ and $Q = (0, \ldots, b)$ in G. The set of masses that are explained by P and Q is the *partial theoretical spectrum*

$$\text{PTS}(P, Q, M) = \{m(\texttt{T}), M - m(\texttt{T}) \mid \texttt{T} \in \text{Pre}(l(P)) \cup \text{Pre}(l(Q))\}.$$

Extending the path ending in v by an edge (v, w) additionally explains the following set of masses:

$$\text{TSe}((v, w), M) = \{m(\texttt{T}) + v, \ M - (m(\texttt{T}) + v) \mid \texttt{T} \in \text{Pre}(l(v, w)), \ m(\texttt{T}) \neq 0\}.$$

Fig. 3. Computation of $T[w, (a, b)]$. Either a path $P = (0, \ldots, w)$ ends with an edge (v, w) with $v \leq a$ (left) or it ends in an edge (v', w) with $v' > a$ (right).

Note that we do not consider the empty string that is in $\text{Pre}(l(v, w))$, because v and $M - v$ have already been explained when extending this path to the vertex v in a previous step of the algorithm.

The following invariant holds in every step of **DeNovo$\varDelta$**: Let P and Q be the two substrings with $m(\mathtt{P}) \leq m(\mathtt{Q})$ and let $P = (0, \ldots, w)$ and $Q = (0, \ldots, a, b)$ be the paths such that $\mathtt{P} = l(P)$ and $\mathtt{Q} = l(Q)$. Then, $a \leq w \leq b$ and all masses that have already been explained in a previous step are in $\text{TSe}((a, b), M)$. That is, for every edge (w, w'), we have that $\text{TSe}((w, w'), M) \cap \text{PTS}(P, Q, M) = \text{TSe}((w, w'), M) \cap \text{TSe}((a, b), M)$. Thus, it is sufficient to not consider masses in $\text{TSe}((a, b), M)$ to compute the set of newly explained masses by extending P. We do not need to remember the complete paths P and Q, but it is sufficient to remember the two last vertices of both paths to update the number of newly explained masses that are in X, respectively not in X:

$$\text{gain}((w, w'), (a, b)) = \left| \left(\text{TSe}((w, w'), M) \setminus \text{TSe}((a, b), M) \right) \cap X \right|$$
$$- \left| \left(\text{TSe}((w, w'), M) \setminus \text{TSe}((a, b), M) \right) \setminus X \right|.$$

We can now formulate a dynamic program that computes a string with mass M that minimizes the symmetric difference: We define a two-dimensional table T with $|V|$ rows and $|E|$ columns, where V denotes the set of vertices and E the multiset of edges of G. The algorithm only considers an entry $T[w, (a, b)]$ if $a \leq w \leq b$ and $w + b \leq M$. This entry $T[w, (a, b)]$ contains the maximal number of explained masses that are in X minus the number of explained masses that are not in X for any two paths $P = (0, \ldots, w)$ and $Q = (0, \ldots, a, b)$, i.e.

$$T[w, (a, b)] = \max_{P, Q} \left\{ \left| \text{PTS}(P, Q, M) \cap X \right| - \left| \text{PTS}(P, Q, M) \setminus X \right| \right\} \quad (1)$$

where the maximum is taken over all paths $P = (0, \ldots, w)$ and all paths $Q = (0, \ldots, a, b)$ in G. We can compute an entry $T[w, (a, b)]$ given the values of all entries $T[x, (c, d)]$ with $x < w$ or $x = w$ and $c < a$ (Fig. 3): Let $P = (0, \ldots, w)$ and $Q = (0, \ldots, a, b)$ be the paths that maximize Eq. (1). Either P ends with an edge (v, w) with $v \leq a$ or it ends with an edge (v', w) with $v' > a$. In the first case, we consider the entry $T[a, (v, w)]$ and add $\text{gain}((a, b), (v, w))$. In the latter case, we add $\text{gain}((v', w), (a, b))$ to the value of $T[v', (a, b)]$. Hence,

$$T[w, (a, b)] = \max \begin{cases} \max\limits_{\substack{(v,w)\in E, \\ v\leq a}} \left\{ T[a, (v, w)] + \text{gain}((a, b), (v, w)) \right\} \\ \\ \max\limits_{\substack{(v',w)\in E, \\ v'>a}} \left\{ T[v', (a, b)] + \text{gain}((v', w), (a, b)) \right\}. \end{cases}$$

DeNovoΔ computes table T by first initializing every entry by $-\infty$. To simplify the notation, we assume that E contains a loop edge $(0,0)$ and set $T[0, (0,0)] = 2$ (an empty string explains 0 and M). The algorithm then considers all vertices v in ascending order and for a vertex v all edges (a, b) with $T[v, (a, b)] \neq -\infty$ in ascending order of a and b. It extends the path ending in v by every outgoing edge of v and updates the corresponding entry in T. Once all entries have been computed, the optimal solution can be reconstructed starting from an entry $T[v, (w, M - v)]$ with maximal value among all $v, w \in V$.

Theorem 1. *Given a peptide mass $M \in \mathbb{R}^+$ and a set $X = \{x_i \in \mathbb{R}^+ \mid i = 1, \dots, k\}$ of fragment masses, algorithm **DeNovoΔ** computes a solution for the de novo sequencing problem.*

Proof. We prove by induction that algorithm **DeNovoΔ** computes the entries of table T correctly. It is easy to see that the entries $T[0, (0, v)]$ for all $(0, v) \in E$ are computed correctly. Assume that all entries $T[w', (a', b')]$ with $w' < w$ or $a' \leq w' = w$ are correct. The entry $T[w, (a, b)]$ is either computed considering an entry $T[a, (v, w)]$ with $v \leq a$ or an entry $T[v', (a, b)]$ with $a < v' < w \leq b$. Both entries are correct by the induction hypothesis. Consider the first case. $T[a, (v, w)] = |\text{PTS}(P', Q, M) \cap X| - |\text{PTS}(P', Q, M) \setminus X|$ for some paths $P' = (0, \dots, a)$ and $Q = (0, \dots, v, w)$. A path P ending in b can be constructed by extending P' with the edge (a, b). It remains to show that

$$\begin{aligned} T[w, (a, b)] &= |\text{PTS}(P', Q, M) \cap X| - |\text{PTS}(P', Q, M) \setminus X| + \text{gain}((a, b), (v, w)) \\ &= |\text{PTS}(P, Q, M) \cap X| - |\text{PTS}(P, Q, M) \setminus X|. \end{aligned}$$

We denote the empty path by $\emptyset$. The set $\text{TSe}((a, b), M) \cap \text{PTS}(P', \emptyset, M)$ is empty, because every mass in $\text{PTS}(P', \emptyset, M)$ is in the interval $[0, a]$ or $[M - a, M]$, but $a < m < M - a$ for every mass $m \in \text{TSe}((a, b), M)$. Moreover, $\text{TSe}((a, b), M) \cap \text{PTS}(\emptyset, Q, M) = \text{TSe}((a, b), M) \cap \text{TSe}((v, w), M)$ due to the fact that $v \leq a \leq w$. Therefore, no mass considered by $\text{gain}((a, b), (v, w))$ has already been considered when computing $T[a, (v, w)]$. We can prove the second case similarly.

Let **S** be an optimal string for the de novo sequencing problem. There are exactly two consecutive prefixes of **S** with masses v and w such that $v \leq M/2 < w$. The entry $T[M - w, (v, w)]$ is equal to $|\text{PTS}(P, Q, M) \cap X| - |\text{PTS}(P, Q, M) \setminus X|$ for some paths $P = (0, \dots, M - w)$ and $Q = (0, \dots, w)$. Concatenating $l(P)$ and the reversed string of $l(Q)$ either results in **S** or in another string **S**$'$ with $|\text{TS}(\mathbf{S}) \triangle X| = |\text{TS}(\mathbf{S}') \triangle X|$, because **S** is an optimal solution. $\square$

```
1  T[v, (a, b)] ← −∞ for all (a, b) ∈ E and v ∈ V
2  T[0, (0, 0)] ← 2
3  for v ∈ V in ascending order do
4  |   foreach (a, b) ∈ E with T[v, (a, b)] ≠ −∞ in ascend. order of a and b do
5  |   |   foreach (v, w) ∈ E with w + b ≤ M do
6  |   |   |   if w ≤ b then
7  |   |   |   |   T[w, (a, b)] ← max ( T[w, (a, b)], T[v, (a, b)] + gain((v, w), (a, b)) )
8  |   |   |   else
9  |   |   |   |   T[b, (v, w)] ← max ( T[b, (v, w)], T[v, (a, b)] + gain((v, w), (a, b)) )
10 |   |   |   end
11 |   |   end
12 |   end
13 end
```

Algorithm **DeNovoΔ**.

Theorem 2. *The time complexity of **DeNovoΔ** is in $\mathcal{O}(|V| \cdot |E| \cdot d \cdot p)$, where d is the maximal out-degree of a vertex in G and p is the maximal length of an edge label.*

Proof. The table T can be initialized in $\mathcal{O}(|V| \cdot |E|)$ time. Then, the algorithm considers for every entry $T[v, (a, b)]$ all outgoing edges of v, i.e. at most d edges. For an edge (v, w) that is labeled with a string S, the time for computing gain() depends linearly on the length of S. Note that G is a multigraph and that there exists an edge from v to w for every permutation of the characters of S. The maximal length of an edge label is p, which is bounded by $\mathcal{O}(M/\mu)$, where μ is the smallest mass of a character in Σ. For every edge, it takes $O(p)$ time to update an entry in lines 7 or 9. Thus, the runtime of **DeNovoΔ** is in $\mathcal{O}(|V| \cdot |E| \cdot d \cdot p)$. □

When considering practical applications, the parameter p depends on the data quality rather than on the size of the input X and M. If we assume p to be a constant, there are only $\mathcal{O}(1)$ edges between two vertices and every vertex has only a constant out-degree. Hence, our algorithm matches the time complexity of Chen's Algorithm [1] unless the length of the edge labels grows asymptotically with the size of the input.

3.2   Computing the k Best Solutions

In this section we briefly sketch how to find the k best solutions for the de novo peptide sequencing problem. We model the table T as a directed acyclic graph [9] and define weighted edges that correspond to extension steps of **DeNovoΔ**. A solution for the de novo sequencing problem then corresponds to a path in this graph and the score of the solution corresponds to the length of the path.

The matrix graph MG is a directed acyclic graph on vertices $V(MG) \subseteq (V \times E)$. A vertex $v_{v,(a,b)}$ represents the entry $T[v, (a, b)]$, where $a \leq v \leq b$ and $v + b \leq M$. Every vertex $v_{v,(a,b)}$ has the following outgoing edges in MG:

$$\big\{(v_{v,(a,b)}, v_{w,(a,b)}) \mid (v,w) \in E, \ w \le b, \ w+b \le M\big\} \ \cup$$
$$\big\{(v_{v,(a,b)}, v_{b,(v,w')}) \mid (v,w') \in E, \ w > b, \ w'+b \le M\big\}.$$

The weight of each of these edges is gain$((v,w),(a,b))$, resp. gain$((v,w'),(a,b))$. Note that the edges defined above correspond exactly to the extension steps in lines 7 and 9 of **DeNovoΔ**. A vertex $v_{v,(a,b)}$ in MG is a *terminal vertex* if $v = M - b$. A path from $v_{0,(0,0)}$ to a terminal vertex represents a string S such that the sum of the edge weights of this path is equal to $|\mathrm{TS}(\mathsf{S}) \triangle X|$. A solution for the de novo sequencing problem corresponds to a longest path from $v_{0,(0,0)}$ to some terminal vertex in MG. The k-th best solution corresponds to the k-th longest path from $v_{0,(0,0)}$ to a terminal vertex in MG.

We can compute the longest paths in MG efficiently, as MG is directed and acyclic. Eppstein's Algorithm [4] computes the k shortest paths connecting a pair of vertices s and t in a directed acyclic graph with n vertices and m edges in $\mathcal{O}(n + m + k)$ time. The algorithm outputs an implicit representation of the paths and the sequence of edges of a path can be listed in time proportional to the length of the path. By negating the edge weights, the algorithm can compute the k longest paths. We can build MG while executing **DeNovoΔ** in time $O\left(|V| \cdot |E| \cdot d \cdot p\right)$, where V is the set of vertices and E the multiset of edges of G, d is the maximal out-degree of a vertex in G and p is the maximal length of an edge label in G. The matrix graph has $\mathcal{O}(|V| \cdot |E|)$ vertices and $\mathcal{O}(|V| \cdot |E| \cdot d)$ edges. Hence, we can find the k best solutions for the de novo peptide sequencing problem in $\mathcal{O}(|V| \cdot |E| \cdot d \cdot p + k)$ time.

4   General De Novo Sequencing Problem

In the previous section we studied the de novo sequencing problem in a simplified version: We assumed that a mass in X is exactly the mass of the corresponding string. In real experiments, a mass in X can have a small offset from the mass of its string as a peptide can split at different chemical bonds between two amino acids and can loose small neutral molecules (e.g. water, ammonia). In this section we study a more general version of the de novo sequencing problem that considers such mass offsets with bounded maximal pairwise difference and present a modified version of **DeNovoΔ** for this problem.

We first formulate the general de novo sequencing problem for a given a set of possible mass offsets. We define the extended theoretical spectrum for a string S as the set of all fragment masses with all possible mass offsets. As the possible offsets for prefixes and suffixes can differ, the extended theoretical spectrum of a string S is not equal to the extended theoretical spectrum of the reversed string of S. Therefore, our modified algorithm **DeNovoΔ_g** for the general de novo sequencing problem needs to distinguish the prefix and the suffix string.

An important difference to the simplified problem is that mass offsets can alter the order of masses in X with respect to the masses of the corresponding strings and computing the newly explained masses by an extension becomes more complicated. While the order of the masses does not change if the maximal difference of two offsets is smaller than the smallest mass μ of a character in Σ, we propose an algorithm that handles maximal difference smaller than $2 \cdot \mu$.

Let O_p and O_s be the sets of all possible mass offsets for a prefix fragment, respectively a suffix fragment. A mass offset can both be positive or negative. The maximal mass offset difference of two sets (O_p, O_s) is $\gamma = \max_{\delta \in (O_p \cup O_s)}(\delta) - \min_{\delta' \in (O_p \cup O_s)}(\delta')$. Two sets (O_p, O_s) of mass offsets are α-basic if $\gamma < \alpha \cdot \mu$. A prefix with mass m explains all masses $(m + \delta)$ for $\delta \in O_p$ and all masses $(M - m + \delta')$ for $\delta' \in O_s$, i.e. $\mathrm{OM}(m, M) = \bigcup_{\delta \in O_p}(m + \delta) \cup \bigcup_{\delta \in O_s}(M - m + \delta)$. The *extended theoretical spectrum* of a string $\mathtt{S}$ is the set of all prefix and suffix masses with all possible offsets $\mathrm{TS}_x(\mathtt{S}) = \bigcup_{\mathtt{T} \in \mathrm{Pre}(\mathtt{S})} \mathrm{OM}(m(\mathtt{T}), m(\mathtt{S}))$.

The general de novo sequencing problem. *Let Σ be a set of characters, with a mass $m(\mathtt{a}) \in \mathbb{R}^+$ for each $\mathtt{a} \in \Sigma$. Given a set $X = \{x_i \in \mathbb{R}^+ \mid i = 1, \ldots, k\}$ of experimentally measured fragment masses, a peptide mass $M \in \mathbb{R}^+$, and 2-basic sets (O_p, O_s) of mass offsets, find a string $\mathtt{S}$ of characters in Σ with $m(\mathtt{S}) = M$ that minimizes $|\mathrm{TS}_x(\mathtt{S}) \triangle X|$.*

In a similar but more complicated fashion than for the de novo sequencing problem, we can compute a solution for the general de novo sequencing problem: The algorithm **DeNovoΔ_g** uses a multigraph $G_x = (V_x, E_x)$ with up to $|O|$ vertices for every mass in X and a multiset of edges that is similarly defined as in the previous section for the multigraph G. The algorithm can compute a solution for the general de novo sequencing problem in time $\mathcal{O}(|V_x| \cdot |E_x| \cdot d \cdot p \cdot |O|)$, where d is the maximal out-degree of a vertex in G_x, p is the maximal length of an edge label, and $|O|$ is the number of possible mass offsets.

5   Experimental Results

We implemented **DeNovoΔ** and studied the quality of its solution in comparison to Chen's algorithm [1,9]. We chose **DeNovoΔ** rather than **DeNovoΔ_g** in our experiments to clearly expose the effect of the symmetric difference in a scoring function: While **DeNovoΔ** computes a string $\mathtt{S}$ that minimizes $|\mathrm{TS}(\mathtt{S}) \triangle X|$, Chen's algorithm aims for maximizing $|\mathrm{TS}(\mathtt{S}) \cap X|$. While we are not primarily interested in runtime differences of both approaches, we observed that both algorithms have very similar performances (usually less than a minute for a single peptide). We are aware that our comparison must necessarily be preliminary in nature, and we plan a more extensive experimental evaluation for the future.

We used the SWATH-MS Gold Standard (SGS) dataset [12] containing measurements of 342 synthesized peptides under different backgrounds and dilutions with annotations that have been manually validated by field experts. We extract for every annotated peptide the measurement at the time of maximum concentration (according to the annotations). Unlike in the usual scenario for de novo sequencing, we thus are not only given the peptide mass M and the set X of fragment masses, but additionally the amino acid string of the analyzed peptide that is necessary to compare the results of the algorithms.

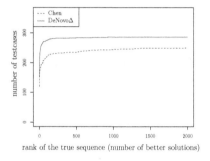
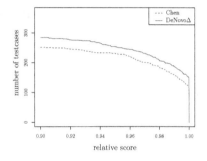

Fig. 4. Comparison of Chen's algorithm (dashed line) and **DeNovoΔ** (solid line) for the SGS dataset (water background, 1x dilution). Left: The complete annotated string is among the 100 best solutions in 219 cases (64 %) for Chen and in 274 cases (80 %) for **DeNovoΔ**. Right: In 252 cases (74 %) the relative score is at least 0.9 for Chen's algorithm and in 285 cases (83 %) for **DeNovoΔ**.

The mass spectrometer not only measures the mass-to-charge ratio of a fragment, but also a signal intensity that reflects the abundance of the fragment. Contamination or measurement errors appear as weak signals and strong signals are usually regarded to be more important. Instead of considering only the size of the sets $TS(S) \cap X$ and $TS(S) \setminus X$, we choose to sum up the intensities of the corresponding signals, just like in [10]. As there are no measured intensities for the masses in $TS(S) \setminus X$, we introduce a parameter $\mathcal{I}$ for penalizing explained masses that are not in X. This parameter is always negative and the choice of its value depends both on the average intensity of measured masses and on how likely it is that a fragment of the analyzed peptide is not measured by the mass spectrometer. This parameter can be defined depending on the type of the fragment (prefix or suffix) and on the mass offset. Setting $\mathcal{I} = 0$ is equivalent to maximizing the intersection $|TS(S) \cap X|$. We used $\mathcal{I} = -2500$ in our experiments. In several tests with different values $\mathcal{I}$ the result appeared to be not very sensitive; other choices led to comparable results. It would be interesting to evaluate thresholding instead of summing up intensities.

We compared the algorithms by generating all solutions with a score of at least 90 % of the maximum score. First, we compared the position of the true, annotated string in the lists of solution of both algorithms. Figure 4 (left) shows in how many cases the true string is among the top x solutions computed by the algorithms. The complete annotated string was among the top 100 strings in 80 % of the analyzed spectra for **DeNovoΔ** and in 64 % of the analyzed spectra for Chen's algorithm. Secondly, we defined a *relative score* that is the score of the true string divided by the score of the best string computed by the algorithm. The relative score is 1 if the true string is the best solution is never smaller than 0.9, because we only computed solutions with a score of at least 90 % of the maximum score. Figure 4 (right) shows a comparison of the results of both algorithms with respect to this score.

6   Discussion and Conclusion

In this paper we propose and study a new formulation of the de novo sequencing problem. Several previous approaches [1,3,7,10] consider the set of masses that are both explained by a string and measured in the experiment. Although it has already been pointed out [3] that penalizing the fact that a explained mass is not measured improves the performance of algorithms for peptide identification, to the best of our knowledge the problem of minimizing the symmetric difference of the set of explained masses and the set of measured masses has not been studied before. We develop a dynamic programming algorithm that can compute both the best and the k best solutions for this new de novo sequencing problem.

Acknowledgments. We would like to thank Tomas Hruz, George Rosenberger, and Hannes Röst for helpful discussions.

References

1. Chen, T., Kao, M.-Y., Tepel, M., Rush, J., Church, G.M.: A dynamic programming approach to de novo peptide sequencing via tandem mass spectrometry. In: Proceedings of the Eleventh Annual ACM-SIAM Symposium on Discrete Algorithms (SODA 2000) (2000). (Conference version of [2])
2. Chen, T., Kao, M.-Y., Tepel, M., Rush, J., Church, G.M.: A dynamic programming approach to de novo peptide sequencing via tandem mass spectrometry. J. Comput. Biol. **8**(3), 325–337 (2001). (Journal version of [1])
3. Dančík, V., Addona, T.A., Clauser, K.R., Vath, J.E., Pevzner, P.A.: De novo peptide sequencing via tandem mass spectrometry. J. Comput. Biol. **6**(3–4), 327–342 (1999)
4. Eppstein, D.: Finding the k shortest paths. SIAM J. Comput. **28**(2), 652–673 (1998)
5. Gabow, H., Maheshwari, S., Osterweil, L.: On two problems in the generation of program test paths. IEEE Trans. Softw. Eng. **SE–2**(3), 227–231 (1976)
6. Hughes, C., Ma, B., Lajoie, G.A.: De novo sequencing methods in proteomics. Proteome Bioinf. **604**, 105–121 (2010)
7. Jeong, K., Kim, S., Pevzner, P.A.: UniNovo: a universal tool for de novo peptide sequencing. Bioinformatics **29**(16), 1953–1962 (2013). (Oxford, England)
8. Kinter, M., Sherman, N.E.: Protein Sequencing and Identication Using Tandem Mass Spectrometry. Wiley-Interscience, New York (2000)
9. Lu, B., Chen, T.: A suboptimal algorithm for de novo peptide sequencing via tandem mass spectrometry. J. Comput. Biol. **10**(1), 1–12 (2003)
10. Ma, B., Zhang, K., Liang, C.: An effctive algorithm for the peptide de novo sequencing from MS/MS spectrum. Comb. Pattern Matching **2676**, 266–277 (2003)
11. Mo, L., Dutta, D., Wan, Y., Chen, T.: MSNovo: a dynamic programming algorithm for de novo peptide sequencing via tandem mass spectrometry. Anal. Chem. **79**(13), 4870–4878 (2007)
12. Röst, H.L., Rosenberger, G., Navarro, P., Gillet, L., Miladinović, S.M., Schubert, O.T., Wolski, W., Collins, B.C., Malmström, J., Malmström, L., Aebersold, R.: OpenSWATH enables automated, targeted analysis of data-independent acquisition MS data. Nat. Biotechnol. **32**(3), 219–223 (2014)

$StreAM\text{-}T_g$: Algorithms for Analyzing Coarse Grained RNA Dynamics Based on Markov Models of Connectivity-Graphs

Sven Jager[1]([✉]), Benjamin Schiller[2], Thorsten Strufe[2], and Kay Hamacher[1,3]

[1] Computational Biology and Simulation, Department of Biology,
TU Darmstadt, Darmstadt, Germany
`jager@bio.tu-darmstadt.de`
[2] Privacy and Data Security, Department of Computer Science,
TU Dresden, Dresden, Germany
`{benjamin.schiller1,thorsten.strufe}@tu-dresden.de`
[3] Department of Physics, Department of Computer Science,
TU Darmstadt, Darmstadt, Germany
`hamacher@bio.tu-darmstadt.de`

Abstract. In this work, we present a new coarse grained representation of RNA dynamics. It is based on cliques and their patterns within adjacency matrices obtained from molecular dynamics simulations. RNA molecules are well-suited for this representation due to their composition which is mainly modular and assessable by the secondary structure alone. Each adjacency matrix represents the interactions of k nucleotides. We then define transitions between states as changes in the adjacency matrices which form a Markovian dynamics. The intense computational demand for deriving the transition probability matrices prompted us to develop $StreAM\text{-}T_g$, a stream-based algorithm for generating such Markov models of k-vertex adjacency matrices representing the RNA. Here, we benchmark $StreAM\text{-}T_g$ (a) for random and RNA unit sphere dynamic graphs. (b) we apply our method on a long term molecular dynamics simulation of a synthetic riboswitch (1,000 ns). In the light of experimental data our results show important design opportunities for the riboswitch.

Keywords: RNA · Markovian dynamics · Dynamic graphs · Molecular dynamics · Coarse graining · Synthetic biology

1 Introduction

The computational design of switchable and catalytic ribonucleic acids (RNA) becomes a major challenge for synthetic biology [22]. So far, available models and simulation tools to design and analyze functionally complex RNA based devices are very limited [8]. Although several tools are available to assess secondary as well as tertiary RNA structure [3], current capabilities to simulate dynamics are still underdeveloped [15] and rely heavily on atomistic molecular dynamics (MD)

© Springer International Publishing Switzerland 2016
M. Frith and C.N.S. Pedersen (Eds.): WABI 2016, LNBI 9838, pp. 197–209, 2016.
DOI: 10.1007/978-3-319-43681-4_16

techniques [13]. RNA structure is largely modular and composed of repetitive motifs [15] that form structural elements such as hairpins and stems based on hydrogen-bonding patterns [30]. Such structural modules play an important role for nano design [18,22].

In order to understand RNA dynamics [1,26] we develop a new method to quantify all possible structural transitions, based on a coarse grained, transferable representation of different module sizes. The computation of Markov State Models have recently become practical to reproduce long-time conformational dynamics of biomolecules using data from MD simulations [9]. To this end, we convert MD trajectories into dynamic graphs and derive the Markovian dynamics in the space of adjacency matrices. Aggregated matrices for each nucleotide represent RNA coarse grained dynamics. However, a full computation is computationally expensive.

To address this challenge we extend *Stream* - a stream based algorithm for counting motifs in dynamic graphs with an outstanding performance of counting motifs in biomolecular trajectories [21]. The extension *StreAM* computes one transition matrix for a single set of vertices or a full set for combinatorial many matrices. To gain insight into global folding and stability, we propose $StreAM\text{-}T_g$: It combines all Markov models for an RNA into a global weighted stochastic transition matrix T_g.

The remainder of this paper is structured as follows: In Sect. 2, we introduce the concept as well as our biological test setup. We describe details in Sect. 3. We present run-time evaluations of our algorithm in Sect. 4 for a synthetic tetracycline (TC) dependent Riboswitch (TC-Aptamer). Finally, we summarize our work in Sect. 5.

2   Our Approach for Coarse Grained Analysis

2.1   Structural Representation of RNA

Predicting the function of complex RNA molecules depends critically on understanding both, their structure as well as their conformational dynamics [14,17]. To achieve the latter we propose a new coarse grained RNA representation and the dynamics in the implied state space at the nucleotide level. For our approach, we start with a MD simulation to obtain a trajectory of the RNA. We reduce these simulated trajectories to nucleotides represented by their ($C3'$) atoms. From there, we represent RNA structure as an undirected graph [11] using each $C3'$ as a vertex and distance dependent interactions as edges [3]. It is well known that nucleotide-based molecular interactions take place between more than one partner [28]. For this reason interactions exist for several edges observable in the adjacency matrix (obtained via a Euclidean distance cut-off) of $C3'$ coordinates at a given time-step. The resulting edges represent, e.g., strong local interactions such as Watson-Crick pairing, Hoogsteen, or $\pi - \pi$-stacking.

Our algorithm estimates adjacency matrix transition rates of a given set of vertices (nucleotides) and builds a Markov model. Moreover, by deriving all Markov models of all possible combinations of vertices, we can reduce them afterwards into a global weighted transition matrix for each vertex representing the ensemble that the vertex/nucleotide is immersed in.

2.2 Dynamic Graphs, Their Analysis and Markovian Dynamics

A *graph* $G = (V, E)$ is an ordered pair of *vertices* $V = \{v_1, v_2, \ldots v_{|V|}\}$ and *edges* E. Here, we only consider *undirected graphs without loops*, i.e., $E \subseteq \{\{v, w\} : v, w \in V, v \neq w\}$. For a subset V' of the vertex set V, we refer to $G^{V'} = (V', E'), E' := \{\{v, w\} \in E : v, w \in V'\}$ as the V'-*induced subgraph* of G.

The *adjacency matrix* $A(G) = A_{i,j}$ (Eq. 1) of a graph G is a $|V| \times |V|$ matrix, defined as follows:

$$A_{i,j} := \begin{cases} 0 : i < j \wedge \{v_i, v_j\} \notin E \\ 1 : i < j \wedge \{v_i, v_j\} \in E \\ \uparrow : \text{otherwise} \end{cases} \tag{1}$$

We denote the set of all adjacency matrices of size k as $\mathscr{A}_k$, with $|\mathscr{A}_k| = 2^{\frac{k \cdot (k-1)}{2}}$. With $concat(A)$, we denote the row-by-row *concatenation* of all defined values of an adjacency matrix A. We define the *adjacency id* of a matrix A as the numerical value of the binary interpretation of its concatenation, i.e., $id(A) = concat(A)_2 \in \mathbb{N}$. We refer to $id(V') := id(A(G^{V'}))$ as the adjacency id of the V'-induced subgraph of G. For example, the concatenation of the adjacency matrix of graph $G_1^{V'}$ (shown in Fig. 1) is $concat(A(G_1^{V'})) = 011011$ and its adjacency id is $id(V') = 011011_2 = 27_{10}$.

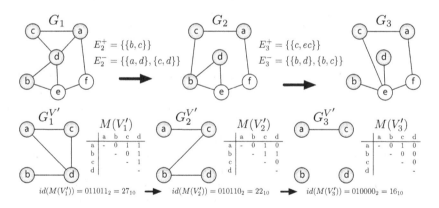

Fig. 1. Example of a dynamic graph and induced subgraphs for $V' = (a, b, c, d)$

As a *dynamic graph* $G_t = (V, E_t)$, we consider a graph whose edge set changes over time. For each point in time $t \in [1, \tau]$, we consider G_t as the *snapshot* or *state* of the dynamic graph at that time. The *transition of a dynamic graph* G_{t-1} to the next state G_t is described by a pair of edge sets which contain the edges added to and removed from G_{t-1}, i.e., (E_t^+, E_t^-). We refer to these changes as a *batch*, defined as follows: $E_t^+ := E_t \backslash E_{t-1}$ and $E_t^- := E_{t-1} \backslash E_t$. The *batch size* is referred as $\delta_t = |E_t^+| + |E_t^-|$.

The *analysis* of dynamic graphs is commonly performed using *stream-* or *batch-based* algorithms. Both output the desired result for each snapshot G_t. Stream-based algorithms take a single update to the graph as input, i.e., the addition or removal of an edge e. Batch-based algorithms take a pair (E_{t+1}^+, E_{t+1}^-) as input. They can always be implemented by executing a stream-based algorithm for each edge addition $e \in E_{t+1}^+$ and removal $e \in E_{t+1}^-$.

The result of analyzing the adjacency id of V' for a dynamic graph G_t is a list $(id_t(V') : t \in [1, \tau])$. We consider each pair $(id_t(V'), id_{t+1}(V'))$ as an *adjacency transition* of V' and denote the *set of all transitions* as $\mathscr{T}(V')$. Then, we define the *local transition matrix* $T_{i,j}(V')$ of V' as a $|\mathscr{A}_k| \times |\mathscr{A}_k|$ matrix which contains the number of transitions between any two adjacency ids over time, i.e., $T_{i,j}(V') := |(i+1, j+1) \in \mathscr{T}(V')|$. From $T_{i,j}(V')$, we can derive a *Markov model* to describe these transitions.

By combining all possible $T_{i,j}(V')$, a specific vertex v is immersed in a subset V', we derive a transition tensor $C_{i,j,l}(V')$ with dimensions $|\mathscr{A}_k| \times |\mathscr{A}_k| \times (k-1)!\binom{|V|}{k-1}$. We define a global weighting parameter w_l, by considering the local distribution weighted by its global distribution of transitions matrices. A global transition matrix T_g is defined as $\sum_l w_l \times C_l(V')$ with the dimensions $|\mathscr{A}_k| \times |\mathscr{A}_k|$.

For a local or global transition matrix the respective dominant eigenvector[1] is called π and represents the stationary distribution attained for infinite (or very long) times. The corresponding conformational entropy of the ensemble of motifs is $H := -\sum_i \pi_i \cdot \log \pi_i$. The change in conformational entropy upon, e.g., binding a ligand is then given as $\Delta H = H_{wt} - H_{complex}$.

2.3 Workflow

MD Simulation Setup. We use a structure of a synthetic tetracycline binding riboswitch (PDB: 3EGZ, chain B, resolution: 2.2 Å, Fig. 2) [32] and perform two simulations: the riboswitch with tetracycline in complex and without tetracycline. As tetracycline binding alters the structural entropy of the molecule [31] our proposed method should be able to detect changes in (local) dynamics due the presence of tetracycline. Both simulations were performed with Gromacs using the charmm27 force field with parameters for synthetic tetracycline derivatives (7CL) [2,5,19]. Simulations were performed at constant temperature (300 K) and pressure (1 bar). The simulation box is filled with Tip3p water, and Mg^{2+} counter ions. After minimization we equilibrate the solvent with fixed RNA for 10 ns, release the RNA and started the simulation with an integration step-size of 1.5 fs. Both all-atom MD simulations last for 1,000 ns.

Graph Transformation. Both MD simulations contain 160,000 snapshots. We generated dynamic graphs $G_t = (V, E_t)$ containing $|V| = 65$ vertices (Table 1),

---

[1] Guaranteed to exist due to the Perron-Frobenius theorem with an eigenvalue of $\lambda = 1$.

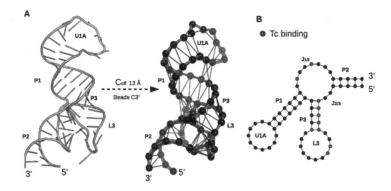

Fig. 2. Structural representation of TC-Aptamer. **A:** TC-Aptamer with a cut-off of 13 Å and using $C3'$ atom for coarse graining reveals edges for dominant WC base-pairings. **B:** Secondary structure representation of TC-Aptamer. Nucleotides participating in TC binding are colored in red. Graphics were created using `Pymol` and `R` [23,29].

each modelling a nucleic $3C'$ (Fig. 2). This resolution is sufficient to represent both small secondary structure elements as well as large quaternary RNA complexes [10,12]. We create undirected edges between two vertices in case their Euclidean cut-off (d) is shorter than $d \in [10,15]$Å (cmp. Table 1).

Markov State Models (MSM) of Local Adjacency Transitions. *StreAM* counts transitions of a k-vertex set with a size of $|\mathscr{A}_k|$ for a given set of V' obtained from a dynamic graph $G_t(V, E_t)$. Afterwards, we can compute respective probabilities resulting in a transition matrix: $T_{i,j}(V') = |(i+1, j+1) \in \mathscr{T}(V')|$. Not all possible states are necessarily visited in a given, finite simulation, although a "missing state" potentially might occur in longer simulations. In order to allow for this, we introduce a minimal pseudo-count [24] of $P_k = \frac{1}{|\mathscr{A}_k|}$.

Global Transition Matrix. Here $C_{i,j,l}(V)$ is the count tensor of transitions between i and j in matrix $\mathscr{T}(V')$. It contains all $T_{i,j}(V')$ a specific vertex is immersed in and due to this it contains all possible information of local markovian dynamcis. $C_{i,j,l}(V)$ is normalized by the count of all transitions of i in all matrices $S_{j,l} = \sum_i C_{i,j,l}(V)$. The global weighting parameter $w_l = \frac{S_{jl}}{\sum_l S_{jl}}$ can be derived by taking all transitions $\mathscr{A}_k$ into account with respect to their probability. For a given set of l transition matrices $T_{i,j}(V')$ we can combine them into a global model:

$$T_{gi,j}(V) = \sum_l \frac{S_{jl}}{\sum_l S_{jl}} \cdot C_{i,j,l}(V). \tag{2}$$

Stationary Distribution and Entropy. As T_g (Eq. 2) is a row stochastic matrix we can compute its dominant eigenvector from a spectral decomposition. It represents a basic quantity of interest: the stationary probability $\boldsymbol{\pi} := (\pi_1, \ldots, \pi_i, \ldots)$ of micro-states i [24]. To this end we used the `markovchain` library in R [27, 29]. For measuring the changes in conformational entropy $H := -\sum_{i=1}^{|\mathscr{A}_k|} \pi_i \cdot \log \pi_i$ upon binding a ligand, we define $\Delta H = H_{wt} - H_{complex}$, form a stationary distribution.

3   Algorithm

StreAM and StreAM$_B$. We compute the adjacency id $id(V')$ for vertices $V' \subseteq V$ in the dynamic graph G_t using the stream-based algorithm *StreAM*, as described in Algorithm 1. Here, $id(V') \in [0, |\mathscr{A}_{|V'|}|]$ is the unique identifier of the adjacency matrix of the subgraph $G^{V'}$. Each change to G_t consists of the edge $\{a, b\}$ and a type to mark it as addition or removal (abbreviated to *add,rem*). In addition to edge and type, *StreAM* takes as input the ordered list of vertices V' and their current adjacency id.

An edge $\{a, b\}$ is only processed by *StreAM* in case both a and b are contained in V'. Otherwise, its addition or removal has clearly no impact on $id(V')$.

Assume $pos(V', a), pos(V', b) \in [1, k]$ to be the positions of vertices a and b in V'. Then, $i = min(pos(V', a), pos(V', b))$ and $j = max(pos(V', a), pos(V', b))$ are the row and column of adjacency matrix $A(G^{V'})$ that represent the edge $\{a, b\}$. In the bit representation of its adjacency id $id(V')$, this edge is represented by the bit $(i-1) \cdot k + j - i \cdot (i+1)/2$. When interpreting this bit representation as a number, an addition or removal of the respective edge corresponds to the addition or subtraction of $2^{k \cdot (k-1)/2 - ((i-1) \cdot k + j - i \cdot (i+1)/2)}$. This operation is performed to update $id(V')$ for each edge removal or addition. In the following, we refer to this position as $e(a, b, V') := \frac{|V'| \cdot (|V'| - 1)}{2} - ((i - 1) \cdot |V'| + j - \frac{i \cdot (i+1)}{2})$.

Data: V', id, $\{a, b\}$, $type \in \{add, rem\}$
begin
   if $a \in V' \wedge b \in V'$;   /* process only relevant edges */
   then
      if $type == add$ **then**
         | $A := A + 2^{e(a,b,V')}$;   /* set corresponding bit to 1 */
      else
         | $A := A - 2^{e(a,b,V')}$;   /* set corresponding bit to 0 */
      end
   end
   return id ;
end

Algorithm 1. *StreAM*: stream-based computation of the adjacency id

---

Furthermore, in Algorithm 2 we show *StreAM$_B$* for the batch-based computation of the adjacency id for vertices V'

Data: V', id_{t-1}, E_t^+, E_t^-
begin
    $id_t(V') := id_{t-1}(V')$;   /* init id with previous one */
    for *all* $\{a,b\} \in E_t^+$ **do**
      | $id_t := StreAM(V', id_t, \{a,b\}, add)$;   /* process addition */
    end
    for *all* $\{a,b\} \in E_t^-$ **do**
      | $id_t := StreAM(V', id_t, \{a,b\}, rem)$;   /* process removal */
    end
    return id_t ;
end

Algorithm 2. *StreAM$_B$*: batch-based computation of the adjacency id

StreAM-T$_g$. We present *StreAM-T$_g$*, an algorithm for the computation of global transition matrices, one particular vertex is participating in, given in Algorithm 3. A full computation with *StreAM-T$_g$* can be divided into the following steps. The first step is the computation of all possible Markov models with *StreAM* from all $\binom{|V|}{k} \cdot k! = \frac{|V|!}{(|V|-k)!}$ combinations, where k is the adjacency size and $|V|$ the number of vertices of G_t. Afterwards, *StreAM-T$_g$* sorts the matrices by vertex id into different sets, each with the size of $\binom{|V|}{k-1} \cdot (k-1)!$. For each vertex, *StreAM-T$_g$* computes a global count tensor C which is normalized by the global distribution of transition states a vertex is immersing in, taking the whole ensemble into account.

Data: T, a
begin
    $C(V) := \{V' \in \mathbb{P}(V) : |V'| = k, a \in V'\}$;   /* C vertex a immersed in */
    $T_g(a) := 0_{|\mathscr{A}_k|,|\mathscr{A}_k|}$;   /* initialize $T_g(a)$ */
    for *all* $V' \in C(V)$ **do**
      | $T_g(a) := T_g(a) + \frac{S(V')}{\sum_{V'' \in C(V)} |S(V'')|} \cdot T(V')$;   /* sum up $T_g(a)$ */
    end
    return $T_g(a)$
end

Algorithm 3. *StreAM-T$_g$(a)* for computing the global transition matrix $T_g(a)$

4 Evaluation

4.1 Objectives

As *StreAM-T$_g$* is intended to analyze large MD trajectories we first measured the speed of *StreAM* for computing a single $\mathscr{T}(V')$ to estimate overall computational resources. With this in mind, we benchmark different G_t with increasing

Table 1. Details of the dynamic graphs obtained from MD simulation trajectories. $|V|$ is the number of vertices, $|E|$ the number of edges and δ_t is the average batch size of a simulation. We convert simulations to unit sphere dynamic graphs with $d \in [10, 15]$Å.

	10 Å	11 Å	12 Å	13 Å	14 Å	15 Å	Rand$_{g1}$	Rand$_{g2}$	Rand$_{g3}$
$|V|$	65	65	65	65	65	65	500	500	500
$|E|$	94	129	189	241	298	353	500	1000	1200
δ_{avg}	6.1	15.6	19.4	18	19.6	23.8	80	100	120

adjacency size k (Table 1). Furthermore, we need to quantify the dependence of computational speed with respect to δ_t. Note, δ_t represents changes in conformations within G_t. For the full computation of T_g, we want to measure computing time in order to benchmark $StreAM\text{-}T_g$ by increasing network size $|V|$ and k for a given system due to exponentially increasing matrix dimesnions $|\mathscr{A}_k| = 2^{\frac{k \cdot (k-1)}{2}}$. We expect due to combinatorial complexity of matrix computation a linear relation between $|V|$ and speed and an exponential relation between increasing k and speed. For the last part, we want to compare Markovian dynamics between both simulations and discuss it with experimental data. We discuss the details in Sects. 4.2 and 4.3. Furthermore, we want to illustrate the biological relevance by applying it to a riboswitch design problem; this is shown in detail in Sect. 4.3.

4.2   Evaluation Setup

All benchmarks were performed on a machine with four *Intel(R) Xeon(R) CPU E5-2687W v2* processors with 3.4 GHz running a Debian operating system. We implemented *StreAM* in Java; all sources are available in a GitHub repository[2]. The final implementation *StreAM-T_g* is integrated in a `Julia` repository[3]. We created plots using the `AssayToolbox` library for R [6,29]. We generate all random graphs using a generator for dynamic graphs[4].

Run-Time Dependencies of StreAM on Adjacency Size. For every dynamic graph $G_t(V, E_t)$, we selected a total number of 100,000 snapshots to measure *StreAM* run-time performance. In order to perform benchmarks with increasing k, we chose randomly nodes $k \in [3, 10]$ and repeated this 500 times for different numbers of snapshots (every 10,000 steps). We determined the slope (speed $\frac{frames}{ms}$) of compute time vs. k for random and MD graphs with different parameters (Table 1).

Run-Time Dependence of StreAM on Batch Size. We measured run-time performance of *StreAM* for the computation of a set of all transitions $\mathscr{T}(V')$

[2] https://github.com/BenjaminSchiller/Stream.

[3] http://www.cbs.tu-darmstadt.de/streAM-Tg.tar.gz.

[4] https://github.com/BenjaminSchiller/DNA.datasets.

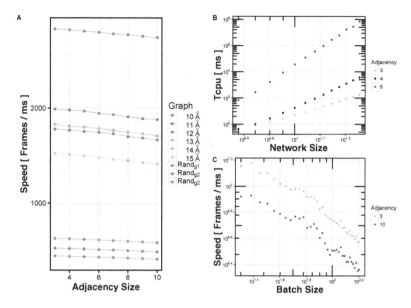

Fig. 3. Run-time performance of *StreAM-T$_g$*. **A**: Speed of computing a set of $\mathscr{T}(V')$ using *StreAM*. **B**: Performance of T_g full computation with increasing network size $|V|$ and different adjacency sizes $k = 3, 4, 5$. **C**: Speed of *StreAM* with increasing batch size for $k = 3, 10$. (Color figure online)

with different adjacency sizes k as well as dynamic networks with increasing batch sizes. To test *StreAM* batch size dependencies, 35 random graphs were drawn with increasing batch size and constant numbers of vertex and edges. All graphs contained 100,000 snapshots and k is calculated from 500 random combinations of vertices.

Run-Time Dependencies of StreAM-T_g on Network Size. We benchmarked the full computation of T_g with different $k \in [3, 5]$ for increasing network sizes $|V|$. Therefore we performed a full computation with *StreAM*. *StreAM-T_g* sorts the obtained transition list, converts them into transition matrices and combines them into a global Markov model for each vertex.

4.3 Run-Time Evaluation

Figure 3B shows computational speeds for each dynamic graph. Speed decreases linearly with a small slope (Fig. 3A). While this is encouraging the computation of transition matrices for $k > 5$ is still prohibitively expensive due to the exponential increase of the matrix dimensions with $2^{\frac{k \cdot (k-1)}{2}}$. For G_t obtained from MD simulations, we observe fast speeds due to small batch sizes (Table 1).

Figure 3B reveals that T_{cpu} increases linearly with increasing $|V|$ and with k exponentially. We restrict the T_g full computation to $k < 5$. In Fig. 3C, speed decreases linearly with δ_t. As δ_t represents the changes between snapshots our

observation has implications for the choice of MD integration step lengths as
well as trajectory granularity.

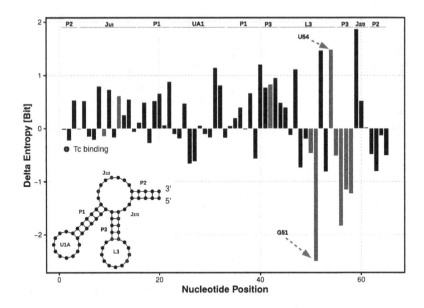

Fig. 4. ΔH for T_g of the native riboswitch and the one in complex with tetracycline
(TC). Nucleotides with TC in complex are colored in red. At the top we annotate the
nucleotides with secondary structure information. (Color figure online)

Application to Molecular Synthetic Biology. For both simulations of
Sect. 2.3, we computed 17,039,360 transition matrices and combined them into
65 global models (one for each vertex of the riboswitch, Eq. 2). To account for
both the pair-interactions and potential stacking effects we focus on $k = 4$-vertex
adjacencies and use dynamic RNA graphs with $d = 13$ Å. One global transition
matrix contains all the transitions a single nucleotide participates in. The stationary distribution and the implied entropy (changes) help to understand the
effects of ligand binding and potential improvements on this (the design problem
at hand). The ΔH obtained are shown in Fig. 4.

A positive value of ΔH in Fig. 4 indicates a loss of conformational entropy
upon ligand binding. Interestingly, the binding loop as well as complexing
nucleotides gain entropy. This is due to the fact of rearrangements between
the nucleotides in spatial proximity to the ligand because 70 % of the accessible
surface area of TC is buried within the binding pocket L3 [32]. Experiments con-
firmed that local rearrangement of the binding pocket are necessary to prevent
a possible release of the ligand [20]. Furthermore crystallographic studies have
revealed that the largest changes occur in L3 upon TC binding [32].

Furthermore, we observe the highest entropy difference for nucleotide G51. Experimental data reveals that G51 crosslinks to tetracycline when the complex is subjected to UV irradiation [4]. These findings suggest a strong interaction with TC and thus a dramatic, positive change in ΔH.

Nucleotides A52 and U54 show a positive entropy difference inside L3. Interestingly, molecular probing experiments show that G51, A52, and U54 of L3 are – in the absence of the antibiotic – the most modified nucleotides [25,32]. Clearly, they change their conformational flexibility upon ligand binding due they direct interaction with the solvent. U54 further interacts with A51, A52, A53 and A55 building the core of the riboswitch [32]. Taken together, these observations reveal that U54 is necessary for the stabilization of L3. A more flexible dynamics (ΔH) will change the configuration of the binding pocket and promotes TC release.

5 Summary, Conclusion, and Future Work

In this study, we demonstrate that *StreAM-T$_g$* fulfills our demands for a method to extract the coarse-grained Markovian dynamics of motifs of a complex RNA molecule. The effects observed in a designable riboswitch could be related to known experimental facts, such as conformational altering caused by ligand binding. Hence *StreAM-T$_g$* derived Markov models in an abstract space of motif creation and destruction. This allows for the efficient analysis of large MD trajectories. Thus we hope to elucidate molecular relaxation timescales, spectral analysis in relation to single-molecule studies, as well as transition path theory in the future. At present, we use it for the design of switchable synthetic RNA based circuits in living cells [7,8].

To broaden the application areas of *StreAM-T$_g$* we will extend it to proteins as well as evolutionary graphs mimicking the dynamics of molecular evolution in sequence space [16].

Acknowledgements. The Authors gratefully acknowledge financial support by the LOEWE project CompuGene of the Hessen State Ministry of Higher Education, Research and the Arts. Parts of this work have also been supported by the DFG, through the Cluster of Excellence cfaed as well as the CRC HAEC.

References

1. Alder, B.J., Wainwright, T.E.: Studies in molecular dynamics. J. Chem. Phys. **31**, 459–466 (1959)
2. Aleksandrov, A., Simonson, T.: Molecular mechanics models for tetracycline analogs. J. Comp. Chem. **30**(2), 243–255 (2009)
3. Andronescu, M., Condon, A., Hoos, H.H., Mathews, D.H., Murphy, K.P.: Computational approaches for RNA energy parameter estimation. RNA **16**(12), 2304–2318 (2010)
4. Berens, C., Thain, A., Schroeder, R.: A tetracycline-binding RNA aptamer. Bioorg. Med. Chem. **9**(10), 2549–2556 (2001)

5. Brooks, B.R., Bruccoleri, R.E., Olafson, B.D., States, D.J., Swaminathan, S., Karplus, M.: Charmm: a program for macromolecular energy, minimization, and dynamics calculations. J. Comp. Chem. **4**(2), 187–217 (1983)
6. Buß, O., Jager, S., Dold, S.-M., Zimmermann, S., Hamacher, K., Schmitz, K., Rudat, J.: Statistical evaluation of HTS assays for enzymatic hydrolysis of β-keto esters. PloS One **11**(1), e0146104 (2016). doi:10.1371/journal.pone.0146104
7. Cameron, D.E., Bashor, C.J., Collins, J.J.: A brief history of synthetic biology. Nat. Rev. Microbiol. **12**(5), 381–390 (2014)
8. Carothers, J.M., Goler, J., Juminaga, D., Keasling, J.D.: Model-driven engineering of RNA devices to quantitatively program gene expression. Science **334**(6063), 1716–1719 (2011)
9. Chodera, J.D., Noé, F.: Markov state models of biomolecular conformational dynamics. Curr. Opin. Struct. Biol. **25**, 135–144 (2014)
10. Deigan, K.E., Li, T.W., Mathews, D.H., Weeks, K.M.: Accurate SHAPE-directed RNA structure determination. PNAS **106**(1), 97–102 (2009)
11. Gan, H.H., Pasquali, S., Schlick, T.: Exploring the repertoire of RNA secondary motifs using graph theory; implications for RNA design. Nuc. Acids Res. **31**(11), 2926–2943 (2003)
12. Hamacher, K., Trylska, J., McCammon, J.A.: Dependency map of proteins in the small ribosomal subunit. PLoS Comput. Biol. **2**(2), 1–8 (2006)
13. Cheatham III, T.E.: Simulation and modeling of nucleic acid structure, dynamics and interactions. Curr. Opin. Struct. Biol. **14**(3), 360–367 (2004)
14. Jonikas, M.A., Radmer, R.J., Laederach, A., Das, R., Pearlman, S., Herschlag, D., Altman, R.B.: Coarse-grained modeling of large RNA molecules with knowledge-based potentials and structural filters. RNA **15**(2), 189–99 (2009)
15. Laing, C., Schlick, T.: Computational approaches to RNA structure prediction, analysis, and design. Curr. Opin. Struct. Biol. **21**(3), 306–318 (2011)
16. Lenz, O., Keul, F., Bremm, S., Hamacher, K., von Landesberger, T.: Visual analysis of patterns in multiple amino acid mutation graphs. In: 2014 IEEE Conference on Visual Analytics Science and Technology (VAST), pp. 93–102 (2014)
17. Manzourolajdad, A., Arnold, J.: Secondary structural entropy in RNA switch (Riboswitch) identification. BMC Bioinform. **16**(1), 133 (2015)
18. Parisien, M., Major, F.: The MC-Fold and MC-Sym pipeline infers RNA structure from sequence data. Nature **452**(7183), 51–55 (2008)
19. Pronk, S., Páll, S., Schulz, R., Larsson, P., Bjelkmar, P., Apostolov, R., Shirts, M.R., Smith, J.C., Kasson, P.M., van der Spoel, D., Hess, B., Lindahl, E.: Gromacs 4.5: a high-throughput and highly parallel open source molecular simulation toolkit. Bioinformatics **29**(7), 845–854 (2013)
20. Reuss, A., Vogel, M., Weigand, J., Suess, B., Wachtveitl, J.: Tetracycline determines the conformation of its aptamer at physiological magnesium concentrations. Biophys. J. **107**(12), 2962–2971 (2014)
21. Schiller, B., Jager, S., Hamacher, K., Strufe, T.: StreaM - a stream-based algorithm for counting motifs in dynamic graphs. In: Dediu, A.-H., Hernández-Quiroz, F., Martín-Vide, C., Rosenblueth, D.A. (eds.) AlCoB 2015. LNCS, vol. 9199, pp. 53–67. Springer, Heidelberg (2015). doi:10.1007/978-3-319-21233-3_5
22. Schlick, T.: Mathematical and biological scientists assess the state of the art in RNA science at an IMA Workshop, RNA in biology, bioengineering, and biotechnology. Int. J. Multiscale Comput. Eng. **8**(4), 369–378 (2010)
23. Schrödinger, L.L.C.: The PyMOL molecular graphics system, version 1.8, November 2015

24. Senne, M., Trendelkamp-schroer, B., Noe, F.: EMMA: A software package for Markov model building and analysis. J. Chem. Theory Comput. **8**(7), 2223–2238 (2012)
25. Hanson, S., Gesine Bauer, B.F., Suess, B.: Molecular analysis of a synthetic tetracycline-binding riboswitch. RNA **11**, 2549–2556 (2005)
26. Shapiro, B.A., Yingling, Y.G., Kasprzak, W., Bindewald, E.: Bridging the gap in RNA structure prediction. Curr. Opin. Struct. Biol. **17**(2), 157–165 (2007)
27. Spedicato, G.A.: Markovchain: discrete time Markov chains made easy (2015), R package version 0.4.3
28. Stombaugh, J., Zirbel, C.L., Westhof, E., Leontis, N.B.: Frequency and isostericity of RNA base pairs. Nucleic Acids Res. **37**(7), 2294–2312 (2009)
29. Team, R.D.C.: R: A Language and Environment for Statistical Computing. R Foundation for Statistical Computing, Vienna, Austria (2008)
30. Tung, C.S.: RNA Structural Motifs. Life Sciences, pp. 1–4 (2002)
31. Wunnicke, D., Strohbach, D., Weigand, J.E., Appel, B., Feresin, E., Suess, B., Muller, S., Steinhoff, H.J.: Ligand-induced conformational capture of a synthetic tetracycline riboswitch revealed by pulse EPR. RNA **17**(1), 182–188 (2011)
32. Xiao, H., Edwards, T.E., Ferré-D'Amaré, A.R.: Structural basis for specific, high-affinity tetracycline binding by an in vitro evolved aptamer and artificial riboswitch. Chem. Biol. **15**(10), 1125–1137 (2008)

Solving Generalized Maximum-Weight Connected Subgraph Problem for Network Enrichment Analysis

Alexander A. Loboda[1], Maxim N. Artyomov[2],
and Alexey A. Sergushichev[1]([✉])

[1] Computer Technologies Department, ITMO University,
Saint Petersburg 197101, Russia
{loboda,alserg}@rain.ifmo.ru
[2] Department of Pathology and Immunology,
Washington University in St. Louis, St. Louis, MO, USA
martyomov@pathology.wustl.edu

Abstract. Network enrichment analysis methods allow to identify active modules without being biased towards *a priori* defined pathways. One of mathematical formulations of such analysis is a reduction to a maximum-weight connected subgraph problem. In particular, in analysis of metabolic networks a generalized maximum-weight connected subgraph (GMWCS) problem, where both nodes and edges are scored, naturally arises. Here we present the first to our knowledge practical exact GMWCS solver. We have tested it on real-world instances and compared to similar solvers. First, the results show that on node-weighted instances GMWCS solver has a similar performance to the best solver for that problem. Second, GMWCS solver is faster compared to the closest analogue when run on GMWCS instances with edge weights.

Keywords: Network enrichment · Maximum weight connected subgraph problem · Exact solver · Mixed integer programming

1 Introduction

Gene set enrichment methods are widely used for the analysis of untargeted biological data such as transcriptomic, proteomic or metabolomic profiles. These methods allow to identify molecular pathways, in a form of gene sets, that have non-random group behaviour in the data. Determining such overenriched pathways provides insights into the data and allows to better understand the considered system.

Network enrichment methods, in opposite to gene set enrichment, do not rely on the predefined gene sets and, thus, allow to identify novel pathways. These methods use network of interacting entities, such as genes, proteins, metabolites, etc. and try to identify the most regulated subnetwork. There are different mathematical formulations of the network enrichment problem, but many of them are NP-hard [1,6,9].

© Springer International Publishing Switzerland 2016
M. Frith and C.N.S. Pedersen (Eds.): WABI 2016, LNBI 9838, pp. 210–221, 2016.
DOI: 10.1007/978-3-319-43681-4_17

Dittrich et al. in [6] suggested a formulation as a maximum-weight connected subgraph (MWCS) problem. Originally, the authors considered node-weighted graph, such that positive weight corresponded to "interesting" nodes and negative weight corresponded to "non-interesting" nodes. The goal was to find a connected graph with the maximal sum of weights of its nodes, which corresponded to an "active module".

Here we consider a slightly different form of MWCS: generalized MWCS (GMWCS), which naturally arises in the studies of metabolic networks [4,11]. In such networks nodes in the graph represent metabolites and edges represent their interconversions via reactions. Compared to MWCS, GMWCS has edges also weighted: the nodes can be scored using metabolomic profiles and the edges can be scored using gene or protein expression profiles.

In recent years, a huge role of metabolic regulation became more and more recognised, especially in a context of immune system [10] and cancer [5]. This warrants the development of effective computational approaches for studying it, such as metabolic network enrichment. The method results in a subnetwork of connected reactions which are hypothesized to be the most important in the considered process. Using such subnetwork one can get a better understanding of the corresponding metabolic regulation and, for example, to infer its critical points [13].

In this paper we describe an exact solver for the node-and-edge-weighted GMWCS problem. First, in Sect. 2 we give formal definitions. Then in Sect. 3 we describe preprocessing steps adapted for the edge-based formulation. In Sect. 4 we show how the instance can be split into three smaller instances. Section 5 is dedicated to a mixed-integer programming (MIP) formulation of the problem. In Sect. 6 we show experimental results of running the solver on real-world instances that appear in GAM web-service and show that it is faster and more accurate than *Heinz* [3] on edge-weighted instances and is similar in performance to *Heinz2* [7] on node-weighted instances.

2   Formal Definitions

Here we consider the Maximum-Weight Connected Subgraph (MWCS) problem for which there are two slightly different formulations. In the most commonly used definition of MWCS only nodes are weighted [2,7]. In this paper we consider problem where edges are weighted too [8]. To remove the ambiguity we call the former problem Simple MWCS (SMWCS) and the latter one Generalized MWCS (GMWCS).

The goal of MWCS problems is to find in a given graph a connected subgraph with the maximal the maximal sum of weights. As a subgraph is connected we can consider connected components of the graph independently. Thus, below we assume that the input graph is connected.

First, we give definition of a Simple Maximum-Weight Connected Subgraph problem.

Definition 1. *Given a connected undirected graph $G = (V, E)$ and weight function $\omega_v : V \to \mathbb{R}$, the Simple Maximum-Weight Connected Subgraph (SMWCS) problem is the problem of finding a connected subgraph $\widetilde{G} = (\widetilde{V}, \widetilde{E})$ with the maximal total weight*

$$\Omega(\widetilde{G}) = \sum_{v \in \widetilde{V}} \omega(v) \to max$$

Second, we define generalized variant of this problem, where both nodes and edges could be weighted.

Definition 2. *Given a connected undirected graph $G = (V, E)$ and a weight function $\omega : (V \cup E) \to \mathbb{R}$, the Generalized Maximum-Weight Connected Subgraph (GMWCS) problem is the problem of finding a connected subgraph $\widetilde{G} = (\widetilde{V}, \widetilde{E})$ with the maximal total weight*

$$\Omega(\widetilde{G}) = \sum_{v \in \widetilde{V}} \omega(v) + \sum_{e \in \widetilde{E}} \omega(e) \to max$$

Now we define a rooted variant of the problem with one of the vertices forced to in a solution. It is used as an auxiliary subproblem of GMWCS.

Definition 3. *Given a connected undirected graph $G = (V, E)$, a weight function $\omega : (V \cup E) \to \mathbb{R}$ and a root node $r \in V$ the Rooted Generalized Maximum-Weight Connected Subgraph (R-GMWCS) problem is the problem of finding a connected subgraph $\widetilde{G} = (\widetilde{V}, \widetilde{E})$ such that $r \in \widetilde{V}$ and*

$$\Omega(\widetilde{G}) = \sum_{v \in \widetilde{V}} \omega(v) + \sum_{e \in \widetilde{E}} \omega(e) \to max$$

El-Kebir and Klau in [7] have shown that MWCS problem is NP-hard. Since MWCS is a special case of GMWCS then GMWCS is also NP-hard. R-GMWCS problem is NP-hard too because any instance of GMWCS problem can be solved by solving an R-GMWCS instance for each node as a root.

Finally, below we use n as a shorthand for the number of nodes $|V|$ and m for the number of edges $|E|$ in the graph G.

3   Preprocessing

We introduce two preprocessing rules adapted from [7] that simplify the problem. These rules make a new graph with a smaller number of vertices and edges in such a way that the GMWCS solution for the original graph can be easily recovered from the GMWCS solution for the simplified graph.

First, we merge groups of close vertices that either none or all of them are in the optimal solution (Fig. 1A). Let $e = (u, v)$ be an edge with $\omega(e) \geq 0$ with simultaneously $\omega(e) + \omega(v) \geq 0$ and $\omega(e) + \omega(u) \geq 0$. In this case if one of the vertices is included in the solution then the edge and the other vertex can also be included without decreasing the total weight. Thus, we can contract edge e into a new vertex w with a weight $\omega(w) = \omega(e) + \omega(u) + \omega(v)$. After the contraction parallel edges between w and some vertex t could appear. In that case we merge all non-negative one into a single edge with weight of the sum of their weights. After that, we remove all edges between w and t except one with the maximal weight. To exhaustively apply the rule in $O(m + kn)$ time, where k is a number of contracted edges, we can use Algorithm 1.

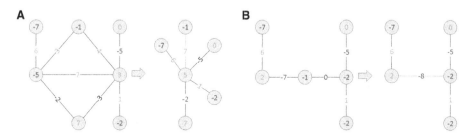

Fig. 1. Applying first rule that contract an edge (A) and second rule that replace negative chain by a single edge (B). New vertices and nodes painted yellow.

Algorithm 1. Edges contraction preprocessing

1: **procedure** CONTRACTEDGES(V, E)
2:     **for all** $e \in E$ **do**
3:         $(u, v) \leftarrow e$
4:         **if** $\omega(u) + \omega(e) < 0$ or $\omega(v) + \omega(e) < 0$ **then**
5:             $e \leftarrow null$
6:         **while** $e \neq null$ **do**
7:             $w \leftarrow contract(e)$
8:             $e \leftarrow null$
9:             **for all** $z \in \delta_w$ **do**
10:                 **if** $\exists$ parallel edges e_1, e_2 between w, z **then**
11:                     **if** $\omega(e_1) \geq 0$ and $\omega(e_2) \geq 0$ **then**
12:                         $merge(e_1, e_2)$
13:                     **else** $remove(\arg\min_{e' \in \{e_1, e_2\}} (\omega(e')))$
14:             **for all** $z \in \delta_w$ **do**
15:                 $e' \leftarrow (z, w)$
16:                 **if** $\omega(u) + \omega(e') \geq 0$ and $\omega(v) + \omega(e') \geq 0$ **then**
17:                     $e \leftarrow e'$

Second, similarly to the previous step, we merge nonpositive chains (Fig. 1B). Let v be a vertex with $deg(v) = 2$ with corresponding incident edges $e_1 = (u, v)$ and $e_2 = (v, w)$. If all three weights $\omega(v)$, $\omega(e_1)$ and $\omega(e_2)$ are nonpositive, then v, e_1 and e_2 could be replaced with a single edge $e = (u, w)$ with a weight $\omega(e) = \omega(v) + \omega(e_1) + \omega(e_2)$. Merging negative chains is implemented in a single pass by iteratively trying to apply the rule for all the nodes. This operation takes $\Theta(n)$ time.

4   Cut Vertex Decomposition

In this section we discuss how a GMWCS instance can be decompose into three smaller problems. The decomposition is based on the idea that biconnected components can be considered separately [7].

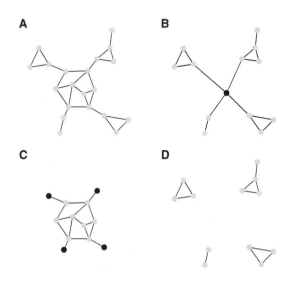

Fig. 2. Input graph and instances spawned by decomposition

Briefly, we have a GMWCS instance as input (Fig. 2A). First, we merge the largest biconnected component into a single vertex with zero weight and solve an R-GMWCS instance for this modified graph and the new vertex as a root (Fig. 2B). Then, we replace each of the components branching from the largest biconnected component by a single vertex with weight equal to the weight of corresponding subgraph in the R-GMWCS solution from the previous step (Fig. 2C). Last, we try to find a subgraph with a greater weight which fully lies in one of the branching components (Fig. 2D).

Formally, let B be a biconnected component of the graph G with the maximal number of vertices. Let C be a set of cut vertices of the graph G that are also contained in B. Let B_c be a component containing c in the graph $G \setminus (B \setminus C)$.

Proposition 1. *Let a subgraph $\widetilde{G}$ of G be an optimal solution of GMWCS for graph G and $\widetilde{G}_c$, $\forall c \in C$, are optimal solutions for R-GMWCS instances for graphs B_c with a root c. In this case, if $\widetilde{G}$ contains a vertex $c \in C$, then we can construct an optimal solution $\widetilde{G}'$ such that: (1) $\widetilde{G}' \cap B = \widetilde{G} \cap B$ and (2) $\widetilde{G}' \cap B_c = \widetilde{G}_c$.*

Proof. Let $\widetilde{B}_c = \widetilde{G} \cap B_c$. We prove that it can be replaced by $\widetilde{G}_c$ without loss of connectivity and optimality. First, $\widetilde{B}_c$ must be connected. Let it be disconnected. Then there is no path between c and some vertex v. Since $\widetilde{G}$ is connected then there is a simple path vc in G. However, by definition of cut vertex, path vc can not contain vertices from $G \setminus B_c$ and, thus it fully lies in B_c, a contradiction. Since $\widetilde{B}_c$ is connected and contains c then it cannot have weight greater than $\widetilde{G}_c$ by construction of $\widetilde{G}_c$.

Now we prove that the replacement keeps the graph connected. Repeating the reasoning from the previous step we can get that $\widetilde{G} \cap B$ must be connected. So, $\widetilde{G}_c$ is connected, $\widetilde{G} \cap B$ connected and both these graphs contain c. Thus, $\widetilde{G}'$ is also connected.   □

This proposition allows us to consider only optimal solutions that either include a vertex from B and in subgraphs B_c are identical to the corresponding R-GMWCS instance or fully lie in some of the subgraphs B_c.

First, for each $c \in C$ we want to know the best solution of the problem for the graph B_c containing vertex c. It is precisely an R-GMWCS instance. For practical reasons, it is better to spawn one instance at this step instead of $|C|$ instances. Let $G^* = \bigcup_{\forall c \in C} B_c$. Then we merge all vertices from C contained in G^* into a single vertex r with $\omega(r) = 0$ and solve R-GMWCS problem for such graph. Let S to be the solution of this instance. To get solution for the graph B_c we replace back r to c in S, and remove all the vertices which are not contained in B_c.

Second, we find best scored subgraph of G that do not lies fully in some of B_c. Let $\widetilde{G}_c$ be the solution of R-GMWCS for graph B_c with root c obtained on the previous step. We obtain a new GMWCS instance by considering the component B and for all $c \in C$ attaching a vertex v with weight $\omega(v) = \Omega(\widetilde{G}_c)$. We solve the resulting instance and then recover a solution for the original problem.

Last, we find all potential solutions that fully lie in B_c for all $c \in C$. For this purpose we spawn one instance for the graph $G^* = \bigcup_{\forall c \in C} B_c$. Clearly that if the solution of the problem for the graph G lies fully in some of B_c then we will find it at this step.

Decomposition of the graph into biconnected components takes $O(n + m)$ time, generating all the three instances also takes linear time, so overall time complexity at this step is $O(n + m)$.

5   Mixed Integer Programming Formulation

Here we describe a MIP formulation of the problem. The GMWCS can be represented as two parts: objective function (weight of the subgraph) that should be maximized and constraints that ensure that the subgraph is connected.

The objective function is linear and can be put into a MIP problem in a straightforward way. However, getting effective linear subgraph connectivity constraints is not trivial. In this section we describe how it can be done. The resulting MIP problem is solved by IBM ILOG CPLEX.

First, we consider a nonlinear formulation of the GMWCS problem, as proposed in [8]. Then, we show how to eliminate nonlinearity and get a linear system. Finally, we introduce extra symmetry-breaking and cuts, which do not impact on the correctness of the formulation, but improve the performance.

5.1 Subgraph Representation

We use one binary variable for each vertex or edge that represent the presence in the subgraph:

1. Binary variable y_v takes the value of 1 iff $v \in V$ belongs to the subgraph.
2. Binary variable w_e takes the value of 1 iff $e \in E$ belongs to the subgraph.

For these variables to be representing a valid subgraph (not necessarily connected) we need to introduce a set of constraints:

$$w_e \leq y_v, \qquad\qquad \forall v \in V, e \in \delta_v. \qquad (1)$$

These constraints state that an edge can be a part of the subgraph, only if both of its endpoints are a part of the subgraph.

5.2 Nonlinear Formulation

The nonlinear formulation of the subgraph connectivity constraints is based on the idea that any connected graph can be traversed from any of its vertices. The output of the traversal can be represented as an arborescence where an arc (v, u) denotes that v has been visited before u. Accordingly, we can ensure connectivity of a subgraph if we can provide an arborescence corresponding to the traversal of this subgraph.

For a given graph $G = (V, E)$, let $S = (V, A)$ be a directed graph, where A is obtained from E by replacing each undirected edge $e = (v, u)$ by two directed arcs (v, u) and (u, v).

Now, we are going to introduce variables that we will use in the formulation and show nonlinear system of constraints, that ensure connectivity of subgraph:

1. Binary variable x_a takes the value of 1 iff $a \in A$ belongs to the arborescence.
2. Binary variable r_v takes the value of 1 iff $v \in V$ is the root of the arborescence.
3. Continuous variable d_v takes the value of n if the path in the arborescence from the root to vertex v contains n vertices. If v does not belong to the solution then value can be arbitrary.

Then we introduce constraints that ensure the validity of an arborescence:

$$\sum_{v \in V} r_v = 1; \tag{2}$$

$$1 \leq d_v \leq n, \qquad \forall v \in V; \tag{3}$$

$$\sum_{(u,v) \in A} x_{uv} + r_v = y_v, \qquad \forall v \in V; \tag{4}$$

$$x_{vu} + x_{uv} \leq w_e, \qquad \forall e = (v, u) \in E; \tag{5}$$

$$d_v r_v = r_v, \qquad \forall v \in V; \tag{6}$$

$$d_u x_{vu} = (d_v + 1)x_{vu}, \qquad \forall (v, u) \in A. \tag{7}$$

Inequality (2) states that there is only one root in the arborescence; (3) is a limitation on the distance between any vertex and the root; (4) states that if a vertex is a part of the subgraph then either it is a root of the arborescence or $deg_{in}(v) = 1$; (5) says that an arc of the arborescence can be in the solution only if the corresponding edge is also in it. Last two inequalities (6) and (7) control correct distances in the arborescence.

Haouari et al. have shown in [8] that this nonlinear system is a correct formulation of GMWCS. That is, the arborescence covers all vertices of the resulting subgraph and the solution can induce this arborescence.

However, inequalities (6) and (7) are not linear and should be replaced, so that the formulation can be represented as a MIP problem.

5.3 Linearization

Nonlinear equations (6) and (7) can be replaced with the following system of linear inequalities:

$$d_v + nr_v \leq n, \qquad \forall v \in V; \tag{8}$$

$$n + d_u - d_v \geq (n + 1)x_{vu}, \qquad \forall (v, u) \in A; \tag{9}$$

$$n + d_v - d_u \geq (n - 1)x_{vu}, \qquad \forall (v, u) \in A. \tag{10}$$

Proposition 2. *Every feasible solution to* (1)–(7) *is also feasible to* (1)–(5), (8)–(10) *and vice versa.*

Proof. First, we prove that (8) is equivalent to (6) in a sense of feasibility of the solution. Since r_v is a binary variable, we can consider two cases. Suppose that $r_v = 1$, then (6) will take the form $d_v = 1$ while (8) will take the from $d_v \leq 1$, and with (3) we have $d_v = 1$. Now suppose that $r_v = 0$, (6) will look $0 = 0$, it means that in this case there is no additional restrictions on variables and (8) will take the form $d_v \leq n$, but system already have such inequality. Thus (6) and (8) are equivalent for both possible values of r_v.

At the second part of the proof we will use the same approach. Here we prove that (7) can be represented as linear inequalities (9) and (10).

1. Let $x_{vu} = 1$. Then after substitution into (7) we have $d_u = d_v + 1$. Then we substitute x_{vu} into (9) and (10)

$$n + d_u - d_v \geq n + 1$$
$$n + d_v - d_u \geq n - 1$$

or, equivalently,

$$d_u \geq d_v + 1$$
$$d_v + 1 \geq d_u$$

or $d_u = d_v + 1$.

2. Let $x_{vu} = 0$. The original nonlinear equation will take the form $0 = 0$. As mentioned above, it means that there is no additional restrictions on variables. We have to show that (9) and (10) also do not add such restrictions. After substitution these inequalities take the form:

$$n + d_u - d_v \geq 0$$
$$n + d_v - d_u \geq 0$$

or $|d_v - d_u| \leq n$. Obviously, variables that hold (3) automatically hold such inequality. Thus, additional restrictions have not be added.   □

5.4 Symmetry-Breaking

It is a common practice to decrease the number of feasible solutions by limiting the number of different but logically equivalent feasible solutions. Such solutions are called symmetric. In our formulation constraints (1)–(5), (8)–(10) allow any arborescence of the graph to show its connectivity. So, in this section we show how to decrease the number of feasible arborescences and thus decrease the search space.

Root Order Rule. First of all, for the unrooted GMWCS problem we force the arborescence root to be a vertex with the maximal weight among present in the subgraph. Corresponding constraint that is added in the MIP instance is:

$$\sum_{v \prec u} r_v \leq 1 - y_u, \qquad\qquad \forall u \in V, \qquad (11)$$

where $v \prec u$ if $\omega(v) < \omega(u)$ or if weights are equal, we use some fixed linear order on vertices.

For the R-GMWCS we set root of the arborescence to be the same as the instance root.

Restricting Traversal. Moreover, connected graph can be traversed from the same vertex in different ways. Similarly to [12], we show how to make infeasible such solutions that could not be reached by a breadth-first search (BFS).

To achieve such form of the arborescence we add constraints:

$$d_v - d_u \leq n - (n-1)w_e, \qquad \forall e = (v,u) \in E; \qquad (12)$$
$$d_u - d_v \leq n - (n-1)w_e, \qquad \forall e = (v,u) \in E. \qquad (13)$$

These constraints state that if an edge e is present the subgraph then the distances to endpoints differ by one.

Proposition 3. *For any connected subgraph G_s of the graph G there exists a solution $(\overline{r}, \overline{y}, \overline{w}, \overline{x}, \overline{d})$ that encodes subgraph G_s and is feasible to (1)–(5), (8)–(10) and (11)–(13).*

Proof. First, for any subgraph G_s we can select any of its vertices, in particular one with the maximal weight, and make a BFS traversal starting from that vertex. As was shown above for any connected subgraph G_s and any its arborescence there is a corresponding encoding $(\overline{r}, \overline{y}, \overline{w}, \overline{x}, \overline{d})$ that satisfy constraints (1)–(5) and (8)–(10). By selection of the vertex with the maximal weight as an arborescence root constraint (11) holds. Constraints (12)–(13) also hold as they directly follow from the BFS ordering.   □

6   Experimental Results

As a testing dataset we used 101 instance generated by Shiny GAM, a web-service for integrated transcriptional and metabolic network analysis [11], based on user-submitted data during its testing phase. In the dataset, there are 38 instances of node-weighted SMWCS and 63 instances of GMWCS. Archive with instances is available at http://genome.ifmo.ru/files/papers_files/WABI2016/ gmwcs/instances.tar.gz. Briefly, node-weighted instances contain about 2200 nodes and 2500 edges and correspond to a network with nodes for both metabolites and reactions which are connected if the metabolite is a substrate or a product of the reaction. Edge-weighted instances contain about 700 nodes and 900 edges. Metabolites and reactions are scored proportionally to logarithm of corresponding differential expression p-values.

For the comparison we selected two other solvers: *Heinz* version 1.68 [6] and *Heinz2* version 2.1 [7]. The first one, *Heinz*, was initially developed for node-weighted SMWCS, but later was adjusted to account for edge weights, however, only acyclic solutions are considered. The second one, *Heinz2*, does not accept edge weights, but works faster than *Heinz* on node-weighted instances.

We ran each of the solver on each of the instances for 10 times with a time limit of 1000 s. *Heinz2* and our GMWCS solver were run using 4 threads. The processor was AMD Opteron 6380 2.5 GHz. A table with the results table are available at http://genome.ifmo.ru/files/papers_files/WABI2016/gmwcs/ results.final.tsv.

6.1 Results for Simple MWCS

The experiments have shown that on the node-weighted instances GMWCS solver has a performance similar to *Heinz2* (Fig. 3A). For 24 instances (63 %) GMWCS is slower than *Heinz2*. However, 32 instances (84 %) were solved by GMWCS within 30 s, compared to 27 (71 %) of *Heinz2*. Moreover, 4 instances were not solved by *Heinz2* in the allowed time of 1000 s compared to only 1 instance for GMWCS.

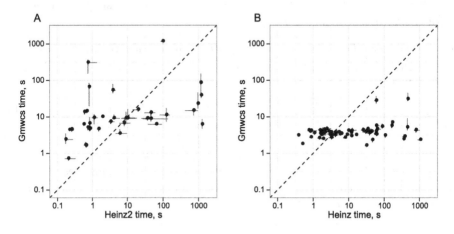

Fig. 3. Comparison of GMWCS with *Heinz2* and *Heinz* solvers on node-weighted (A) and node-and-edge-weighted (B) instances. The points represent median times of 10 runs on one instance. Horizontal and vertical grey lines represent the second minimal and the second maximal times. For convenience a small random noise was added to the median values of more then 950 s.

6.2 Results for Generalized MWCS

For the edge-weighted GMWCS instances GMWCS solver was able to find optimal solutions within 10 s all instances except two, while it took for *Heinz* more than 10 s to solve 30 of the instances (48 %) (Fig. 3B). Moreover, only 35 instances (56 %) had an acyclic solution, accordingly, 28 instances were not solved to GMWCS-optimality by *Heinz*.

7   Conclusion

Network analysis approaches are being actively developed for analyzing biological data. From the mathematical point of view this usually correspond to NP-hard problems. Here we described an exact practical solver for a particular formulation of generalized maximum weight connected subgraph problem that naturally arises in metabolic networks. We have tested the method on the real-world data and have shown that the developed solver is similar in performance to an existing solver *Heinz2* on a simple MWCS instances and works better and more accurately compared to *Heinz* on the edge-weighted instances. The implementation is freely available at https://github.com/ctlab/gmwcs-solver.

Funding. This work was supported by Government of Russian Federation [Grant 074-U01 to A.A.S., A.A.L.].

References

1. Alcaraz, N., Pauling, J., Batra, R., Barbosa, E., Junge, A., Christensen, A.G.L., Azevedo, V., Ditzel, H.J., Baumbach, J.: KeyPathwayMiner 4.0: condition-specific pathway analysis by combining multiple omics studies and networks with cytoscape. BMC Syst. Biol. **8**(1), 99 (2014)
2. Álvarez-Miranda, E., Ljubić, I., Mutzel, P.: The maximum weight connected subgraph problem. In: Jünger, M., Reinelt, G. (eds.) Festschrift for Martin Grötschel, pp. 245–270. Springer, Heidelberg (2013)
3. Beisser, D., Brunkhorst, S., Dandekar, T., Klau, G.W., Dittrich, M.T., Müller, T.: Robustness and accuracy of functional modules in integrated network analysis. Bioinformatics **28**(14), 1887–1894 (2012). (Oxford, England)
4. Beisser, D., et al.: Integrated pathway modules using time-course metabolic profiles and EST data from Milnesium tardigradum. BMC Syst. Biol. **6**, 72 (2012)
5. Cairns, R.A., Harris, I.S., Mak, T.W.: Regulation of cancer cell metabolism. Nat. Rev. Cancer **11**(2), 85–95 (2011)
6. Dittrich, M.T., Klau, G.W., Rosenwald, A., Dandekar, T., Müller, T.: Identifying functional modules in protein-protein interaction networks: an integrated exact approach. Bioinformatics **24**(13), i223–i231 (2008). (Oxford, England)
7. El-Kebir, M., Klau, G.W.: Solving the maximum-weight connected subgraph problem to optimality (2014). arXiv:1409.5308
8. Haouari, M., Maculan, N., Mrad, M.: Enhanced compact models for the connected subgraph problem and for the shortest path problem in digraphs with negative cycles. Comput. Oper. Res. **40**(10), 2485–2492 (2013)
9. Ideker, T., Ozier, O., Schwikowski, B., Siegel, A.F.: Discovering regulatory and signalling circuits in molecular interaction networks. Bioinformatics **18**(Suppl 1), S233–S240 (2002). (Oxford, England)
10. Mathis, D., Shoelson, S.E.: Immunometabolism: an emerging frontier. Nat. Rev. Immunol. **11**(2), 81 (2011)
11. Sergushichev, A., Loboda, A., Jha, A., Vincent, E., Driggers, E., Jones, R., Pearce, E., Artyomov, M.: GAM: a web-service for integrated transcriptional and metabolic network analysis. Nucleic Acids Res. (2016). http://nar.oxfordjournals.org/citmgr?type=bibtex&gca=nar%3Bgkw266v1
12. Ulyantsev, V., Zakirzyanov, I., Shalyto, A.: BFS-based symmetry breaking predicates for DFA identification. In: Dediu, A.-H., Formenti, E., Martín-Vide, C., Truthe, B. (eds.) LATA 2015. LNCS, vol. 8977, pp. 611–622. Springer, Heidelberg (2015)
13. Vincent, E.E., et al.: Mitochondrial phosphoenolpyruvate carboxykinase regulates metabolic adaptation and enables glucose-independent tumor growth. Mol. Cell **60**(2), 195–207 (2015)

A Natural Encoding of Genetic Variation in a Burrows-Wheeler Transform to Enable Mapping and Genome Inference

Sorina Maciuca, Carlos del Ojo Elias, Gil McVean, and Zamin Iqbal[✉]

Wellcome Trust Centre for Human Genetics, University of Oxford,
Roosevelt Drive, Oxford OX3 7BN, UK
{sorina.maciuca,gil.mcvean,zamin.iqbal}@well.ox.ac.uk,
carlos.delojoelias@ndm.ox.ac.uk

Abstract. We show how positional markers can be used to encode genetic variation within a Burrows-Wheeler Transform (BWT), and use this to construct a generalisation of the traditional "reference genome", incorporating known variation within a species. Our goal is to support the inference of the closest mosaic of previously known sequences to the genome(s) under analysis. Our scheme results in an increased alphabet size, and by using a wavelet tree encoding of the BWT we reduce the performance impact on rank operations. We give a specialised form of the backward search that allows variation-aware exact matching. We implement this, and demonstrate the cost of constructing an index of the whole human genome with 8 million genetic variants is 25 GB of RAM. We also show that inferring a closer reference can close large kilobase-scale coverage gaps in *P. falciparum*.

Keywords: Pan-genome · Burrows-Wheeler Transform · FM-index · Genome

1 Introduction

Genome sequencing involves breaking DNA into fragments, identifying substrings (called "reads"), and then inferring properties of the genome. Recently, it has become possible to study within-species genetic variation on a large scale [6,7], where the dominant approach is to match substrings to the canonical "reference genome" which is constructed from an arbitrary individual. This problem ("mapping") has been heavily studied (see [5]) and the Burrows-Wheeler Transform (BWT) [2] underlies the two dominant mappers [3,4]. Mapping reads to a reference genome is a very effective way of detecting genetic variation caused by single character changes (SNPs - single nucleotide polymorphisms). However, this method becomes less effective the further the genome differs from the reference. This is an important problem to address since, in many organisms, biologically relevant genomic regions are highly diverse.

© Springer International Publishing Switzerland 2016
M. Frith and C.N.S. Pedersen (Eds.): WABI 2016, LNBI 9838, pp. 222–233, 2016.
DOI: 10.1007/978-3-319-43681-4_18

For a given species, our goal is to build a compact representation of the genomes of N individuals, which we call a Population Reference Genome (PRG). This data structure facilitates the following inference: we take as input sequence data from a new sample, an estimate of how many genomes the sample contains and their relative proportions - e.g. a normal human sample would contain 2 genomes in a 1:1 ratio, a bacterial isolate would contain 1 genome and a malaria sample might contain 3 genomes in the ratio 10:3:1. We would then infer the sequence of the underlying genomes. In this paper we describe a method for encoding genetic variation designed to enable this approach.

Genomes evolve mainly via two processes - mutation (changing a few characters) and recombination (either two chromosomes exchange a chunk of DNA, or one chromosome copies a chunk from another). Thus once we have seen many genomes of a given species, a new genome is likely to look like a mosaic of genomes we have seen before. If we can infer a close mosaic, we have found a "personalised reference genome", and reads are more likely to match exactly. This approach was first described in [8], applied to the human MHC region. However their implementation was quite specific to the region and would not scale to the whole genome. Valenzuela *et al.* [1] have also espoused a find-the-closest-reference approach.

Other "reference graph" methods have been published [9–11], generally approaching just the alignment step. Siren *et al.* developed a method (GCSA [10]), with construction costs for a whole human genome (plus mutations) of more than 1 TB of RAM. Huang *et al.* [11] developed an FM-index [13] encoding of a reference genome-plus-variation ("BWBBLE") by extending the genetic alphabet to encode single-character variants with new characters and then concatenating padded indel variants to the end of the reference genome. We do something similar, but treat all variation in an equivalent manner, and retain knowledge of allelism naturally. While completing this paper, the preprint for GCSA2 was published [12], which drops RAM usage of human genome index construction to <100 GB at the cost of >1 TB of disk I/O.

We show below how to encode a set of genomes, or a reference plus genetic variation, in an FM-index which naturally distinguishes alternate alleles. We extend the well known BWT backward search and show how read-mapping can be performed in a way that allows reads to cross multiple variants, allowing recombination to occur naturally. Our data structure supports bidirectional search (which underlies the Super Maximal Exact Match algorithms of bwa-mem [3]), but currently we have only implemented exact matching. We use empirical datasets to demonstrate low construction cost (human genome) and the value of inferring a personalised reference in *P. falciparum*.

2   Background: Compressed Text Indexes

Burrows-Wheeler Transform. The Burrows-Wheeler Transform (BWT) of a string is a reversible permutation of its characters. The BWT of a string $T = t_1 t_2 \dots t_n$ is constructed by sorting its n cyclic shifts $t_1 t_2 \dots t_n, t_2 \dots t_n t_1, \dots,$

$t_n t_1 \ldots t_{n-1}$ in lexicographic order. The matrix obtained is called the Burrows-Wheeler Matrix (BWM) and the sequence from its last column is the BWT.

Suffix Arrays. The suffix array of a string T is an array of integers that provides the starting position of T's suffixes, after they have been ordered lexicographically. Formally, if $T_{i,j}$ is the substring $t_i t_{i+1} \ldots t_j$ of T and SA is the suffix array of T, then $T_{SA[1],n} < T_{SA[2],n} < \ldots < T_{SA[n],n}$. It is related to the BWT, since looking at the substrings preceding the terminating character \$ in the BWM rows gives the suffixes of T in lexicographical order.

Backward Search. Any occurrence of a pattern P in text is a prefix for some suffix of T, so all occurrences will be adjacent in the suffix array of T since suffixes starting with P are sorted together in a SA-interval. Let $C[a]$ be the total number of occurrences in T of characters smaller than a in the alphabet. If P' is a suffix of the query P and $[l(P'), r(P'))$ is its corresponding SA-interval, then the search can be extended to aP' by calculating the new SA-interval:

$$l(aP') = C[a] + rank_a(BWT, l(P') - 1) \tag{1}$$

$$r(aP') = C[a] + rank_a(BWT, r(P)), \tag{2}$$

where the operation $rank_a(S, i)$ returns the number of occurrences of symbol a in $S[1, i]$. The search starts with the SA-interval of the empty string, $[1, n]$ and successively adds one character of P in reverse order. When the search is completed, it returns a SA-interval $[l, r)$ for the entire query P. If $r > l$, there are $r - l$ matches for P and their locations in T are given by $SA[i]$ for $l \le i < r$. Otherwise, the pattern does not exist in T. If the C-array and the ranks have already been stored, the backward search can be performed in $O(|P|)$ time in strings with DNA alphabet.

Wavelet Trees. Rank queries scale linearly with the alphabet size by default. The wavelet tree [14] is a data structure designed to store strings with large alphabets efficiently and provide rank calculations in logarithmic time. The tree is defined recursively: take the lexicographically ordered alphabet, split it into 2 equal halves; in the string corresponding to the current node (start with the original string at root), replace the first half of letters with 0 and the other half with 1; the left child node will contain the 0-encoded symbols and the right child node will contain the 1-encoded symbols, preserving their order from the original string; re-apply the first step for each child node recursively until the alphabet left in each node contains only one or two symbols (so a 0 or 1 determines which symbol it is).

To answer a rank query over the original string with large alphabet, repeated rank queries over the bit vectors in the wavelet tree nodes are used to locate the subtree that contains the leaf where the queried symbol is non-ambiguously encoded.

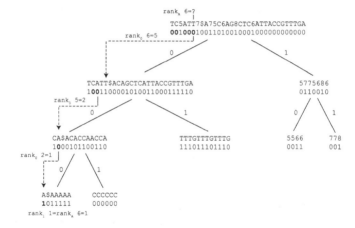

Fig. 1. Wavelet tree encoding of a string that is the same as the BWT in Fig. 4. Calculating the rank of the marked "A" is performed by repeated rank() calls moving down the binary tree until the alphabet remaining is just 2 characters. Note that only the bit vectors are stored in the tree, the corresponding strings are only shown here for clarity.

The rank of the queried symbol in this leaf is equal to its rank in the original string. The number of rank queries needed to reach the leaf is equal to the height of the tree, i.e. $\log_2 |\Sigma|$ if we let Σ be the set of symbols in the alphabet. Computing ranks over binary vectors can be done in constant time, so a rank query in a wavelet tree-encoded string has complexity $O(\log_2 |\Sigma|)$.

3   Encoding a Variation-Aware Reference Structure

3.1   Terminology

A *variant site* or *site* is a region of the chromosome where there are a number of alternative options for what sequence can be present. These alternatives are termed *alleles* and might be as short as a single character, or could be many hundreds of characters long. A *pan-genome* refers to a representation (with unspecified properties) of a number (greater than 1) of genomes within a species. A Population Reference Graph is an encoding of a pan-genome that enables matching of sequence data to the datastore, inference of nearest mosaic with the appropriate ploidy, and then discovery of new variants not present in the PRG.

3.2   PRG Encoding

We use a PRG conceptually equivalent to a directed, acyclic, partial order graph, that is generated from a reference sequence and a set of alternative sequences at given variation sites. The graph is linearised into a long string over an alphabet extended with new symbols marking the variants, for which the FM-index can be constructed. We call this string the *linear PRG*.

```
Ref   CAAGGCTAT--ACCTACT
Alt1  CAAGGTTATTTACCTGCT    ➡    CAAGG5CTAT6TTATTT6C5ACCT7A8G7CT
Alt2  CAAGGC-----ACCTACT
```

Fig. 2. A simple PRG linearised according to our encoding. The first site has 3 alleles, which do not here look at all similar, and the second is a SNP.

Building this data structure requires multiple steps.

1. Corresponding regions of shared sequence between the input genomes must be identified. These must be of size k at least (where k is pre-defined), and act as anchors.
2. For any site between two anchor regions, the set of possible alleles/haplotypes must be determined, but do not need to be aligned. Indels are supported by haplotypes of different lengths.
3. Each variation site is assigned two unique numeric identifiers, one even and one odd, which we call variation markers. The odd identifiers will mark variation site boundaries and will sometimes be referred to as site markers. The even identifiers will mark alternative allele boundaries and will sometimes be referred to as allele boundary markers.
4. For each variation site, its left anchor is added to the linear PRG, followed by its odd identifier. Then each sequence coming from that site, starting with the reference sequence, is successively added to the linear PRG, followed by the even site identifier, except the last sequence, which is followed by the odd identifier.
5. Convert the linear PRG to integer alphabet ($A \rightarrow 1$, $C \rightarrow 2$, $G \rightarrow 3$, $T \rightarrow 4$, variation site identifiers $\rightarrow 5,6,...$)
6. The FM-index (suffix array, BWT, wavelet tree over BWT) of the linear PRG is constructed and we will call this the vBWT.

An illustration of these steps on a toy example is given in Fig. 2.

Importantly, *the markers force the ends of alternative sequences coming from the same site to be sorted together in a separate block in the Burrows-Wheeler matrix, even if they do not have high sequence similarity.* Therefore, alternative alleles from each site can be queried concurrently.

3.3 Graph Structure: Constraints

We show in Fig. 3(a) two sequences which differ by 3 SNPs and give two graph encodings in Fig. 3(b) and (c). Both represent the sequence content equally well, and we allow both. In Fig. 3(d) we have an example where a long deletion lies "over" two other alleles. We would encode this in our PRG as shown in Fig. 3(e). This works but results in many alternate alleles. An alternative would be to allow "nested" variation, where variants lie on top of other alleles, as shown in Fig. 3(f). This could be encoded in our system, but we do not allow it for our initial implementation, as it would potentially impact mapping speed.

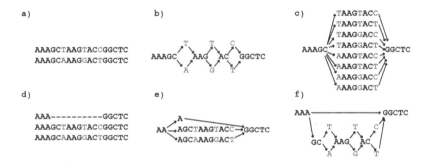

Fig. 3. PRG graph structure. The sequences shown in Fig. 3(a) could be represented either as 3 separate mutations (shown in (b)), or enumerated as 8 small haplotypes, shown in (c). Both are supported by our encoding. Similarly, the sequences in (d) could be represented in our implementation as shown in (e). However, we do not support "nesting" of alleles, as shown in (f).

4   Variation-Aware Backward Search in vBWT

In this section, we present a modified backward search algorithm for exact matching against the vBWT that is aware of alternative sequence paths. Our implementation leverages the succinct data structures library SDSL [18] and is incorporated in our software called **gramtools**.

When reads align to the non-variable part of the PRG or when an allele is long enough to enclose the entire read, the usual backward search algorithm can be used. Otherwise, when the read must cross variation site junctions, site identifiers and some alternative alleles must be ignored by the search. This means a read can align to multiple substrings of the linear PRG that may not be adjacent in the BWM, so the search can return multiple SA-intervals. We give pseudocode in Algorithm 1 below, and outline the idea in Fig. 4.

At each step in the backward search, before extending to the next character, we need to check whether the current matched read substring is preceded by a variation marker anywhere in the linear PRG. A scan for symbols larger than 4 ("range_search_2d" in the pseudocode) must be performed within the range given by the current SA-interval. With a wavelet tree this range search can be done in $O(d \log(|\Sigma|/d))$ time, where d is the size of the output. If a variation marker is found and it is an odd number, the read is about to cross a site boundary. The suffix array can be queried to find the position of the two odd numbers (start/end of the site) in the linear PRG.

If the search cursor is next to the start of the site, it is just the site marker that needs to be skipped, so the SA-interval (size 1) of the suffix starting with that marker needs to be added to the set of intervals that will be extended with the next character in the read. If the search cursor is next to the end of a site, all alternative alleles from that site need to be queried. Their ends are sorted together in the BWM because of the markers, so they can be queried concurrently by adding the SA-interval of suffixes starting with all numbers marking that site (even and odd).

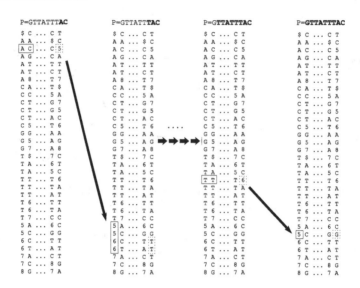

Fig. 4. Backward search across the vBWT of the linear PRG in Fig. 2. We start at the right-hand end of the read GTTATTTAC, with the character C, and as we extend we hit the character 5, signalling the start or end of a variation site. We check the suffix array to get the coordinate in the linear PRG, and find it is the end. Therefore, the read must now continue into one of the alleles, signalled by the number 6. Continuing in this manner (the shorter arrows signify multiple intermediate steps not shown) we are able to align across the site.

If the variation marker found is even, the read is about to cross an allele boundary, which means its current suffix matches the beginning of an alternative allele and the read is about to walk out of a site, so the search cursor needs to jump to the start of site. The odd markers corresponding to that site can be found in the first column of the BWM, and then querying the suffix array decides which one marks the start of site. The SA-interval (size 1) for the BWM row starting with this odd marker is recorded. Once the check for variation markers is finished and all candidate SA-intervals have been added, each interval can be extended with the next character in the read by using Eqs. 1 and 2.

5   Experiments

5.1   Construction Cost: The Human Genome

We constructed a PRG from the human reference genome (GRC37 without "alt" contigs) plus the 1000 genomes final VCF (12 GB in size) [6]. We excluded variants without specified alleles, and those with allele frequency below 5 % (rare variation offers limited benefit - our goal is to maximise the proportion of reads mismatching the graph by at most 1 SNP). If two variants occurred at consecutive bases, they were merged, and all haplotypes enumerated. If the VCF

Algorithm 1. Variation-aware backward search

Input: pattern $P[1, m]$ and FM-index of PRG in integer alphabet
Output: list of SA intervals corresponding to matches of P

```
 1: l ← C(P[m])                                            ▷ l for left
 2: r ← C(P[m] + 1)                                        ▷ r for right
 3: i ← m
 4: SA_int = {[l, r]}                                      ▷ list of SA intervals
 5: Extra_int = ∅                                          ▷ Extra intervals
 6: while i > 1 and SA_int ≠ ∅ do
 7:     for all [l, r] ∈ SA_int do
 8:         M ← WT.range_search_2d(l, r − 1, 5, |Σ|)       ▷ find variation site markers
 9:         for all (idx, num) ∈ M do                      ▷ idx ∈ [l, r), num ∈ [5, |Σ|]
10:             if num%2 = 0 then
11:                 odd_num = num − 1
12:             else
13:                 odd_num = num
14:             if SA[C(odd_num)] < SA[C(odd_num) + 1] then
15:                 start_site ← C(odd_num), end_site ← C(odd_num) + 1
16:             else
17:                 start_site ← C(odd_num) + 1, end_site ← C(odd_num)
18:             if num%2 = 1 and SA[idx] = SA[site_end] + 1 then
19:                 Extra_int = Extra_int ∪ {[C(num), C(num + 2)]}
20:             else
21:                 Extra_int = Extra_int ∪ {[C[start_site], C[start_site] + 1]}
22:     i ← i − 1
23:     SA_int = SA_int ∪ Extra_int
24:     for all [l, r] ∈ SA_int do
25:         l = C(P[i]) + rank_{P[i]}(BWT, l − 1)
26:         r = C(P[i]) + rank_{P[i]}(BWT, r)
```

---

contained two consecutive records which overlapped, the second was discarded. This resulted in a dataset of 7.4 million SNPs and 978000 indels. We give construction costs in Table 1, along with comparative figures for BWBBLE with identical input.

For comparison, GCSA took over 1 TB of RAM building chromosomes separately and pruning the graph in high diversity regions. GCSA2 reduces the memory footprint to below 128 GB RAM, running in 13 h with 32 cores, and using over 1 Tb of I/O to fast disk. Our vBWT construction has a lower memory cost than GCSA, GCSA2 and BWBBLE, is faster than GCSA/GCSA2, has no (significant) I/O burden, but is significantly slower than BWBBLE.

Table 1. FM-index construction costs and final data structure size for human reference genome plus 1000 genomes variants

Software	Peak memory (GB)	Time (h/min)	Final/in-use memory (GB)
vBWT	25	4h24m	17.5
BWBBLE	60	1h5m	12

5.2 Inferring a Closer Reference Genome

P. falciparum is a haploid parasite that undergoes recombination. It has an unusual genome that contains more indels than SNPs [15]. The gene MSP3.4 is known to have two diverged lineages at high frequencies in multiple populations from across the world. The lineages differ by around 1 SNP every 3 bases over a 500 bp region (the DBL domain) of the gene. We constructed a catalog of MSP3.4 variation from Cortex [16] variant calls from 650 *P. falciparum* samples and built a PRG just for that chromosome. We show in Fig. 5 the density of variants and number of alleles.

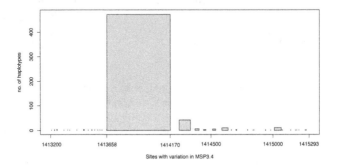

Fig. 5. Histogram of number of alleles at each site in MSP3.4 plotted above the chromosome coordinate.)

We aligned Illumina 76 bp reads from a well-studied sample that was not used in graph construction (named 7G8) to the PRG using backward search (exact matching, which took 3 mins), and collected counts on the number of reads supporting each allele. At each site we chose the allele with the highest coverage to extract the path through the graph with maximum support - this was our graph-inferred personalised reference for this sample. We then mapped the reads (using bwa_mem [17]) independently to the reference and to the inferred genome. As can be seen in Figs. 6 and 7, our method gives dramatically better pileup results over the MSP3.4 gene.

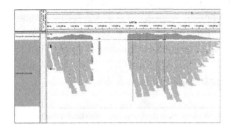
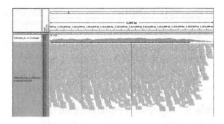

Fig. 6. Mapping reads from sample 7G8 to *P. falciparum* 3D7 reference genome results in a gap covering the DBL domain

Fig. 7. Mapping reads from sample 7G8 to our vBWT-inferred genome removes the gap, leaving isolated variants easy to detect with standard methods

5.3 Simulations, Usability, Future Performance Improvements

We took 44,439 *P. falciparum* SNPs and indels called with Cortex from a single genetic cross (7G8xGB4) [15] and created a whole-genome PRG, and simulated 10,000 reads from one random haplotype. All reads were 150 bp, error-free. We precalculate a hash of the SA intervals corresponding to all 9-mers in the PRG that overlap a variation site. This one-time precalculation was done in 1 h 43 min using 25 threads. In order to avoid unfairly slowing BWBBLE we constrained it to do exact matching only. All experiments were performed single-threaded on a machine with 64 processors Intel Xeon CPU E5-4620 v2 @ 2.60 GHz and 1 TB of memory.

Table 2. Simulation results

Software	Step	Time (sec)	Speed (reads/sec)
Gramtools	Map and Infer reference	1051	9.5
Gramtools	bwa map to ref	0.861	11614
BWBBLE	align	2.6	3846

Results are shown in Table 2. Gramtools mapping speed is notably slower than BWBBLE, although it is usable for megabase sized genomes - a 30x whole-genome dataset for *P. falciparum* would take 5.8 h using 24 cores. However, the output is directly usable and interpretable by any bioinformatician - a reference genome close to the sample, and a standard SAM file. By comparison, BWBBLE outputs a SAM file with respect to an artificial reference with indels appended at the end - to use this in a normal pipeline requires software development and innovation.

There are a number of performance improvements we can make. We store an integer array that allows us to determine if a position in the PRG is in a site, and if so, which allele; this is naively encoded (in std::vector). For the human example, this costs us around 12 GB of RAM. This array, which contains a zero at every non-variable site in the chromosome, could be stored much more compactly. More significantly, there is one significant speed improvement which we have yet to implement - precalculating and storing an array of ranks at marker positions across the BWT - just as in a standard FM-index. This is not normally done for large alphabets, but we can store only for A,C,G,T.

6 Discussion

We have described a whole-genome scale implementation of a PRG designed to enable inference of a within-graph mosaic reference close to that of a new individual, followed by discovery of novel variation as a "delta" from that. As with any reference graph approach, there is an implicit coupling between mapping and graph structure (for handling alternate alleles). By placing positional markers, we are able to ensure that alternate alleles sort together in the BWT matrix, allowing mapping across sites and recombination. For haploids we naturally infer a personalised reference genome. For other ploidies, our implementation readily lends itself to "lightweight alignment" [19–21] followed by an HMM to infer haplotypes, followed by full MEM-based graph alignment.

Software. Our software, gramtools, is available here: http://github.com/iqbal-lab/gramtools, and scripts for reproducing this paper are here: https://github.com/iqbal-lab/paper-2016-vBWT-gramtools-WABI.

Acknowledgments. We would like to thank Jacob Almagro-Garcia, Phelim Bradley, Rayan Chikhi, Simon Gog, Lin Huang, Jerome Kelleher, Heng Li, Gerton Lunter, Rachel Norris, Victoria Popic, and Jouni Siren for discussions and help. We thank the SDSL developers for providing a valuable resource.

References

1. Valenzuela, D., Valimaki, N., Pitkanen, E., Makinen, V.: On enhancing variation detection through pan-genome indexing. Biorxiv. http://dx.doi.org/10.1101/021444
2. Burrows, M., Wheeler, D.J.: A block sorting lossless data compression algorithm. Digital Equipment Corporation, Tech. Rep. **124** (1994)
3. Li, H., Durbin, R.: Fast and accurate short read alignment with Burrows-Wheeler Transform. Bioinformatics **25**(14), 1754–1760 (2009)
4. Langmead, B., Salzberg, S.: Fast gapped-read alignment with Bowtie 2. Nat. Meth. **9**(4), 357–359 (2012)
5. Reinert, K., Langmead, B., Weese, D., et al.: Alignment of next-generation sequencing reads. Annu. Rev. Genomics Hum. Genet. **16**, 13–51 (2015)
6. The 1000 Genomes Project Consortium: A global reference for human genetic variation. **526**, pp. 68–74 (1000)
7. Ossowski, S., Schneeberger, K., Clark, R.M., et al.: Sequencing of natural strains of Arabidopsis thaliana with short reads. Genome Res. **18**, 2024–2033 (2008)
8. Dilthey, A., Cox, C., Iqbal, Z., et al.: Improved genome inference in the MHC using a population reference graph. Nat. Genet. **47**, 682–688 (2015)
9. Schneeberger, K., Hagmann, J., Ossowski, S., et al.: Simultaneous alignment of short reads against multiple genomes. Genome Biol. **10**, R98 (2009)
10. Siren, J., Valimaki, N., Makinen, V.: Indexing graphs for path queries with applications in genome research. IEEE/ACM Trans. Comput. Biol. Bioinform. **11**(2), 375–388 (2014)
11. Huang, L., Popic, V., Batzoglou, S.: Short read alignment with populations of genomes. Bioinformatics **29**(13), i361–i370 (2013)
12. Siren, J.: Indexing Variation Graphs. arXiv:1604.06605
13. Ferragina, P., Manzini, G.: Opportunistic datastructures with applications. In: Proceedings of the 41st Symposiumon Foundations of Computer Science (FOCS 2000), pp. 390–398. IEEE Computer Society, Los Alamitos (2000)
14. Grossi, R., Gupta, A., Vitter, J.: High-order entropy-compressed text indexes. In: Proceedings of the 14th Annual ACM-SIAM Symposium on Discrete Algorithms, pp. 841850. Society for Industrial and Applied Mathematics (2003)

15. Miles, A., Iqbal, Z., Vauterin, P., et al.: Genome variation and meiotic recombination in Plasmodium falciparum: insights from deep sequencing of genetic crosses (2015). Biorxiv http://dx.doi.org/10.1101/024182
16. Iqbal, Z., Caccamo, M., Turner, I., et al.: De novo assembly and genotyping of variants using colored de Bruijn graphs. Nat. Genetics **44**, 226–232 (2012)
17. Li, H.: Aligning sequence reads, clone sequences, assembly contigs with BWA-MEM. arXiv:1303.3997
18. Gog, S., Beller, T., Moffat, A., et al.: From theory to practice: plug and play with succinct data structures. In: 13th International Symposium on Experimental Algorithms, (SEA 2014), pp. 326–337 (2014)
19. Patro, R., Mount, S.M., Kingsford, C.: Sailfish enables alignment-free isoform quantification from RNA-seq reads using lightweight algorithms. Nat. Biotechnol. **32**, 462–464 (2014)
20. Srivastava, A., Sarkar, H., Gupta, N., Patro, R.: RapMap: a rapid, sensitive and accurate tool for mapping RNA-seq reads to transcriptomes. Bioinformatics **32**(12), i192–i200 (2016)
21. Bray, N., Pimentel, H., Melsted, P., Pachter, L.: Near-optimal probabilistic RNA-seq quantification. Nat. Biotechnol. **34**, 525–527 (2016)

Inferring Population Genetic Parameters: Particle Filtering, HMM, Ripley's K-Function or Runs of Homozygosity?

Svend V. Nielsen[✉], Simon Simonsen, and Asger Hobolth

Aarhus University, Bioinformatics Research Centre, C.F. Møllers Allé 8,
8000 Aarhus C, Denmark
svn@cs.au.dk

Abstract. The coalescent with recombination is widely accepted as the key model to understand genetic diversity within a species. Many theoretical properties of the model are well understood, but formulating and implementing efficient inference methods remains a challenge. A major breakthrough has been to approximate the coalescent with recombination by a Markov chain along the sequences. Here we describe a new tool, RECJumper, for inference in the Markov approximated coalescent model. Previous methods are often based on a discretisation of the tree space and hidden Markov models. We avoid the discretisation by using particle filtering, and compare several proposal distributions. We also investigate runs of homozygosity, and introduce a new summary statistics from spatial statistics: Ripley's K-function. We find that (i) choosing an appropriate proposal distribution is crucial to obtain satisfactory behaviour in particle filtering, (ii) tree space discretisation in HMM-methodology is non-trivial and the choice can influence the results, and (iii) Ripley's K-function is a much more informative statistics than runs of homozygosity for recombination rate estimation.

1 Introduction

Consider the black piecewise constant function in Fig. 1(a). The function provides the coalescence times $x = \{x_i\}_{1 \leq i \leq N}$ for two genomic sequences (e.g. from sequencing a single diploid individual) in positions indexed by $i, i = 1, \ldots, N$. The coalescence times are unknown (hidden), and we instead observe the binary mutation pattern $y = \{y_i\}_{1 \leq i \leq N}$. Sites with large coalescence times are more likely to experience a mutation than sites with short coalescence times; a simple mutation model assigns probability

$$p(y_i = 1 | x_i = t) = 1 - \exp(-\theta t) \tag{1}$$

for being heterozygote (and $\exp(-\theta t)$ for being homozygote). Recombination events are responsible for the jumps in the coalescence path, and in general the dependence structure between coalescence times is very complex [8]. Fortunately the Markov assumption is a good approximation [13]. We observe that

© Springer International Publishing Switzerland 2016
M. Frith and C.N.S. Pedersen (Eds.): WABI 2016, LNBI 9838, pp. 234–245, 2016.
DOI: 10.1007/978-3-319-43681-4_19

the Markov approximated coalescent with recombination and mutation is a state space model where $\{x_i\}_{1\le i\le N}$ is the latent process and $\{y_i\}_{1\le i\le N}$ is the measurement data. State space models have been analysed in financial econometrics for more than a decade (e.g. [2]), and an efficient class of algorithms is particle filtering.

Particle filtering is an importance sampling method. Recall that the main idea in importance sampling is to simulate from a distribution $q(x)$ (possibly depending on y), assign each simulation a weight $w(x) = f(x)p(x)/q(x)$, and approximate the mean, $E[f(X)]$, by the weighted sample average. For example if the quantity of interest is the likelihood $p(y)$ we have

$$p(y) = \int_x p(x,y)dx = \int_x \frac{p(x)p(y|x)}{q(x)}q(x)dx = \int_x w(x)q(x)dx \approx \frac{1}{n}\sum_{j=1}^n w(z_j),$$

where z_j is a sample from q. The challenge in particle filtering is to formulate proposal distributions that are easy to simulate *and* close to the optimal proposal distribution $p(x|y)$.

Figure 1(a) shows three coalescence paths from a naive but very fast proposal distribution, and Fig. 1(b) shows their corresponding weights $w(x)$. The distribution of weights in Fig. 1(b) shows that the price for the fast proposal distribution is very high: Almost all coalescence paths have a very low weight and are therefore not useful for subsequent analysis. This unfortunate situation is avoided in Figs. 1(c) and 1(d) where we have used the proposal function from RECJumper. Very many RECJumper coalescent paths have a reasonable high weight.

Particle filtering is far from the only way to infer population genetic parameters. Another popular method is to discretise the state space so that only a discrete and finite number of coalescent times are possible. Most known is perhaps the PSMC [11] that uses the Sequential Markov Coalescent (SMC) model [13] on two DNA sequnces. ARGweaver [16] and MSMC [17] are extensions of the PSMC to more than two sequences. The CoalHMM [12] is an implementation with the Simonsen Churchill model [18] for two sequences. The advantage of a discrete and finite, latent state space is that the classical HMM algorithms apply. The challenge is to formulate a reasonable discretisation procedure. How many bins and where to place them?

In Fig. 2 we show the results of a simulation study. We simulated 100 data sets from the model with a recombination rate of $\rho = 0.1$, a fixed mutation rate $\theta = 0.1$ in a genomic segment of size $N = 20,000$ base pairs. We then estimated the recombination rate using particle filtering and an HMM with a state space of size 20. We observe that they yield similar results. This is expected since they are both approximations to the same integral.

In Fig. 2 we have also included two more parameter estimation procedures. The first is based on runs of homozygosity. Harris and Nielsen [9] use this statistics for demographic inference, but we observe that actually a lot of power is lost by summarizing the data this way. The second is based on Ripley's K-function. In spatial statistics the runs of homozygosity (often called the nearest neighbour

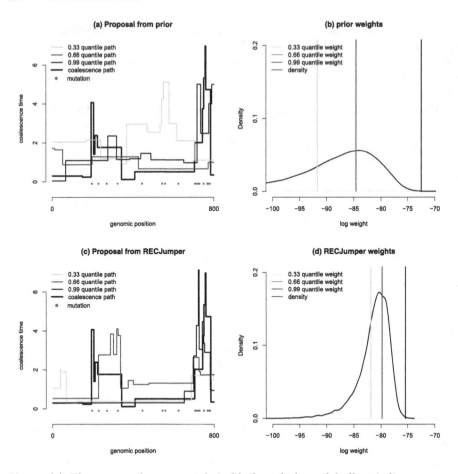

Fig. 1. (a) The true coalescence path is black and the red bullets indicate mutations (the observed sequence). The coloured lines are different paths from the proposal distributions. Paths that are likely to produce the mutation pattern get a high weight. (b) Distributions of weights from the prior distribution. (c) The proposals from RECJumper ($h = 100$). (d) The variance of the prior weights is larger than the variance of the RECJumper weights. (Color figure online)

function) is only seldom used. Ripley's K-function is a much more popular summary of a point pattern. Ripley's K-funcion gives the mean number of points $K(r)$ within a distance r from a typical point. It is straight forward to determine an empirical estimate of Ripley's K-function. In Fig. 2 we observe that Ripley's K-function is a very useful statistics. We emphasize that this observation also has important consequences for simulation-based procedures such as Approximate Bayesian Computation [1] where the data is summarized in terms of simple summary statistics.

Our paper is organized as follows. In Sect. 2 we describe in detail the state space model. In particular we provide the probability of a new coalescent height,

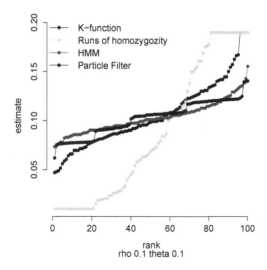

Fig. 2. The four methods have each produced an estimate of the recombination rate, ρ, based on 100 simulations of size 20,000 bases from the state space model. This plot visualises their distributions by sorting all the estimates. The K-function produces better estimates of ρ than runs of homozygosities whereas HMM and particle filtering performs even better. The 'steps' of the particle filter estimates are due to the grid of driver values of ρ.

and the density for a new height conditional on the old height. We work with the most general recombination model for two loci and two sequences. In Sect. 3 we describe particle filtering and RECjumper, and in Sect. 4 we consider the HMM framework. Section 5 is concerned with Ripley's K-function and runs of homozygosity. Our paper ends with a general discussion of the various methods.

2   The State Space Model

A state space model is fully specified by its *transition probabilities* $p(x_i|x_{i-1})$ and its *emission probabilities* $p(y_i|x_i)$. Therefore, only the joint distribution of the coalescence times at two adjacent genomic positions is needed in order to specify the model. Below we let $s = x_{i-1}$ and $t = x_i$ denote the left and right coalescence times. The coalescence times are determined by the Simonsen-Churchill model [18] which is a continuous time Markov chain. The states are the ancestry of two pairs of loci from two sequences; see Fig. 3. The model is given a careful treatment in the textbooks [6,19], and we only use the following theorem which provides the transition probabilities.

Theorem 1. *Let Λ denote the 8×8 rate matrix for the states in the Simonsen-Churchill model (see Fig. 3(c)). The conditional probability of no change from the left to the right tree is*

$$P(T = s|S = s) = e^s[e^{\Lambda s}]_{11}, \tag{2}$$

and the conditional density $\pi(t|s)$ of T given $S = s$ and given $T \neq S$ is

$$\pi(t|s) = \begin{cases} e^{-(s-t)} \frac{[e^{\Lambda t}]_{12} + [e^{\Lambda t}]_{13}}{e^{-s} - [e^{\Lambda s}]_{11}} & \text{for } t < s, \\ e^{-(t-s)} \frac{[e^{\Lambda s}]_{12} + [e^{\Lambda s}]_{13}}{e^{-s} - [e^{\Lambda s}]_{11}} & \text{for } t > s. \end{cases} \tag{3}$$

Proof. See Lemma 2 of Hobolth and Jensen [10].

A simplification of the above model, called the SMC model has a simpler structure and will also be used in the particle filter. In this model we have (see Hobolth and Jensen [10], p. 52, bottom right)

$$P^{\text{SMC}}(T = s|S = s) = e^{-\rho s} \tag{4}$$

and

$$\pi^{\text{SMC}}(t|s) = \begin{cases} \frac{\rho(e^{-\rho t} - e^{-t})}{(1-\rho)(1-e^{-\rho s})} & \text{for } t < s \\ \frac{\rho e^{-(t-s)}(e^{-\rho s} - e^{-s})}{(1-\rho)(1-e^{-\rho s})} & \text{for } t > s. \end{cases} \tag{5}$$

For more discussions of sequential Markov chains for two loci, two sequences we refer to Wilton et. al. [21].

3 Particle Filtering

Particle filtering is a statistical method that can be used to improve a specific type of importance sampler [4]. The goal is to simulate from a distribution $p(\cdot)$ such that a low variance estimate of $E_p[f(X)]$ can be constructed. This is achieved by simulating particles $z_1, \ldots, z_n$ from any distribution, q, called the proposal distribution, which satisfies that every possible sample under p is also possible under q. The particles are each assigned a weight $w(z) = f(z)p(z)/q(z)$. In this study $f(x) = p(y|x)$ such that the weighted particles

$$(z_1, w(z_1)), \ldots, (z_n, w(z_n))$$

constitute a sample from $p(x|y)$ in the sense that $p(y) = E_p[f(Z)] \approx \frac{1}{n} \sum_{j=1}^{n} w(z_j)$. Having n weighted particles is not as powerful as having n regular samples because some particles may be insignificant due to a very low weight. This problem is called sample degeneracy. Furthermore, all simulated particles could, in principle, be made useless by a future particle that has a much higher weight. Under the assumption that no such future particle exists, we will interpret the approximation to the Effective Sample Size

$$\text{ESS}(w_1, \ldots, w_n) = \frac{\left(\sum_{j=1}^{n} w_j\right)^2}{\sum_{j=1}^{n} w_j^2}$$

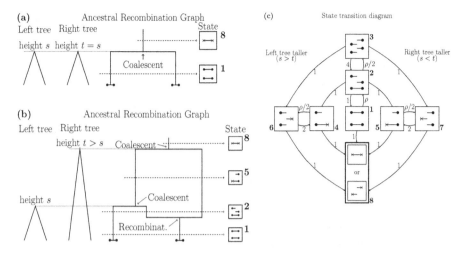

Fig. 3. (a) Realisation of the continuous time Markov Chain where no recombination between the two sites occur. In the present both pairs of loci are linked (—) yet not coalesced with each other (•) meaning that the chain is positioned in state 1. At time point s the chain jumps directly from state 1 to state 8 where both loci are coalesced($\times$). (b) A recombination occurs which unlinks the two loci. They subsequently find different coalescence times. (c) All possible transitions are shown with their corresponding transition rates. The corresponding rate matrix is denoted Λ.

as the actual number of samples from p. It is evident that the choice of $q(x)$ affects the weights and thereby the ESS. In the case $q \propto p(x,y)$ all the weights will be constant and ESS will be the highest attainable value, n. This suggests that choosing q close to $p(x|y)$ will lead to a high ESS.

An importance sampler is susceptible to particle filtering methods if p and q can be decomposed as follows

$$p(x_{1:N}, y_{1:N}) = p(x_1)p(x_2|x_1) \cdots p(x_N|x_{N-1})p(y_1|x_1) \cdots p(y_N|x_N) \quad (6)$$
$$q(x_{1:N}|y_{1:N}) = q(x_1|y_{1:N})q(x_2|x_1, y_{1:N}) \cdots q(x_N|x_{N-1}, y_{1:N}). \quad (7)$$

The natural way to simulate from q is to simulate x_1 from $q(x_1|y_{1:N})$, then x_2 from $q(x_2|x_1, y_{1:N})$ and so forth. If done this way, conditions (6) and (7) allow calculation of preliminary weights after i steps

$$w(x_{1:i}) = \frac{p(x_1)p(x_2|x_1) \cdots p(x_i|x_{i-1})p(y_1|x_1)p(y_2|x_2) \cdots p(y_i|x_i)}{q(x_1|y_{1:N})q(x_2|x_1, y_{1:N}) \cdots q(x_i|x_{i-1}, y_{1:N})}.$$

Preliminary weights make it possible to gauge the final weight of a particle before the particle is fully produced. If a particle turns out to yield low preliminary weights, we would like to discard it so that we do not waste computing power on a particle that will most likely be insignificant. In addition we duplicate the preliminarily high-weighted particles. This is done through resampling which removes the problem of sample degeneracy completely but introduces sample impoverishment. Sample impoverishment is the dependencies between the particles and it consists of two issues: (i) A particle having a low preliminarily weight could recover as more coordinates are simulated. Removing such a comeback particle through resampling will give insufficient diversity within the sample, (ii) The first coordinates of the particles will eventually converge to one value if there are enough resamples. The last coordinates of the particles will not have this issue because they are only resampled a few times.

Issue (ii) is normally solved using smoothing and in particular the smoothing algorithm called Forward-Backward Recursions. Issue (i) is a more fundamental problem and its remedies are often costly in terms of computations. The simplest remedies are increasing the number of particles, choosing a good proposal distribution and fine-tuning the positions of resampling. The optimal resampling positions depend both on the target distribution and proposal distribution, but will not be explored in this study.

One of the simplest proposal distributions in particle filtering is the prior as proposed in the early literature [7]

$$q(x_i|x_{i-1}, y_{1:N}) = p(x_i \mid x_{i-1}).$$

In this paper we use the distribution $p^{\mathrm{SMC}}(x_i|x_{i-1})$ specified in (4) and (5). The advantage is that it is fast and easy to simulate. On the other hand it does not use the data y so we expect a lot of particles with low weight. Some realisations from this proposal are shown in the upper panel of Fig. 1.

We formulated a more informed choice of proposal distribution

$$q(x_i|y_{1:N}, x_{i-1}) = \tilde{p}^{\mathrm{SMC}}(x_i|x_{i-1}, y_{i:(i+h)}),$$

for some lag h. Our desire was to simulate from the distribution $p^{\mathrm{SMC}}(x_i \mid x_{i-1}, y_{i:(i+h)})$. Its distribution is fully specified by the emission probabilities in (1), transition probabilities in (4) and (5), and the state space model assumption. It was, however, not computationally feasible without making some approximations. Therefore, we increased the forgetfulness of the latent Markov chain and we substituted a binomial distribution with a Poisson distribution. This reduced model is denoted $\tilde{p}^{\mathrm{SMC}}$. The algorithm first simulates the genomic distance d to the next recombination from $\tilde{p}^{\mathrm{SMC}}(d|x_{i-1}, y_{i:(i+h)})$ and then draws x_i from $\tilde{p}^{\mathrm{SMC}}(x_i|x_{i-1}, y_{i:(i+h)}, d)$. The effect of the forgetfulness assumption seems to be that fewer extreme values are simulated.

When x_i is simulated, the algorithm forgets the previous d and simulates a new d to generate x_{i+1}. A natural extension of this procedure is to use the simulated d by setting $x_i = x_{i+1} = \cdots = x_{i+d-1}$ and then continue the algorithm at x_{i+d}. This speeds up the algorithm yet only decreases the accuracy slightly. We call this faster version RECJumper. Similarly the prior can also be made faster by simulating the next recombination event and jumping to its position. This faster version will simply be called the prior proposal.

We compared the two proposal distributions by looking at how the ratio between ESS and time consumption depends on sequence length in Fig. 4. The proposals were informed of the true values $(\rho, \theta) = (0.1, 0.02)$. RECJumper is superior for sequences that are longer than 400 base pairs when h is big. For large h RECjumper simulates particles more slowly but closer to the true distribution, $p(x|y)$.

If the distribution p is parametrised in terms of a parameter, ρ, it makes sense to talk about the likelihood. Remembering that the Particle Filter produces weighted samples $(z_j, w(z_j))_{i=1,\ldots,n}$ from p_{ρ_0}, enough samples will justify the importance sampling approximation to the likelihood function

$$L(\rho) \approx \sum_{j=1}^{n} w(z_j) \frac{p_\rho(z_j)}{p_{\rho_0}(z_j)}. \tag{8}$$

In Fig. 2 we show 100 estimates using this method.

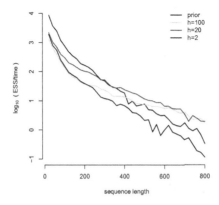

Fig. 4. The measure of samples per seconds is ESS divided per time. Proposals from prior and RECJumper with different lags are shown. For sequences of length 450 and above, the prior is inferior to the more informed versions of RECJumper.

4   Hidden Markov Model

When there is only a finite number of states in a state space model, it becomes a Hidden Markov Model. In this framework powerful algorithms exists. The Forward (or Backward) algorithm calculates the likelihood exactly and quickly [5], and hidden paths are also simulated quickly [3].

The established coalescent HMM methods obtain a finite number of states by discretising the state space model. Discretisation is done by dividing the time axis into a number of intervals and letting the hidden states be the intervals in which the coalescence time falls. Unfortunately, discretising a state space model does not preserve the dependence structure of a state space model. It does not follow that the hidden states of the discretised state space model form a Markov chain nor that the observed data are independent conditioned on the hidden states.

Therefore, in addition to the usual loss of power when discretising, discretisation introduces a bias from the intended model. These disadvantages are alleviated by applying a finer discretisation at the cost of more computations. Nevertheless, the choice of number and shape of intervals in the discretisation influences the results as demonstrated in Fig. 5.

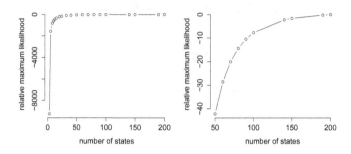

Fig. 5. A typical relationship between the maximum likelihood and the number of intervals for a HMM-based coalescence model. The same graph is plotted on both plots but the axis are different. The HMM converges towards an optimum as the number of intervals increases. However, it requires 200+ states to make the number of intervals insignificant in a comparison of maximum likelihood values.

5   Ripleys K-Function

A popular summary function in point processes theory is Ripley's K-function. Let θ be the intensity of events. In our case an event is a mutation, and θ is the mutation rate per base pair. The mutation pattern can be seen as a point process by letting the points be the indices of mutations. On this point process Ripley's K function is

$$K(r) = \frac{1}{\theta} E \left[\#\{s : |s - t| \le r, Y_s = 1\} \mid Y_t = 1 \right]$$

or, equivalently, the relative number of mutations at most distance r away from a position with a mutation. The discreteness of our point process allows us to make sense of the 'derivative' of $K(r)$

$$\kappa(r) = \frac{1}{\theta} E \left[\#\{s : |t - s| = r, Y_s = 1\} \mid Y_t = 1 \right]$$

which we will use in the following.

Another descriptive summary in spatial statistics is the nearest neighbour function. The nearest neighbour function is the probability distribution of the distance from a typical point to the nearest neighbouring point. The nearest neighbour function is less popular than Ripley's K-function in spatial statistics because it has less power to discriminate between point pattern models [20]. However, in population genetics the nearest neighbour function (or 'runs of homozygosity' or 'distance between segregating sites') is very popular [9], and Ripley's K-function is seldom used. In Fig. 6(a) we show $\kappa(r)$ as a function of ρ and θ. Note that the curves converge to $\theta/(1+\theta)$ and that the behaviour for small r is determined by the recombination rate. In Fig. 6(b) we show the distribution of runs of homozygozity as a function of ρ and θ.

To assess the power of the descriptive statistics, we make an estimator which minimises the χ^2-distance between the observed and the theoretical statistics. In Fig. 2 we demonstrate that Ripley's K-function is a much more powerful summary for parameter estimation than nearest neighbour.

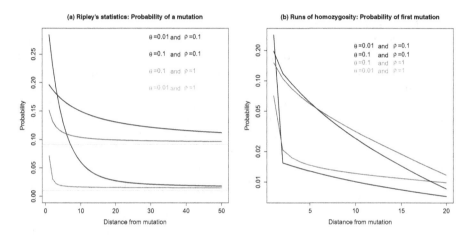

Fig. 6. On (a) and (b) the summary functions, the unnormalised derivative of Ripley's K-function and runs of homozygosity, are plotted for different parameter values. The starting position of the K-statistics depends on the recombination rate ρ and it slowly converges to the mutation rate θ. For runs of homozygosity, the difference between different recombination rates are not as profound as for K-functions.

6 Conclusion

The particle filter is an alternative method for inference about population genetic parameters. Asymptotically, all allowed proposal distributions converge, but the time consumption can be detrimental in applications. In our simulation study, the RECJumper particle filter is significantly better than the prior particle filter when the sequence segments are larger than 400 base pairs. The set-up of resampling positions and number of particles determines whether simulating segments of length larger than 400 base pairs is advantageous. It could be enough to sample segments shorter than 400 base pairs and then stitching them together with particle filtering methods. Best practice might also differ from dataset to dataset which makes it convenient to have both proposals; prior and RECJumper. Our particle filter is slower than the HMM when it comes to evaluating the likelihood $p(y)$ at a precision necessary to estimate the recombination rate on two sequences.

Generally, there are two strategies for calculating an integral numerically. One is to discretise the function so that the integral of the discretised function can be calculated exactly while another strategy is to make a Monte Carlo estimate which converges towards the correct integral. In this study we estimated the integral $L(\rho) = \int p_\rho(y|x)p_\rho(x)\,\mathrm{d}x$ using the latter strategy in contrast to the widespread HMM based methods applying the former. The discretisation method normally struggles in higher dimensions like the HMM methods do with many sequences. If time has to be divided into 200 intervals, the speed will be even slower.

Besides estimating a constant recombination rate, the model can also be extended to estimation of varying recombination rate, and varying mutation rate. It could also be used to estimate variability in population sizes as Palacios and Wakeley [15] successfully did with simulated coalescent paths under a constant population size model.

The summary statistics investigated proved to display a big difference in power of estimating the recombination rate. The variance of the estimator based on Ripley's K-function was significantly lower than that of runs of homozygosity in the set-up with a constant-sized, panmictic population. In this study the theoretical values of the K-function and runs of homozygosity were calculated using the Simonsen-Churchill model. These could also be estimated empirically from simulations avoiding the Markov assumption.

The K-function can also be used for posterior predictive checks where a fitted model is tested by comparing datasets simulated from the fitted model with the actual dataset. The comparison is through a summary statistic on the datasets and a transformation of the K-function could be such a summary statistic. The global rank envelope test [14] is one way to make the transformation.

References

1. Beaumont, M.A., Zhang, W., Balding, D.J.: Approximate Bayesian computation in population genetics. Genetics **162**(4), 2025–2035 (2002)
2. Cappé, O., Moulines, E., Rydén, T.: Hidden Markov Models. Springer, New York (2004)
3. Cawley, S.L., Pachter, L.: HMM sampling and applications to gene finding and alternative splicing. Bioinformatics **19**(suppl. 2), 36–41 (2003)
4. Doucet, A., Johansen, A.M.: A tutorial on particle filtering and smoothing: fifteen years later. Handb. Nonlinear Filtering **12**(3), 656–704 (2009)
5. Durbin, R., Eddy, S.R., Krogh, A., Mitchison, G.: Biological Sequence Analysis: Probabilistic Models of Proteins and Nucleic Acids. Cambridge University Press, Cambridge (1998)
6. Durrett, R.: Probability Models for DNA Sequence Evolution. Springer Science & Business Media, Berlin (2008)

7. Gordon, N.J., Salmond, D.J., Smith, A.F.M.: Novel approach to nonlinear/non-gaussian bayesian state estimation. IEE Proc. F - Radar Sign. Process. **140**(2), 107–113 (1993)
8. Griffiths, R.C., Marjoram, P.: Ancestral inference from samples of DNA sequences with recombination. J. Comput. Biol. **3**(4), 479–502 (1996)
9. Harris, K., Nielsen, R.: Inferring demographic history from a spectrum of shared haplotype lengths. PLoS Genet. **9**(6), 1–20 (2013)
10. Hobolth, A., Jensen, J.L.: Markovian approximation to the finite loci coalescent with recombination along multiple sequences. Theoret. Popul. Biol. **98**, 48–58 (2014)
11. Li, H., Durbin, R.: Inference of human population history from individual whole-genome sequences. Nature **475**(7357), 493–496 (2011)
12. Mailund, T., Halager, A.E., Westergaard, M., Dutheil, J.Y., Munch, K., Andersen, L.N., Lunter, G., Prüfer, K., Scally, A., Hobolth, A., Schierup, M.H.: A new isolation with migration model along complete genomes infers very different divergence processes among closely related great ape species. PLoS Genet. **8**(12), 1–19 (2012)
13. McVean, G.A., Cardin, N.J.: Approximating the coalescent with recombination. Philos. Trans. Roy. Soc. B: Biol. Sci. **360**(1459), 1387–1393 (2005)
14. Myllymki, M., Mrkvika, T., Grabarnik, P., Seijo, H., Hahn, U.: Global envelope tests for spatial processes. J. Roy. Stat. Soc.: Ser. B (Statistical Methodology) (2016)
15. Palacios, J.A., Wakeley, J., Ramachandran, S.: Bayesian nonparametric inference of population size changes from sequential genealogies. Genetics **201**(1), 281–304 (2015)
16. Rasmussen, M.D., Hubisz, M.J., Gronau, I., Siepel, A.: Genome-wide inference of ancestral recombination graphs. PLoS Genet. **10**(5), 1–27 (2014)
17. Schiffels, S., Durbin, R.: Inferring human population size and separation history from multiple genome sequences. Nat. Genet. **46**(8), 919–925 (2014)
18. Simonsen, K.L., Churchill, G.A.: A Markov chain model of coalescence with recombination. Theor. Popul. Biol. **52**(1), 43–59 (1997)
19. Wakeley, J.: Coalescent Theory: An Introduction. W. H. Freeman, San Francisco (2009)
20. Wiegand, T., He, F., Hubbell, S.P.: A systematic comparison of summary characteristics for quantifying point patterns in ecology. Ecography **36**(1), 92–103 (2013)
21. Wilton, P.R., Carmi, S., Hobolth, A.: The SMC' is a highly accurate approximation to the ancestral recombination graph. Genetics **200**(1), 343–355 (2015)

A Graph Extension of the Positional Burrows-Wheeler Transform and Its Applications

Adam M. Novak[1]([✉]), Erik Garrison[2], and Benedict Paten[1]([✉])

[1] Genomics Institute, University of California Santa Cruz,
Santa Cruz, CA 95064, USA
{anovak,benedict}@soe.ucsc.edu
[2] Wellcome Trust Sanger Institute, Cambridge, UK
eg10@sanger.ac.uk

Abstract. We present a generalization of the Positional Burrows-Wheeler Transform, or PBWT, to genome graphs, which we call the gPBWT. A genome graph is a collapsed representation of a set of genomes described as a graph. In a genome graph, a haplotype corresponds to a restricted form of walk. The gPBWT is a compressible representation of a set of these graph-encoded haplotypes that allows for efficient subhaplotype match queries. We give efficient algorithms for gPBWT construction and query operations. We describe our implementation, showing the compression and search of 1000 Genomes data.

As a demonstration, we use the gPBWT to quickly count the number of haplotypes consistent with random walks in a genome graph, and with the paths taken by mapped reads; results suggest that haplotype consistency information can be practically incorporated into graph-based read mappers.

1 Introduction

The PBWT is a compressible data structure for storing haplotypes that provides an efficient search operation for subhaplotype matches [2]. Implementations, such as BGT (https://github.com/lh3/bgt), can be used to compactly store and query thousands of samples. The PBWT can also allow existing haplotype-based algorithms to work on much larger collections of haplotypes than would otherwise be practical [4]. In the PBWT, each site (corresponding to a genetic variant) is a binary feature and the sites are totally ordered. The input haplotypes to the PBWT are binary strings, with each element in the string indicating the state of a site. In the generalization we present, each input haplotype is a walk in a general bidirected graph. This allows haplotypes to be partial (they can start and end at arbitrary nodes) and to traverse arbitrary structural variation. It does not require the sites (nodes in the graph) to have a biologically relevant ordering to provide compression. However, despite these generalizations, the core data structures are similar, the compression still exploits genetic linkage and the haplotype matching algorithm is essentially the same.

© Springer International Publishing Switzerland 2016
M. Frith and C.N.S. Pedersen (Eds.): WABI 2016, LNBI 9838, pp. 246–256, 2016.
DOI: 10.1007/978-3-319-43681-4_20

2 Definitions

We define $G = (V, E)$ as a **genome graph** in a bidirected formulation [5,6]. Each node in V has a DNA-sequence label; a left, or $5'$, **side**; and a right, or $3'$, side. Each edge in E is a pairset of sides. The graph is not a multigraph: only one edge may connect a given pair of sides and thus only one **self-loop** can be present on any given side.

We consider all the sides in the graph to be (arbitrarily) ordered relative to one another. We also define the idea of the **opposite** of a side s, with the notation $\bar{s}$, meaning the side of s's node which is not s (i.e. the left side of the node if s is the right side, and the right side of the node if s is the left side). Finally, we use the notation $n(s)$ to denote the node to which a side s belongs.

Within the graph G, we define the concept of a **thread**, which can be used to represent a haplotype or haplotype fragment. A thread t on G is a reversible nonempty sequence of sides, such that for $0 \leq i < N$ sides t_{2i} and t_{2i+1} are opposites of each other, and such that G contains an edge connecting every pair of sides t_{2i} and t_{2i+1}. In other words, a thread is a walk through the sides of the graph that alternates traversing nodes and traversing edges and which starts and ends with nodes. Note that a thread is reversible: exactly reversing the sequence of sides making up a thread produces an equivalent thread. We call a thread traversed in a certain direction an **orientation**.

We consider G to have associated with it a collection of **embedded** threads, denoted as T. We propose an efficient storage and query mechanism for T given G.

Our high-level strategy is to store T by grouping together threads that have recently visited the same sequences of sides, and storing in one place the next sides that those threads will visit. As with the Positional Burrows-Wheeler Transform, used to store haplotypes against a linear reference, and the ordinary Burrows-Wheeler transform, we consider the recent history of a thread to be a strong predictor of where the thread is likely to go next [2]. By grouping together the next side data such that nearby entries are likely to share values, we can use efficient encodings (such as run-length encodings) and achieve high compression.

More concretely, our approach is as follows. We call an instance of side in a thread a **visit**; a thread may visit a given side multiple times. Consider all visits of threads in T to a side s where the thread arrives at s either by traversing an edge incident to s (and not by traversing $n(s)$) or by beginning at s. For each such visit, take the sequence of sides coming before this arrival at s in the thread and reverse it, and then sort the visits lexicographically by these sequences of sides, breaking ties by an arbitrary global ordering of the threads. Then, for each visit, look two steps ahead in its thread (past s and $\bar{s}$), and note what side comes next (or the null side if the thread ends). After repeating for all the sorted visits to s, take all the noted sides in order and produce the array $B_s[]$ for side s. An example $B[]$ array and its interpretation are shown in Fig. 1. (Note that, throughout, arrays are indexed from 0 and can produce their lengths trivially upon demand.)

Each unoriented edge $\{s, s'\}$ in E has two orientations (s, s') and (s', s). Let $c()$ be a function of these oriented edges, such that for an oriented edge (s, s'), $c(s, s')$ is the smallest index in $B_{s'}[]$ of a visit of s' that arrives at s' by traversing $\{s, s'\}$. Note that, because of the global ordering of sides and the sorting rules defined for $B_{s'}[]$ above, $c(s_0, s')$ will be less than or equal to $c(s_1, s')$ for $s_0 < s_1$ both adjacent to s'.

For a given G, we call the combination of the $c()$ function and the $B[]$ arrays a **graph Positional Burrows Wheeler Transform (gPBWT)**. We submit that a gPBWT is sufficient to represent T, and, moreover, that it allows efficient counting of the number of threads in T that contain a given new thread as a subthread. Figure 2 and Table 1 give a worked example.

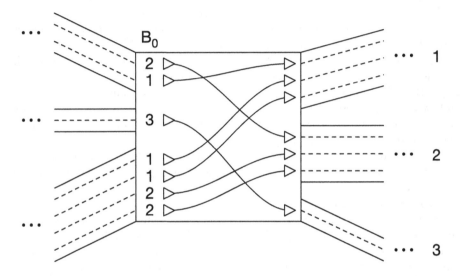

Fig. 1. An illustration of the $B_0[]$ array for a single side numbered 0. Threads visiting this side may enter their next nodes on sides 1, 2, or 3. The $B_0[]$ array records, for each visit of a thread to side 0, the side on which it enters its next node. This determines through which of the available edges it should leave the current node. Because threads tend to be similar to each other, they are likely to run in "ribbons" of multiple threads that both enter and leave together. These ribbons cause the $B_s[]$ arrays to contain runs of identical values, which may be compressed.

3   Extracting Threads

To reproduce T from G, and the gPBWT, consider each side s in G in turn. Establish how many threads begin (or, equivalently, end) at s by taking the minimum of $c(x, s)$ for all sides x adjacent to s. If s has no incident edges, take the length of $B_s[]$ instead. Call this number b. Then, for i running from 0 to b, exclusive, begin a new thread at $n(s)$ with the sides $[s, \bar{s}]$. Next, we traverse from $n(s)$ to the next node. Consult the $B_s[i]$ entry. If it is the null side, stop traversing, yield the thread, and start again from the original node s with the

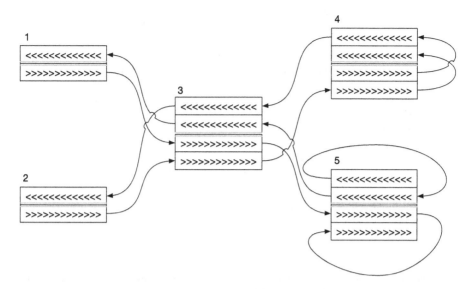

Fig. 2. A diagram of a graph containing two embedded threads. The graph consists of nodes $[1, 2, 3, 4, 5]$, with sides $[1L, 1R, 2L, 2R, \ldots]$, connected by edges $[1R, 3L]$, $[2R, 3L]$, $[3R, 4L]$, $[3R, 5L]$, $[4R, 4R]$, and $[5R, 5L]$. Embedded threads travel on the right-hand side of the nodes they are traveling through. Each thread here corresponds to a pair of "lanes" running in opposite directions. Visits are ordered from top to bottom, with "lanes" for lesser visits above those for greater ones. the "lanes" on the top half of each node are ordered in correspondence with the $B_s[]$ entries for the right side of the node, and those on the bottom half are ordered in correspondence with the $B_s[]$ entries for the left side of the node. The threads shown here are $[1L, 1R, 3L, 3R, 5L, 5R, 5L, 5R]$ and $[2L, 2R, 3L, 3R, 4L, 4R, 4R, 4L]$.

Table 1. $B_s[]$ and $c()$ values for the embedding of threads illustrated in Fig. 2.

Side	$B_s[]$ Array	Edge	$c(s,t)$ count
1L	[3L]	$c(1R, 3L)$	0
1R	[null]	$c(2R, 3L)$	1
2L	[3L]	$c(3R, 4L)$	1
2R	[null]	$c(3R, 5L)$	0
3L	[5L, 4L]	$c(4R, 4R)$	0
3R	[2R, 1R]	$c(5R, 5L)$	1
4L	[4R, 4R]	$c(3L, 1R)$	0
4R	[3R, null]	$c(3L, 2R)$	0
5L	[5L, null]	$c(4L, 3R)$	0
5R	[5R, 3R]	$c(5L, 3R)$	1

Algorithm 1. Algorithm for extracting threads from a graph.

---

function STARTING_AT(*Side*, *G*, *B*[], *c*())
    ▷ Count instances of threads starting at *Side*.
    ▷ Replace by an access to a partial sum data structure if appropriate.
    if *Side* has incident edges **then**
        return $c(s, Side)$ for minimum s over all sides adjacent to *Side*.
    else
        return LENGTH($B_{Side}[]$)
function RANK(*b*[], *Index*, *Item*)
    ▷ Count instances of *Item* before *Index* in *b*[].
    ▷ Replace by RANK of a rank-select data structure if appropriate.
    $Rank \leftarrow 0$
    for all index i in *b*[] **do**
        if $b[i] = Item$ **then**
            $Rank \leftarrow Rank + 1$
    return *Rank*
function WHERE_TO(*Side*, *Index*, *B*[], *c*())
    ▷ For thread visiting *Side* with *Index* in the reverse prefix sort order, get the corresponding sort index of the thread for the next side in the thread.
    return $c(\overline{Side}, B_{Side}[Index]) + $ RANK($B_{Side}[], Index, B_{Side}[Index]$)
function EXTRACT(*G*, *c*(), *B*[])
    ▷ Extract all oriented threads from graph *G*.
    for all Side s in G **do**
        $TotalStarting \leftarrow$ STARTING_AT($s, G, B[], c()$)
        for all i in $[0, TotalStarting)$ **do**
            $Side \leftarrow s$
            $Index \leftarrow i$
            $Thread \leftarrow [s, \overline{s}]$
            $NextSide \leftarrow B_{Side}[Index]$
            while $NextSide \neq$ null **do**
                $Thread \leftarrow Thread + [NextSide, \overline{NextSide}]$
                $Index \leftarrow$ WHERE_TO($Side, Index, B[], c()$)
                $Side \leftarrow NextSide$
                $NextSide \leftarrow B_{Side}[Index]$
            yield *Thread*

---

next i value less than b. Otherwise, traverse to side $s' = B_s[i]$. Calculate the arrival index i' as $c(\overline{s}, s')$ plus the number of entries in $B_s[]$ before entry i that are also equal to s'. This gives the index in $\overline{s'}$ of the thread being extracted. Then append s' and $\overline{s'}$ to the growing thread, and repeat the traversal process with $i \leftarrow i'$ and $s \leftarrow s'$, until the end of the thread is reached.

This process will enumerate all threads in the graph, and will enumerate each such thread twice (once from each end). The threads merely need to be deduplicated (such that two enumerated threads produce one actual thread, as the original collection of embedded threads may have had duplicates) in order to produce the collection of embedded threads T. Pseudocode for thread extraction is shown in Algorithm 1.

4   Succinct Storage

For the case of storing haplotype threads specifically, we can assume that, because of linkage, many threads in T are identical local haplotypes for long runs, diverging from each other only at relatively rare crossovers or mutations. Because of the reverse prefix sorting of the visits to each side, successive entries in the $B[]$ arrays are thus quite likely to refer to locally identical haplotypes, and thus to contain the same value for the side to enter the next node on. Thus, the $B[]$ arrays should benefit from run-length compression. Moreover, since (as will be seen below) one of the most common operations on the $B[]$ arrays will be expected to be rank queries, a succinct representation, such as a collection of bit vectors or a dynamic wavelet tree, would be appropriate. To keep the alphabet of symbols in the $B[]$ arrays small, it is possible to replace the stored sides for each $B[]$ with numbers referring to the nodes adjacent to $\bar{s}$.

We note that, for contemporary variant collections (e.g. the 1000 Genomes Project), the underlying graph G may be very large, while there may be relatively few threads (of the order of thousands) [1]. Implementers should thus consider combining multiple $B[]$ arrays into a single data structure to minimize overhead.

5   Embedding Threads

A trivial construction algorithm for the gPBWT is to independently construct $B_s[]$ and $c(s, s')$ for all sides s and oriented edges (s, s') according to their definitions above. However, this would be very inefficient. Here we present an efficient algorithm for gPBWT construction, in which the problem of constructing the gPBWT is reduced to the problem of embedding an additional thread.

Each thread is embedded by embedding its two orientations, one after the other. To embed a thread orientation $t = [t_0, t_1, \ldots t_{2N}, t_{2N+1}]$, we first look at node $n(t_0)$, entering by t_0. We insert a new entry for this visit into $B_{t_0}[]$, lengthening the array by one. The location of the new entry is near the beginning, before all the entries for visits arriving by edges, with the exact location determined by the arbitrary order imposed on thread orientations. Thus, its addition necessitates incrementing $c(s, t_0)$ by one for all oriented edges (s, t_0) incident on t_0 from sides s in G. If no other order of thread orientations suggests itself, the order created by their addition to the graph will suffice, in which case the new entry can be placed at the beginning of $B_{t_0}[]$. We call the location of this entry k. The value of the entry will be t_2, or, if t is not sufficiently long, the null side, in which case we have finished.

If we have not finished the thread, we first increment $c(s, t_2)$ by one for each side s adjacent to t_2 and after t_1 in the global ordering of sides. This updates the $c()$ function to account for the insertion into $B_{t_2}[]$ we are about to make. We then find the index at which the next visit in t ought to have its entry in $B_{t_2}[]$, given that the entry of the current visit in t falls at index k in $B_{t_0}[]$. This is given by the same procedure used to calculate the arrival index when extracting threads, denoted as $WHERE\_TO(t_1, k)$ (see Algorithm 1). Setting k to this value, we

Algorithm 2. Algorithm for embedding a thread in a graph.

---

procedure INSERT($b[]$, *Index*, *Item*)
  ▷ Insert *Item* at *Index* in $b[]$.
  ▷ Replace by INSERT of a rank-select-insert data structure if appropriate.
  LENGTH($b[]$) ← LENGTH($b[]$) + 1 ▷ Increase length of the array by 1
  for all i in (*Index*, LENGTH($b[]$) − 1], descending **do**
    $b[i] ← b[i − 1]$
  $b[Index] = Item$

procedure INCREMENT_C(*Side*, *NextSide*, $c()$)
  ▷ Modify $c()$ to reflect the addition of a visit to the edge (*Side*, *NextSide*).
  for all side s adjacent to *NextSide* in G **do**
    if $s >$ *Side* in side ordering **then**
      $c(s, NextSide) ← c(s, NextSide) + 1$

procedure EMBED(t, G, $B[]$, $c()$)
  ▷ Embed an oriented thread t in graph G.
  ▷ Call this twice to embed it for search in both directions.
  $k ← 0$ ▷ Index we are at in $B_{t_{2i}}[]$
  for all i in $[0, $ LENGTH(t)$/2)$ **do**
    if $2i + 2 <$ LENGTH(t) **then**
      ▷ The thread has somewhere to go next.
      INSERT($B_{t_{2i}}[], k, t_{2i+2}$)
      INCREMENT_C($t_{2i+1}, t_{2i+2}, c()$)
      $k ←$ WHERE_TO($t_{2i}, k, B[], c()$)
    else
      INSERT($B_{t_{2i}}[], k, $ null)

---

can then repeat the preceding steps to embed t_2, t_3, etc. until t is exhausted and its embedding terminated with a null-side entry. Pseudocode for this process is shown in Algorithm 2.

Assuming that the $B[]$ array information is both indexed for $O(log(n))$ rank queries and stored in such a way as to allow $O(log(n))$ insertion and update (in the length of the array n), this insertion algorithm is $O(N \cdot log(N + E))$ in the length of the thread to be inserted (N) and the total length of existing threads (E). Inserting M threads of length N will take $O(M \cdot N \cdot log(M \cdot N))$ time.

6   Counting Occurrences of Subthreads

The generalized PBWT data structure presented here preserves some of the original PBWT's efficient haplotype search properties [2]. The algorithm for counting all subthread instances in T of a new thread orientation t runs as follows.

We define f_i and g_i as the first and past-the-last indexes for the range of visits of threads in T to side t_{2i}, ordered as in $B_{t_{2i}}[]$.

For the first step of the algorithm, f_0 and g_0 are initialized to 0 and the length of $B_{t_0}[]$, respectively, so that they select all visits to node $n(t_0)$, seen as entering through t_0. On subsequent steps, f_{i+1} and g_{i+1}, are calculated from f_i and g_i merely by applying the $WHERE\_TO()$ function (see Algorithm 1). We calculate $f_{i+1} = WHERE\_TO(t_{2i}, f_i)$ and $g_{i+1} = WHERE\_TO(t_{2i}, g_i)$.

This process can be repeated until either $f_{i+1} \geq g_{i+1}$, in which case we can conclude that the threads in the graph have no matches to t in its entirety, or until t_{2N}, the last entry in t, has its range f_N and g_N calculated, in which case $g_N - f_N$ gives the number of occurrences of t as a subthread in threads in T. Moreover, given the final range from counting the occurrences for a thread t, we can count the occurrences of any longer thread that begins with t, merely by continuing the algorithm with the additional entries in the longer thread.

Assuming that the $B[]$ arrays have been indexed for $O(1)$ rank queries, the algorithm is $O(N)$ in the length of the subthread t to be searched for, and has a runtime independent of the number of occurrences of t. Pseudocode is shown in Algorithm 3.

Algorithm 3. Algorithm for searching for a subthread in the graph.

---

function COUNT(t, G, $B[]$, $c()$)
    ▷ Count occurrences of subthread t in graph G.
    $f \leftarrow 0$
    $g \leftarrow$ LENGTH($B_{t_0}[]$)
    for all i in $[0,$ LENGTH(t)$/2 - 1)$ **do**
        $f \leftarrow$ WHERE_TO($t_{2i}, f, B[], c()$)
        $g \leftarrow$ WHERE_TO($t_{2i}, g, B[], c()$)
        if $f \geq g$ **then**
            return 0
    return $g - f$

---

7   Results

The gPBWT was implemented within xg, the succinct graph indexing component of the **vg** variation graph toolkit [3]. Due to the succinct data structure libraries employed, efficient integer vector insert operations were not possible, and so a batch construction algorithm, applicable only to directed acyclic graphs, was implemented. A modified release of **vg**, which can be used to replicate the results shown here, is available from https://github.com/adamnovak/vg/releases/tag/gpbwt-paper.

The modified **vg** was used to create a genome graph for human chromosome 22, using the 1000 Genomes Phase 3 VCF on the hg19 assembly, embedding information about the correspondence between VCF variants and graph elements [1]. Note that the graph constructed from the VCF was directed and acyclic; it described only substitutions and indels, with no structural variants, and thus was amenable to the the batch gPBWT construction algorithm. Next, haplotype information for the 5,008 haplotypes stored in the VCF was imported and stored in a gPBWT-enabled xg index for the graph, using the batch construction algorithm mentioned above. In cases where the VCF specified self-inconsistent haplotypes (for example, a haplotype with a G to C SNP and a G to GAT insertion at the same position), they were broken apart at the inconsistent positions. The xg indexing and gPBWT construction process took 25 h and 45 min using a single indexing thread on an Intel Xeon X7560 running at 2.27 GHz, and consumed 344 GB of memory. The high memory usage was a result of the decision to retain the entire data set in memory in an uncompressed format during construction. However, the resulting xg index was 662 MB on disk, of which 573 MB was used by the gPBWT. Information on the 5,008 haplotypes across the 1,103,547 variants was thus stored in about 1.7 bits per phased diploid genotype in the succinct self-indexed representation, or 0.018 bits per haplotype base. Extrapolating linearly from the 51 megabases of chromosome 22 to the entire 3.1 gigabase human reference genome, a similar index of the entire 1000 Genomes dataset would take 40 GB, with 35 GB devoted to the gPBWT. This is well within the storage and memory capacities of modern computer systems.

Random Walks. To evaluate query performance, 1 million random walks of 100 bp each were simulated from the graph. To remove walks covering ambiguous regions, walks that contained two or more N bases in a row were eliminated, leaving 686,897 random walks. The number of haplotypes in the gPBWT index consistent with each walk was then determined, taking 81.30 s in total using a single query thread on the above-mentioned Xeon system. The entire operation took a maximum of 685 MB of memory, indicating that the on-disk index did not require significant expansion during loading to be usable. Overall, the gPBWT index required 118 microseconds per count operation on the 100 bp random walks. It was found that 317,681 walks, or 46 %, were not consistent with any haplotype in the graph. The distribution of of the number of haplotypes consistent with each random walk is visible in Fig. 3.

Read Mapping. To further evaluate the performance of the query implementation, 1000 Genomes Low Coverage Phase 3 reads for NA12878 that were mapped in the official alignment to chromosome 22 were downloaded and re-mapped to the chromosome 22 graph, using the xg/GCSA2-based mapper in vg, allowing for up to a single secondary mapping per read. The reads which mapped with scores of at least 90 points out of a maximum of 101 points (for a perfectly-mapped 101 bp read) were selected (so filtering out alignments highly like to be erroneous), and broken down into primary and secondary mappings. The number of haplotypes in the gPBWT index consistent with each read's path through the graph was calculated (Fig. 3). For 1,509,672 primary mappings, the count operation took 226.36 s in total, or 150 microseconds per mapping, again using 685 MB of memory. It was found that 13,918 of these primary mappings, or 0.9 %, and 1,280 of 57,115 secondary mappings, or 2.2 %, were not consistent with any haplotype path in the graph. These read mappings, despite having reasonable edit based scores, may represent rare recombinations, but the set is also likely to be enriched for spurious mappings.

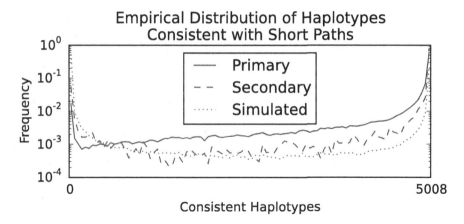

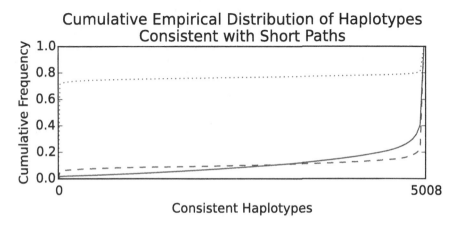

Fig. 3. Distribution (top) and cumulative distribution (bottom) of the number of 1000 Genomes Phase 3 haplotypes consistent with short paths in the chromosome 22 graph. Primary mappings of 101 bp reads with scores of 90 out of 101 or above ($n = 1,509,672$) are the solid blue line. Secondary mappings meeting the same score criteria ($n = 57,115$) are the dashed green line. Simulated 100 bp random walks in the graph without consecutive N characters ($n = 686,897$) are the dotted red line. Consistent haplotypes were counted using the gPBWT support added to **vg** [3].

8   Discussion

We have introduced the gPBWT, a graph based generalization of the PBWT. We have demonstrated that a gPBWT can be built for a substantial genome graph (all of human chromosome 22 and the associated chromosome 22 substitutions and indels in 1000 Genomes). Using this data structure, we have been able to quickly determine that the haplotype consistency rates of random walks and primary and secondary read mappings differ substantially from each other, and based on the observed distributions we hypothesize that consistency with very few haplotypes can be a symptom of a poor alignment. A sophisticated analysis of haplotype consistency rate distributions could thus improve alignment scoring (although from a variant calling perspective it would be incorrect to independently penalize each read with the full cost of a required recombination).

In the present experiment, we have examined only relatively simple variation: substitutions and short indels. More complex variation, like large inversions and translocations, which would have induced cycles in our genome graphs, was both absent from the 1000 Genomes data set we used and unsupported by the optimized DAG-based construction algorithm which we implemented. We expect that complex structural variation is well suited to representation as a genome graph, so supporting it efficiently should be a priority for a serious practical gPBWT construction implementation.

Extrapolating from our results on chromosome 22, we predict that a whole-genome gPBWT could be constructed for all 5,008 haplotypes of the 1000 Genomes data and stored in the main memory of a contemporary computer. Looking forward, this combination of genome graph and gPBWT could potentially enable efficient mapping not just to one reference genome or collapsed genome graph, but simultaneously to a large set of genomes related by a genome graph.

Acknowledgements. We would like to thank Richard Durbin for inspiration and David Haussler for his extremely helpful comments on the manuscript.

References

1. 1000 Genomes Project Consortium, et al.: A global reference for human genetic variation. Nature, **526**(7571), 68–74 (2015)
2. Durbin, R.: Efficient haplotype matching and storage using the positional Burrows-Wheeler transform (PBWT). Bioinformatics **30**(9), 1266–1272 (2014)
3. Garrison, E.: vg: the variation graph toolkit (2016). https://github.com/vgteam/vg/blob/80e823f5d241796f10b7af6284e0d3d3d464c18f/doc/paper/main.tex
4. Lunter, G.: Fast haplotype matching in very large cohorts using the Li and Stephens model. bioRxiv (2016). http://biorxiv.org/content/early/2016/04/12/048280
5. Medvedev, P., Brudno, M.: Maximum likelihood genome assembly. J. Comput. Biol. **16**(8), 1101–1116 (2009)
6. Paten, B., Novak, A., Haussler, D.: Mapping to a reference genome structure, April 2014, ArXiv e-Prints. http://arxiv.org/abs/1404.5010

Compact Universal k-mer Hitting Sets

Yaron Orenstein[1], David Pellow[2], Guillaume Marçais[3], Ron Shamir[2($\boxtimes$)],
and Carl Kingsford[3($\boxtimes$)]

[1] Computer Science and Artificial Intelligence Laboratory,
Massachusetts Institute of Technology, Cambridge, MA, USA
yaronore@mit.edu
[2] Blavatnik School of Computer Science, Tel-Aviv University, Tel-aviv, Israel
{dpellow,rshamir}@tau.ac.il
[3] School of Computer Science, Carnegie Mellon University, Pittsburgh, PA, USA
{gmarcais,carlk}@cs.cmu.edu

Abstract. We address the problem of finding a minimum-size set of
k-mers that hits L-long sequences. The problem arises in the design of
compact hash functions and other data structures for efficient handling
of large sequencing datasets. We prove that the problem of hitting a
given set of L-long sequences is NP-hard and give a heuristic solution
that finds a compact universal k-mer set that hits *any* set of L-long
sequences. The algorithm, called DOCKS (design of compact k-mer sets),
works in two phases: (i) finding a minimum-size k-mer set that hits every
infinite sequence; (ii) greedily adding k-mers such that together they
hit all remaining L-long sequences. We show that DOCKS works well
in practice and produces a set of k-mers that is much smaller than a
random choice of k-mers. We present results for various values of k and
sequence lengths L and by applying them to two bacterial genomes show
that universal hitting k-mers improve on minimizers. The software and
exemplary sets are freely available at acgt.cs.tau.ac.il/docks/.

1   Introduction

Inspired by Grabowski and Raniszewski's sampled suffix array using minimizers [1], we consider the following problem involving covering strings by selecting short k-mer substrings:

Problem 1. Given integers k and L, find a smallest set U_{kL} of k-mers such that any string of length L or longer must contain at least one k-mer from U_{kL}.

The set U_{kL} is called a *universal* set of hitting k-mers, and we call each k-mer in the set *universal*. Such a set has a number of applications in speeding up genomic analyses since it can often be used in places where minimizers have been used in the past [2]. For example:

1. **Hashing for read overlapping.** A naïve read overlapper must test $O(n^2)$ pairs of reads to see whether they overlap. If we require an overlap of length L, any pair of reads with such an overlap must contain a k-mer from set

© Springer International Publishing Switzerland 2016
M. Frith and C.N.S. Pedersen (Eds.): WABI 2016, LNBI 9838, pp. 257–268, 2016.
DOI: 10.1007/978-3-319-43681-4_21

U_{kL} in this overlapped region. By bucketing reads into bins according to the universal k-mers they contain, we need only test pairs of reads in the same bucket. The number of buckets is limited by $|U_{kL}|$.

2. **Sparse suffix arrays.** A sparse suffix array of a string S saves memory by storing an index for only every sth position in S [1,3]. To query a sparse suffix array for string q, we perform at most s queries starting from indices $0, \ldots, s - 1$ in q; one of these queries will intersect a position stored in the suffix array. Using U_{kL}, we can instead store only positions in S that start with a k-mer in U_{kL}. Any query with $|q| \geq L$ must contain one of these selected k-mers and will be matched when searching the suffix array.

3. **Bloom filters to speed up sequence search.** Bloom filters have been used to speed up sequence search by storing k-mers present in a read set for quick testing [4]. In current implementations, all k-mers present in a read set are stored in these filters. If, instead, only the set of k-mers in a U_{kL} is stored, any window of length $\geq L$ is still guaranteed to contain one of these representative queries, potentially reducing the size of Bloom filters that must be maintained.

Minimizers have been used for some of these and similar applications [5–7]. Minimizers are the lexicographically first k-mer within a window of length L, which were introduced by Roberts et al. [2] for genome assembly. MSP [8] compresses k-mers by hashing them to their 4-mer minimizer to efficiently construct a de Bruijn graph for assembly. SparseAssembler [9] represents the de Bruijn graph using only every g-th k-mer in the sequence (and has also been implemented using minimizers). Kraken [10] uses minimizers to speed up database queries for k-mers during metagenome sequence classification. The Locally Consistent Parsing (LCP) [11] provides the concept of "core substrings" which, like minimizers, are guaranteed to be shared by long enough identical strings. The SCALCE software package [12] uses core substrings to compress DNA sequences.

A universal set U_{kL}, if it can be found, has a number of advantages over minimizers for these applications. First, the set of minimizers for a given collection of reads may be as dense as the complete set of k-mers, whereas we show below that U_{kL} is often smaller by a factor of k. Second, for any k and L, the set of universal k-mers needs to be computed only once and not recomputed for every dataset. Third, the hash buckets, sparse suffix arrays, and Bloom filters created for different datasets will contain a comparable set of k-mers if they are sampled according to U_{kL}. The universal set of k-mers also has the advantage over dataset-specific sets because one does not need to look at all the reads before deciding on the k-mers to use, and one does not need to build a dataset-specific de Bruijn graph to select covering k-mers.

The need for faster and more memory efficient genomic analysis methods is rapidly increasing as fast as the size and depth of sequencing data is increasing. The NIH Sequence Read Archive, for example, contains over 3.5 petabytes of sequence data and is growing at a fast pace. Increased use of RNA-seq in many conditions and in clinical settings leads to high processing burdens. Metagenomic sampling at increasing depth to quantify and assemble microbes from environmental samples leads to even larger sequencing datasets. New ideas in indexing, data structures, and algorithms are essential to keep computational pace with this data generation. The minimizer idea has been extremely successful in reducing computational requirements. The universal set of k-mers proposed here will lead to further improvements in speed and memory.

The problem is also of theoretical interest as it can be rephrased as an equivalent problem on the complete (original) de Bruijn graph (see Definition 1 below). This is the viewpoint we take for most of this article:

Problem 2. Given a de Bruijn graph D_k of order k and an integer L, find a smallest set of vertices U_{kL} such that any path in D_k of length $L - k$ passes through at least one vertex of U_{kL}.

A solution to this problem may reveal additional hidden structure contained within the class of de Bruijn graphs.

We show that the problem of finding a minimum-size k-mer set that hits every string in a given set of L-long strings is NP-hard, further motivating the need for a universal k-mer set. We provide a heuristic called DOCKS that is based on the combination of three ideas. First, we use a decycling algorithm [13] to convert a complete de Bruijn graph into a directed acyclic graph (DAG) by removing a minimum number of k-mers, building off an implementation by Knuth [14]. We then supply a novel dynamic program to score remaining k-mers by the number of remaining length-ℓ paths that they hit. Finally, we use that dynamic program in a greedy heuristic to select the additional k-mers and produce a small universal set $\hat{U}_{kL}$, which we show empirically to often be close to the optimal size. Our use of a greedy heuristic is motivated by providing a proof that finding a small ℓ-path cover in a graph G is NP-hard even when G is a DAG.

DOCKS provides the first practical solution to the identification of universal sets of k-mers. The software is freely available on acgt.cs.tau.ac.il/docks/, as are universal sets of k-mers over a range of values of L and k. We report on the size of the universal k-mer hitting set produced by DOCKS and demonstrate on two datasets that we can better cover sequences with a smaller set of k-mers than is possible using minimizers. Our results also provide a starting point for additional theoretical investigation of these path coverings of de Bruijn graphs.

2 Definitions

Definition 1 (de Bruijn Graph). *A de Bruijn graph of order k over alphabet Σ is a directed graph in which every vertex has an associated label (a string over Σ) of length k (k-mer) and every edge has an associated label of length $k + 1$. There are exactly $|\Sigma|^k$ vertices in a de Bruijn graph, each representing a unique k-mer. If an edge (u, v) has a label l, then the label of u must be a k-prefix of l and the label of v must be a k-suffix of l.*

A *complete* de Bruijn graph contains all possible edges, which represent together all $(k + 1)$-mers over Σ. Every path in a de Bruijn graph represents a sequence. A path $v_0, e_0, v_1, e_1, v_2, \ldots, v_n$ of length n spells a sequence s of length $n + k$ such that the label of v_i occurs in s starting at position i for all $0 \leq i \leq n$, and the label of e_i occurs in s starting at position i for all $0 \leq i \leq n - 1$. Note that vertices may repeat in a path.

We will need two bits of terminology involving k-mers intersecting and not intersecting sequences over an alphabet Σ:

Definition 2 (hits). *K-mer w hits sequence S if $w \subseteq S$, i.e. w appears as a contiguous substring in string S. K-mer set X hits sequence S if $\exists w \in X$ s.t. $w \subseteq S$. Denote $hit(w, L) = \{S \in \Sigma^L \mid w \subseteq S\}$ for k-mer w and length L. Denote $hit(X, L) = \underset{w \in X}{\cup} hit(w, L)$.*

Definition 3 (avoids). *Sequence S avoids k-mer w if $w \nsubseteq S$. Sequence S avoids k-mer set X if $\forall w \in X, w \nsubseteq S$. Denote $avoid(w, L) = \Sigma^L \setminus hit(w, L)$ for k-mer w and similarly $avoid(X, L) = \Sigma^L \setminus hit(X, L)$ for k-mer set X.*

3   Methods

It is not known how to efficiently find a minimum universal (k, L)-hitting set. As we show in Sect. 4, the corresponding problem when restricted to a given set of input sequences is NP-hard (Sect. 4.1). Here, we give a practical heuristic to find small (but non-optimal) universal k-mer sets. This algorithm works in two steps: first it finds and removes a minimum-size k-mer set hitting all infinite sequences, and then it removes additional k-mers to hit all remaining L-long sequences. We now describe these two steps in detail.

3.1   Finding a Minimum k-mer Set Hitting All Infinite Sequences

The problem of finding a minimum-size k-mer set hitting all infinite sequences is known in the literature as finding an 'unavoidable set' of constant length [15]. This is a set of words of the same length k that hits any infinite word (but finite words may avoid the set). The problem of finding an unavoidable set for a given k can be solved in time polynomial in the output size [15]. The original algorithm is due to Mykkeltveit [13]. Its running time is $O(kM(k))$, where $M(k)$ is the size of the minimum unavoidable set. $M(k)$ converges to $|\Sigma|^k/k$ (an exact formula is given in Sect. 5.1, Eq. 5), so the running time is $O(|\Sigma|^k)$.

An unavoidable set of constant length k is equivalent to a set of vertices in a complete de Bruijn graph of order k whose removal turn it into a DAG. Each k-mer in the set corresponds to a vertex, and the removal of vertices from every cycle guarantees that no infinite sequence is represented as a path in the graph. This set is known as a *decycling set*.

3.2   A Greedy Algorithm to Hit All Remaining L-long Sequences

Unfortunately, finding an unavoidable set is not enough, as there may be L-long sequences that avoid that set. Thus, we need additional k-mers to hit those.

If we consider the graph formulation, after removal of the unavoidable set from the graph, we are left with a directed acyclic graph, which may contain $(L - k)$-long paths representing L-long sequences. We need to remove additional vertices, so that there is no path of length $\ell = L - k$. The problem of finding a minimum-size set of vertices that hit all ℓ-long paths in a directed acyclic graph is NP-hard, as we prove in Subsect. 4.2. Therefore, we give a heuristic algorithm.

Our algorithm is based on the greedy algorithm for the minimum hitting set [16]. We define the *hitting number* $T_\ell(v)$ of a vertex v as the number of paths of length ℓ that contain it. The main observation is that we can calculate the hitting number of each vertex efficiently using dynamic programming. The solution is based on calculating the number of paths of length i that terminate at vertex v, and the number of paths of length i that start at vertex v, for all $v \in V$ and $0 \le i \le \ell$. Then, the number of ℓ-long paths through v is directly computable from these values by breaking any path into a i-long path ending at v and a $(\ell - i)$-long path starting at v, for all possible values of i. We set $\ell = L - k$ to get the hitting number of each vertex.

Specifically, let $G' = (V', E')$ be the directed acyclic graph, after removing the decycling set. Denote by D and F matrices of size $|V'| \times (\ell+1)$, where $D(v, i)$ is the number of i-long paths in G' starting at vertex v and $F(v, i)$ is the number of i-long paths ending at vertex v.

The calculation of D and F is as follows:

$$D(v, 0) = F(v, 0) = 1, \forall v \in V' \tag{1}$$

$$D(v, i) = \sum_{(v,u)\in E'} D(u, i - 1) \tag{2}$$

$$F(v, i) = \sum_{(u,v)\in E'} F(u, i - 1) \tag{3}$$

To get the number of ℓ-long paths vertex v participates in, we sum:

$$T_\ell(v) = \sum_{i=0}^{\ell} F(v, i) \times D(v, \ell - i) \tag{4}$$

The running time is proportional to the sum of all vertex degrees (which is $\Theta(|E|)$) times ℓ, giving a running time of $O(|\Sigma|^k \cdot \ell)$ for $\ell = L - k$.

3.3   The Complete DOCKS Algorithm

To get the complete algorithm, we combine the two steps. First, we find a decycling set in a complete de Bruijn graph of order k and remove it from the graph. Then, we repeatedly remove a vertex v with the largest hitting number $T_\ell(v)$ until there are no ℓ-long paths, recomputing $T_\ell(u)$ for all remaining u after each removal. This is summarized below (Algorithm DOCKS).

Finding the decycling set takes $O(|\Sigma|^k)$, as the size of the set is $\Theta(|\Sigma|^k/k)$ and the running time for finding each k-mer is $O(k)$ [13]. In the second phase,

---

Algorithm 1. DOCKS: Find a small k-mer set hitting all L-long sequences

---
1: Generate a complete de Bruijn graph G of order k, set $\ell = L - k$.
2: Find a decycling vertex set X using Mykkeltveit's algorithm.
3: Remove all vertices in X from graph G, resulting in G'.
4: **while** there are still paths of length ℓ **do**
5:    Calculate the number of starting and ending i-long paths at each vertex, for
        $0 \leq i \leq \ell$.
6:    Calculate the hitting number for each vertex.
7:    Remove a vertex with maximum hitting number from G', and add it to set X.
8: **end while**
9: Output set X.

---

each iteration calculates the hitting number of all vertices using dynamic programming in time $O(|\Sigma|^k L)$. The number of iterations is $1 + p$, where p is the number of vertices removed. Thus, the total running time is dominated by steps 4–8 and is $O((1 + p)|\Sigma|^k L)$.

4   Complexity

4.1   NP-hardness of MINIMUM (k, L)-HITTING SET

The problem of finding a dataset-specific hitting set is NP-hard, further motivating the need for the design of a universal k-mer set:

MINIMUM (k, L)-HITTING SET
INSTANCE: Set S of L-long sequences over Σ and k.
VALID SOLUTION: Set X of k-mers s.t. $S \subseteq hit(X, L)$.
GOAL: Minimize $|X|$.

We prove that MINIMUM (k, L)-HITTING SET is NP-hard. For simplicity, we study the problem on DNA alphabet, but it can be easily generalized to any finite alphabet Σ. We show a reduction from HITTING SET [17]. While the problems look similar, HITTING SET is not a special case of the other since in HITTING SET the subsets are arbitrary, while in MINIMUM (k, L)-HITTING SET problem each subset is made of overlapping k-mers.

Theorem 1. *MINIMUM (k, L)-HITTING SET is NP-hard.*

Proof. Given an input to HITTING SET, a set S of subsets of $E = \{e_1 \dots e_n\}$, we generate an input to MINIMUM (k, L)-HITTING SET problem as follows: Denote by m the size of the maximum cardinality set, i.e. $m = \max_{S_i \in S} |S_i|$. We choose $\ell = \lceil \log_2(\max(m, n)) \rceil$, $L = 3\ell m$ and $k = 2\ell$. We map each set $S_i \in S$ to a ℓ-long binary representation of i, where instead of bits we use nucleotides C and G. We map each element $e_j \in E$ to a ℓ-long binary representation of j, where instead of bits we use nucleotides A and T. We call these representations the set's $\{C, G\}$-representation and the element's $\{A, T\}$-representation and denote them by $f_{CG}(S_i)$ and $f_{AT}(e_j)$.

We generate a sequence set T, which is the input to MINIMUM (k, L)-HITTING SET. For each set $S_i \in S$ we generate a sequence that contains all of its elements' $\{A, T\}$-representations, each appearing twice consecutively and buffered by the set's $\{C, G\}$-representation. Formally, for the set $S_i = \{e_{i_1}, \ldots, e_{i_{|S_i|}}\}$ we create the sequence: $T_i :=$ $(\prod_{j=1}^{|S_i|} f_{AT}(e_{i_j}) \cdot f_{AT}(e_{i_j}) \cdot f_{CG}(S_i)) \cdot (f_{AT}(e_{i_1}) \cdot f_{AT}(e_{i_1}) \cdot f_{CG}(S_i))^{m-|S_i|}$ (here $\prod$ indicates concatenation). The new instance T is $\{T_1, \ldots, T_{|S|}\}$.

Denote by T^{OPT} an optimal solution to MINIMUM (k, L)-HITTING SET. If a k-mer contains a complete $\{A, T\}$-representation w, then the element $f_{AT}^{-1}(w)$ is in the optimal solution to HITTING SET. If a k-mer contains a complete $\{C, G\}$-representation w, then any element from the set $f_{CG}^{-1}(w)$ can be part of the optimal solution. The running time of the reduction is bounded by $O(|S| \times L)$ to generate the input sequence set T. In terms of m and n the running time is $O(|S| \cdot m \cdot (\log(m) + \log(n)))$.

We now prove the correctness of the reduction. We start with proving several properties of the solution.

Lemma 1. *A k-mer that contains a complete $\{A, T\}$-representation w can be replaced by k-mer ww to produce a hitting set of the same cardinality.*

Proof. The k-mer contains a complete $\{A, T\}$-representation w. Thus, it can only hit sequences that contain w. Since the sequences were constructed to contain two adjacent $\{A, T\}$-representations per element, and since this representation is unique, k-mer ww hits the same set of sequences.   □

Lemma 2. *A k-mer that contains a complete $\{C, G\}$-representation can be replaced by a k-mer that contains two adjacent occurrences of any $\{A, T\}$-representation from this sequence to produce a hitting set of the same cardinality.*

Proof. A $\{C, G\}$-representation is unique to each sequence. Thus, it can only hit one sequence, and replacing it by any other k-mer from that sequence preserves the hitting properties of the set.   □

We now prove the two sides of the reduction:

1. MINIMUM (k, L)-HITTING SET $\Rightarrow$ HITTING SET: all L-long sequences in T are hit by k-mers in T^{OPT}. By Lemmas 1 and 2 we can transform any hitting set to a hitting set of the same cardinality, but containing only k-mers over $\{A, T\}$. These correspond to elements in an optimal solution of HITTING SET. Assume contrary that there is a smaller solution U to HITTING SET. Then, the set $\{f_{AT}(w) \cdot f_{AT}(w) \mid w \in U\}$ hits all sequences in the k-mer hitting problem, and by that producing a smaller solution, contrary to its optimality.
2. HITTING SET $\Rightarrow$ MINIMUM (k, L)-HITTING SET: denote S^{OPT} an optimal solution to HITTING SET. Then, a set of k-mers $\{f_{AT}(w) \cdot f_{AT}(w) \mid w \in S^{OPT}\}$ is an optimal solution to MINIMUM (k, L)-HITTING SET.

Assume contrary that there is a smaller solution U to MINIMUM (k, L)-HITTING SET. By Lemmas 1 and 2 there is a solution composed of k-mers over $\{A, T\}$. The set of element $\{f_{AT}^{-1}(w_{1:k/2}) \mid w \in U\}$ is a smaller hitting set in HITTING SET, contrary to its optimality.   □

4.2 NP-hardness of MINIMUM ℓ-PATH COVER IN A DAG

Our heuristic to find U_{kL} searches for a minimum ℓ-path cover in the DAG created after removing a decycling set (Sect. 3.1). We show now that this problem is in general NP-hard (by a reduction from VERTEX COVER [17]) — motivating our use of a greedy heuristic to solve this subproblem.

MINIMUM ℓ-PATH VERTEX COVER IN A DAG
INSTANCE: A directed acyclic graph $G = (V, E)$ and integer ℓ.
VALID SOLUTION: Vertex set X s.t. $G' = (V \setminus X, E)$ contains no ℓ-long paths.
GOAL: Minimize $|X|$.

Theorem 2. *MINIMUM ℓ-PATH COVER IN A DAG is NP-hard.*

Proof. Given a graph $G = (V, E)$ as input to VERTEX COVER, we construct an instance to MINIMUM ℓ-PATH COVER IN A DAG as follows. We first remove from G any vertices that are incident to self-loop edges, since these must be part of any vertex cover. We then transform the remaining graph into a DAG by arbitrarily ordering the vertices of G, and directing the edges from lower-index to higher-index vertices. Since there are no self-loops, the result is a DAG $D = (V, A)$. The input to the ℓ-path cover is $I = (D, 1)$. The running time of the reduction is linear in the size of the graph.

A set of vertices $U \subseteq V$ is a vertex cover in G iff it intersects every edge in E. But this is true iff it hits every path of length 1 in D. Hence, U is a minimum vertex cover iff it is a minimum 1-path cover in D.   □

5   Results

5.1   A Theoretical Lower Bound for the Number of k-mers

For a given k-mer w, its conjugacy class is the set of k-mers obtained by rotation of w. Conjugacy classes form cycles in the de Bruijn graph and form a partition of the k-mers. The number of conjugacy classes over all k-mers is given by [15]

$$C(|\Sigma|, k) = \sum_{i=1}^{k} |\Sigma|^{\gcd(i,k)} / k. \tag{5}$$

A decycling set necessarily contains a k-mer in each conjugacy class. Golomb's conjecture, proved by Mykkeltveit [13], states that the smallest decycling set has cardinality $C(|\Sigma|, k)$. Consequently, a (k, L)-hitting set has a size $\geq C(|\Sigma|, k)$.

Table 1 reports $L_{max} = \ell + k$, the length of the longest sequence in a complete de Bruijn graph after the decycling set is removed. The length of sequences avoiding the decycling set is too long for most applications. Additional k-mers must be selected to obtain a hitting set for smaller longest paths.

Table 1. Maximum length of longest sequence avoiding an unavoidable set for different k. For each value k, a decycling set was removed from a complete de Bruijn graph, and the length L_{max} of the longest sequence, represented as a longest path, was calculated.

k	2	3	4	5	6	7	8	9	10	11	12	13	14
L_{max}	5	11	20	45	70	117	148	239	311	413	570	697	931

5.2 Computational Results

We implemented and ran DOCKS over a range of k and L: $5 \leq k \leq 9$ with $20 \leq L \leq 200$, in increments of 10. These are typical values used for minimizers of longer k-mers and read lengths of short read sequences. We also implemented two random procedures that we compare to as baselines. One, termed "random", removes random vertices until no $\ell = L - k$ paths remains. The second, termed "decycling+random" (DR), first removes a minimum-size decycling set and then randomly removes vertices until no path of length $\ell = L - k$ exists. In both cases checking the termination condition is done by first testing if there are any cycles, and if there are no cycles, computing the maximum-length path, which takes linear time in a DAG.

The results are summarized in Fig. 1. Our method outputs a set of k-mers that is much smaller than both random procedures. The results also show that there is a significant benefit in removing a minimum-size decycling set first and then additional vertices if we wish to hit all ℓ-long paths, as the random procedure that starts from the complete graph performs far worse than the one that is applied to the graph after removing an optimal decycling set. Note that random sometimes removes the same number for different values of L, since by the time it gets an acyclic graph, only short paths remain. As expected, the ratio compared to the lower bound decreases with L. It is easier to hit longer sequences as they contain more k-mers.

Table 2. Running times of the DOCKS algorithm for different k and L values. The user run time is in seconds (s) or minutes (m).

k/L	100	110	120	130	140	150	160	170	180	190	200
7	0.7 s	0.5 s	0.4 s	0.4 s	0.4 s	0.4 s	0.4 s	0.4 s	0.4 s	0.4 s	0.4 s
8	11.1 s	7.6 s	4.3 s	2.6 s	1.3 s	0.7 s	0.7 s	0.7 s	0.7 s	0.7 s	0.7 s
9	8.8 m	6.9 m	5.4 m	4.2 m	3.2 m	2.5 m	1.8 m	1.4 m	1.0 m	0.7 m	0.5 m

Table 2 reports the running times for different values of k and L. For all instances, the dominant running time is of the second step, greedily finding an ℓ-path cover. This computation needs to be done only once per (k, L) pair. Running times were benchmarked on a single CPU of a 12-CPU Intel Xeon X5680 (3.33 GHz) machine with 47 GB 1333 MHz RAM.

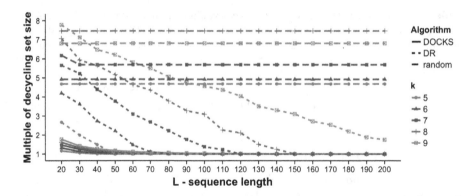

Fig. 1. Performance of DOCKS. For different combinations of k and L we ran DOCKS and two random procedures over the DNA alphabet. The results are shown in comparison to the size of the decycling set. When the ratio is 1, all the sequences avoiding the decycling set were of length shorter than L. DR: decycling+random.

5.3   Comparison to Minimizers on Bacterial Genomes

Although the number of universal hitting k-mers for a given path length can be a significant proportion of all k-mers (around $|\Sigma|^k/k$), the actual number of k-mers hitting a given sequence set is much less, even less than the number of minimizers. In Table 3, we compare the distribution of the universal hitting k-mers and the minimizers in two bacterial genomes. *Acetobacter tropicalis* (RefSeq NZ_CP011120) has a genome of 2.8 Mbp and a GC content of 47.8 %. *Caulobacter vibriodes* (RefSeq NC_002696) is larger at 4.0 Mbp and has a higher GC content of 67.2 %. For each genome, we computed the number of minimizers using $k = 8$ and a window length of 100. Also, for each window of 100 bases we found a k-mer from the set $\hat{U}_{kL}$ for $k = 8, L = 100$, computed by DOCKS. Each such window is guaranteed to contain at least one universal k-mer, and usually more than one. In each window, we select only one of the universal k-mers, the smallest one in lexicographic order. In addition, we measured the distances between the selected k-mers (minimizers or universal k-mers) and computed the mean and standard deviation of the distances. Using universal hitting k-mers instead of minimizers gives a smaller set of selected k-mers, which is also sparser in the sequence and more evenly distributed.

Table 3. Comparison of the number of selected minimizers and universal k-mers, for $k = 8, L = 100$, and their distribution, in bacterial genomes. We report the mean distance ($\pm$std) between positions at which consecutive selected k-mers appear in the sequence.

	Minimizers		Universal k-mers	
	Selected	Mean distance	Selected	Mean distance
Acetobacter	3119	44.1 ± 33.6	2439	50.8 ± 29.2
Caulobacter	7315	47.2 ± 31.0	4585	51.2 ± 28.4

6 Conclusion

In this work, we presented the DOCKS algorithm, which generates a compact set of k-mers that together hit all L-long DNA sequences. DOCKS's good performance can be attributed to its two components. It first optimally removes a minimum-size set that hits all infinite sequences, which takes care of most L-long sequences. It then greedily removes vertices that hit remaining L-long sequences. Its feasibility stems from the first step, which runs in time $O(k)$ times the size of the output, and the second step, which uses dynamic programming to bound the running time to be quadratic in the output size times L.

A limitation of our approach is its heuristic nature, which does not guarantee any ratio over the optimal solution. Unfortunately, as we show, the general problem of finding a minimum (k, L)-hitting set is NP-hard. On top of that, even after removing a decycling edge set, the problem of finding a minimum set that hits all L-long sequences in a directed acyclic graph is NP-hard.

Some problems from this work remain open. First, is the problem of the universal (k, L)-hitting set polynomial in $O(|\Sigma|^k)$? The size of the output $\Theta(|\Sigma|^k/k)$ is doubly exponential in the size of the input (the parameters k and L), but the computational complexity remains open. Second, is the problem of ℓ-path cover in a DAG polynomial in the special case of directed acyclic subgraphs of de Bruijn graphs? Third, since the dominant run time is the second phase, which re-calculates the vertex hitting numbers on each iteration, can we update this number more efficiently after the removal of one vertex? Fourth, is there a tight upper bound on the number p of vertices that will be removed by the greedy heuristic? Fifth, can we give an upper bound or a tighter lower bound on the size of U_{kL}? Sixth, is the ℓ-path cover problem polynomial for $L > 1$?

In conclusion, we demonstrated the ability of DOCKS to generate compact sets of k-mers that hit all L-long sequences. These k-mer sets can be generated once for any desired value of k and L and then used easily for many different purposes. For example, there is a set of only 700 6-mers out of a total of 4096 that hits every sequence longer than 70 bases — a typical read length for many sequencing experiments — enabling efficient binning of reads. These sets of k-mers could improve many of the applications that use minimizers, as we showed that they are both smaller and more evenly distributed across typical sequences.

Acknowledgments. R.S. was supported in part by the Israel Science Foundation as part of the ISF-NSFC joint program 2015–2018. D.P. was supported in part by a Ph.D. fellowship from the Edmond J. Safra Center for Bioinformatics at Tel-Aviv University. This research is funded in part by the Gordon and Betty Moore Foundation's Data-Driven Discovery Initiative through Grant GBMF4554 to C.K., by the US National Science Foundation (CCF-1256087, CCF-1319998) and by the US National Institutes of Health (R01HG007104). C.K. received support as an Alfred P. Sloan Research Fellow. Part of this work was done while Y.O., R.S. and C.K. were visiting the Simons Institute for the Theory of Computing.

References

1. Grabowski, S., Raniszewski, M.: Sampling the suffix array with minimizers. In: Iliopoulos, C., Puglisi, S., Yilmaz, E. (eds.) SPIRE 2015. LNCS, vol. 9309, pp. 287–298. Springer, Heidelberg (2015)
2. Roberts, M., Hayes, W., Hunt, B.R., Mount, S.M., Yorke, J.A.: Reducing storage requirements for biological sequence comparison. Bioinformatics **20**, 3363–3369 (2004)
3. Karkkainen, J., Ukkonen, E.: Sparse suffix trees. In: Cai, J.-Y., Wong, C.K. (eds.) COCOON 1996. LNCS, vol. 1090, pp. 219–230. Springer, Heidelberg (1996)
4. Solomon, B., Kingsford, C.: Fast search of thousands of short-read sequencing experiments. Nat. Biotechnol. **34**, 300–302 (2016)
5. Movahedi, N.S., Forouzmand, E., Chitsaz, H.: De novo co-assembly of bacterial genomes from multiple single cells. In: 2012 IEEE International Conference on Bioinformatics and Biomedicine (BIBM), pp. 1–5 (2012)
6. Deorowicz, S., Kokot, M., Grabowski, S., Debudaj-Grabysz, A.: KMC 2: fast and resource-frugal k-mer counting. Bioinformatics **31**(10), 1569–1576 (2015). Oxford Univ Press
7. Chikhi, R., Limasset, A., Jackman, S., Simpson, J.T., Medvedev, P.: On the representation of de Bruijn graphs. J. Comput. Biol. **22**, 336–352 (2015)
8. Li, Y., Kamousi, P., Han, F., Yang, S., Yan, X., Suri, S.: Memory efficient minimum substring partitioning. In: Proceedings of the VLDB Endowment, vol. 6, pp. 169–180. VLDB Endowment (2013)
9. Ye, C., Ma, Z.S., Cannon, C.H., Pop, M., Douglas, W.Y.: Exploiting sparseness in de novo genome assembly. BMC Bioinform. **13**, S1 (2012)
10. Wood, D.E., Salzberg, S.L.: Kraken: ultrafast metagenomic sequence classification using exact alignments. Genome Biol. **15**, R46 (2014)
11. Sahinalp, S.C., Vishkin, U.: Efficient approximate and dynamic matching of patterns using a labeling paradigm. In: 37th Annual Symposium on Foundations of Computer Science, Proceedings, pp. 320–328 (1996)
12. Hach, F., Numanagi, I., Alkan, C., Sahinalp, S.C.: SCALCE: boosting sequence compression algorithms using locally consistent encoding. Bioinformatics **28**, 3051–3057 (2012)
13. Mykkeltveit, J.: A proof of Golomb's conjecture for the de Bruijn graph. J. Comb. Theory Ser. B **13**, 40–45 (1972)
14. Knuth, D.E.: Unavoidable2 (2003). http://www-cs-faculty.stanford.edu/uno/programs/unavoidable2.w
15. Champarnaud, J.M., Hansel, G., Perrin, D.: Unavoidable sets of constant length. Int. J. Algebra Comput. **14**, 241–251 (2004)
16. Chvatal, V.: A greedy heuristic for the set-covering problem. Math. Oper. Res. **4**, 233–235 (1979)
17. Karp, R.M.: Reducibility among combinatorial problems. In: Jünger, M., Liebling, T.M., Naddef, D., Nemhauser, G.L., Pulleyblank, W.R., Reinelt, G., Rinaldi, G., Wolsey, L.A. (eds.) 50 Years of Integer Programming 1958–2008, pp. 219–241. Springer, Heidelberg (2010)

A New Approximation Algorithm for Unsigned Translocation Sorting

Lianrong Pu, Daming Zhu$^{(\boxtimes)}$, and Haitao Jiang

School of Computer Science and Technology, Shandong University, Jinan, China
lianrong.pu@gmail.com, {dmzhu,htjiang}@sdu.edu.cn

Abstract. Translocation has long been learned as a basic operation to rearrange genomes. Signed translocation sorting can be solved in polynomial time. Unsigned translocation sorting turns to be NP-Hard and Max-SNP-Hard. The best known algorithm by now for unsigned translocation sorting can achieve a performance ratio 1.408. In this paper, we propose a new approximation algorithm for unsigned translocation sorting, which can achieve a performance ratio 1.375 in polynomial time.

Keywords: Algorithm · Complexity · Approximation · Genome · Rearrangement · Translocation

1 Introduction

Rearrangement accounts for the stories of gene order variations in a genome, and can be formalized into such basic operations as reversal, translocation, transposition, etc. [15]. *Rearrangement sorting* of genomes asks to find a shortest sequence of rearrangements that transforms one genome into another, which can be used to trace the evolutionary path between genomes.

Hannenhalli and Pevzner showed that reversal sorting can be solved in polynomial time for signed linear genomes [8]. Subsequent work improved the running time of their algorithm and simplified the underlying theory [2,11]. On the other hand, reversal sorting for unsigned linear genomes is NP-Hard [3]. Unsigned reversal sorting can be approximated to a performance ratio 1.375 [2]. More related algorithmic developments of rearrangement sorting can be consulted in [12,13], as well as [16] for the rearrangement sorting's applications in computational genomics.

There has been known translocation operations happen in living genomes to drive their evolutions [6]. The signed translocation sorting was first studied by Kececioglu and Ravi [10]. An algorithm with the time complexity $O(n^3)$ was first devised by Hannenhalli [7], where n is the gene number of the genomes. Then several improved algorithms have been proposed to reduce the time complexity. Zhu and Ma first improved the algorithm to run in $O(n^2 log n)$ time [20]. Wang et al. again improved it to run in $O(n^2)$ [18]. The algorithm of Ozery-Flato and Shamir with a time complexity $O(n^{\frac{3}{2}}\sqrt{\log n})$ is the fastest seemingly [17].

© Springer International Publishing Switzerland 2016
M. Frith and C.N.S. Pedersen (Eds.): WABI 2016, LNBI 9838, pp. 269–280, 2016.
DOI: 10.1007/978-3-319-43681-4_22

Translocation sorting of unsigned genomes turns to be NP-hard and Max-SNP-Hard [19]. Kececcioglu and Ravi first proposed a 2-approximation algorithm for this problem [10]. Cui et al. improved the performance ratio to 1.75 [4], and further to $1.5 + \epsilon$ [5]. The most recent approach for unsigned translocation sorting was proposed by Jiang et al., whose algorithm can achieve a performance ratio of $1.408 + \epsilon$ [9].

In this paper, we present a new approximation algorithm for unsigned translocation sorting, which can achieve a performance ratio of 1.375. The unsigned translocation sorting can be approximated by decomposing a breakpoint graph with respect to two unsigned genomes into a set of alternating cycles, which equivalently, assigns signs to those genes in the unsigned genomes. We will first decompose a breakpoint graph into a set of disconnected components, such that each component can be decomposed into alternating cycles alone, then decompose each component into as many cycles as those needed to approximate the unsigned translocation sorting to 1.375.

2   Preliminaries for Translocation Sorting

A *genome* is a set of chromosomes, where a *chromosome* is a sequence of genes. Usually, a gene is *signed* and represented by an integer with a sign '+' or '−' before it to indicate its direction, or *unsigned* where its direction is unknown, and represented by just an integer. A genome is *signed* if all its genes are signed, and *unsigned*, otherwise. A chromosome in a signed (resp. unsigned) genome is also *signed* (resp. *unsigned*). Each gene in a genome will be represented by an unique integer. Thus, a genome with n genes admits to have the gene set $\{1, 2, ..., n\}$.

Let $X = [x_1, x_2, ..., x_m]$ be a chromosome in a genome. A *segment* within X is a consecutive subsequence of X, where we refer to the first and the last genes of the segment as the *ends* of it. A segment within X is *signed* (resp. *unsigned*), if X is signed (resp. unsigned). Let $I = [x_i, ..., x_j]$ be a signed (resp. unsigned) segment within a chromosome, then $-I = [-x_j, ..., -x_i]$ (resp. $[x_j, ..., x_i]$). Two segments or chromosomes, say X and Y, are *congruent*, if either $X = Y$ or $X = -Y$. Two genomes are congruent if their chromosome sets are the same, provided that two congruent chromosomes are the same. It will be abbreviated as $A = B$ for A and B to be congruent.

Let $X = [x_1, x_2, ..., x_m]$ and $Y = [y_1, y_2, ..., y_n]$ be two chromosomes. A reciprocal translocation on X and Y first breaks X into two non-empty segments $X_1 = [x_1, ..., x_{i-1}], X_2 = [x_i, ..., x_m]$, and breaks Y into two non-empty segments $Y_1 = [y_1 ..., y_{j-1}], Y_2 = [y_j, ..., y_n]$, then using one of the two ways, reconstructs two new chromosomes from X_1, X_2 and Y_1, Y_2. A *prefix-prefix translocation* connects X_1 with Y_2 and Y_1 with X_2, creates two new chromosomes X_1Y_2 and Y_1X_2. A *prefix-suffix translocation* connects X_1 with $-Y_1$ and $-X_2$ with Y_2, creates two new chromosomes $X_1 - Y_1$ and $-X_2Y_2$. For a translocation ρ on two chromosomes in genome A, we denote by $A \bullet \rho$ the genome transformed from A due to ρ acting on it. Then the translocation sorting problem is,

Instance: Two genomes A and B, which have the same gene sets.

Question: Find a sequence of translocations ρ_1, ρ_2, ..., ρ_t, such that $A \bullet \rho_1 \bullet \rho_2 \bullet \cdots \bullet \rho_t = B$, and t is minimized.

The *translocation distance* between A and B is the minimum number of translocations that transform A into B.

Two genes are *adjacent in a genome* if in the genome, they are adjacent on a chromosome. Let A and B be two signed genomes each of which has the gene set $\{1, ..., n\}$. A *breakpoint graph* with respect to A and B can be constructed as follows. For each gene x in A, we set two vertices $l(x)$, $r(x)$. For two adjacent genes x, y in A where x is on the left side of y, we set a black edge $b(x, y) = (r(x), l(y))$. For two adjacent genes x, y in B, we set a gray edge ,

$$g(x,y) = \begin{cases} (r(x), l(y)), & \text{if } x \text{ and } y \text{ have the same signs as they have in } A; \\ (r(x), r(y)), & \text{if } x \text{ has the same sign as it has in } A,\ y \text{ doesn't}; \\ (l(x), l(y)), & \text{if } y \text{ has the same sign as it has in } A,\ x \text{ doesn't}; \\ (l(x), r(y)), & \text{if neither } x \text{ nor } y \text{ has the same sign as it has in } A. \end{cases}$$

Then the breakpoint graph with respect to A and B is $G_s(A, B) = (V, E_b \cup E_g)$, where, $V = \{l(x), r(x) \mid x \in \{1, 2, ..., n\}\}$, $E_b = \{b(x, y) \mid x, y \text{ are adjacent in } A\}$, $E_g = \{g(x, y) \mid x, y \text{ are adjacent in } B\}$.

In $G_s(A, B)$, each connected component is either a single vertex, or a cycle, in which each vertex is incident with a black and a gray edge. A cycle in $G_s(A, B)$ is usually known as an *alternating cycle*. An alternating cycle in $G_s(A, B)$ is referred to as an *i-cycle*, if it contains i black (resp. gray) edges. An *i-cycle* is *long* if $i \geq 2$, and *even* (resp. *odd*), if i is even (resp. odd).

Note for an arbitrary segment I, I is congruent to $-I$. A segment, say $[x_i, x_{i+1}, ..., x_j]$ of at least three genes within some chromosome in A, is a *sub-permutation* (abbr. *SP*) in A with respect to B, if there exists a segment $[x_i, permutation(x_{i+1}, ..., x_{j-1}), x_j]$ within some chromosome in B and $permutation(x_{i+1}, ..., x_{j-1}) \neq [x_{i+1}, ..., x_{j-1}]$.

Let I be a segment within a chromosome in A. An SP in A with respect to B is *within* I, if it is a consecutive subsequence of I. An SP is *internal* for I, if it is within I, and not congruent to I. A *minimal sub-permutation* (abbr. *minSP*) is a sub-permutation for which no internal SP exists.

An SP in A with respect to B is an *even isolation*, if all the $minSP$s in A occur within it, and are totalized even. Let $I = [x_i, x_{i+1}, ..., x_j]$ be an SP in A with respect to B, $IN(I) = \{r(x_i), l(x_j)\} \cup \{l(x_k), r(x_k) \mid i < k < j\}$. Then no edge of $G_s(A, B)$ has one end in $IN(I)$ and the other end in $V(G_s(A, B)) - IN(I)$. Thus, I will also be mentioned as the subgraph of $G_s(A, B)$ induced by the vertices in $IN(I)$. Moreover, an SP, $minSP$, or even isolation in A with respect to B will be said *in* $G_s(A, B)$.

Let b, c, s be the numbers of black edges, cycles and $minSP$s in $G_s(A, B)$ respectively. It has been shown in [7] that the translocation distance $d_s(A, B)$ between A and B can be formulated by,

Lemma 1. $d_s(A, B) = b - c + s + f$, where $f = 1$ if s is odd, $f = 2$ if $G_s(A, B)$ has an even isolation, $f = 0$ otherwise.

Let A and B be two unsigned genomes each of which has the gene set $\{1, ..., n\}$. A breakpoint graph with respect to A and B (abbr. $G(A, B)$) can be constructed as $G(A, B) = (V, E_b \cup E_g)$, where $V = \{1, ..., n\}$, $E_b = \{(i, j) \mid i, j$ are adjacent in $A\}$, $E_g = \{(i, j) \mid i, j$ are adjacent in $B\}$, and all edges in E_b (resp. E_g) are dyed with black (resp. gray). A vertex is *within* a chromosome if its corresponding gene is within that chromosome. A black (resp. gray) edge is *within* a chromosome in A (resp. B) if both ends of it is within that chromosome. A black (resp. gray) edge is *left* (resp. *right*) incident with a vertex in $G(A, B)$, if its other end is on the left (resp. right) side of the vertex within a chromosome in A (resp. B).

In $G(A, B)$, each vertex is incident with either a black and a gray edge or two black and two gray edges. This allows us to decompose $G(A, B)$ into a set of alternating cycles, and equivalently, to assign signs to the genes in A, if those genes in B have been assigned. A gene is assigned *positively* (resp. *negatively*) if it gets a sign $+$ (resp. $-$). By default, all genes in B are assigned positively, and for an arbitrary chromosome $[y_l, ..., y_r]$ in B, $y_{i+1} = y_i + 1$, $l \leq i \leq r - 1$.

To decompose $G(A, B)$ into alternating cycles, each four degree vertex, say x in $G(A, B)$, has to be split into $l(x)$ and $r(x)$, where $l(x)$ is on the left side of $r(x)$; then respectively, those two black edges left and right incident with x in $G(A, B)$ are made incident with $l(x)$ and $r(x)$. Moreover, those two gray edges left and right incident with x in $G(A, B)$ are made incident with $l(x)$ and $r(x)$ (resp. $r(x)$ and $l(x)$), if and only if that gene corresponding to x in A is assigned positively (resp. negatively).

Assigning a gene in A corresponding to a vertex x, refers to splitting x into $l(x)$ and $r(x)$, turning the black edges left and right incident with x in $G(A, B)$ to be incident with $l(x)$ and $r(x)$, then turning the two gray edges incident with x to be incident with $l(x)$ and $r(x)$ according to the sign the gene has been assigned.

Once $G(A, B)$ is decomposed into one whose connected components are alternating cycles, those $minSP$s and even isolations in this graph can be found in linear time [1]. Thus we can use Lemma 1 to approximate the translocation distance between A and B. For short, a *cycle decomposition* of $G(A, B)$ directly refers to the graph produced by a cycle decomposition of $G(A, B)$. Let $\chi(G(A, B))$ be the set of all cycle decompositions for $G(A, B)$. The translocation distance between A and B can be specialized as $d(A, B) = min\{ b(G_s) - c(G_s) + s(G_s) + f(G_s) \mid G_s \in \chi(G(A, B)) \}$, where $b(G_s)$, $c(G_s)$, $s(G_s)$ are the black edge, cycle, and $minSP$ numbers of G_s respectively, $f(G_s)$ indicates the same as f in Lemma 1.

3   A Sufficient Condition

Let A and B be two unsigned genomes, while $G(A, B)$ the breakpoint graph with respect to A and B. A segment $[x_i, x_{i+1}, ..., x_j]$ of at least three genes within a chromosome in A, is a *candidate sub-permutation* (abbr. CSP) with respect to B, if there exists a segment $[x_i, permutation(x_{i+1}, ..., x_{j-1}), x_j]$ within a chromosome in B and $permutation(x_{i+1}, ..., x_{j-1}) \neq [x_{i+1}, ..., x_{j-1}]$.

Let I be a segment within a chromosome in A. A CSP in A with respect to B is *internal* for I, if it is within I, and not congruent to I. A *minimal* candidate sub-permutation (abbr. $minCSP$) is a CSP for which no internal CSP exists. A CSP or $minCSP$ in A with respect to B is also mentioned to be *in* $G(A, B)$.

Let G_s be a cycle decomposition of $G(A, B)$. Then a gene in A must have been assigned *depending on* G_s. A CSP in $G(A, B)$ *acts as* an SP in a cycle decomposition of $G(A, B)$, if those genes in the CSP along with their signs assigned depending on G_s form an SP in that cycle decomposition of $G(A, B)$.

Since $f \leq 2$ by Lemma 1, we will give up the contribution of f, aim to find a cycle decomposition of $G(A, B)$, such that if it has b black edges, c alternating cycles, and s $minSPs$, then $b - c + s \leq 1.375\, d(A, B)$. Therefore, we refer to $b - c + s$ as the *translocation distance snapshot* (abbr. *TD-snapshot*) a cycle decomposition of $G(A, B)$ with b black edges, c cycles, and s $minSPs$ has. Thus a cycle decomposition of $G(A, B)$ is *optimal*, if its TD-snapshot is minimized over all cycle decompositions of $G(A, B)$. The *TD-snapshot* of $G(A, B)$ is the TD-snapshot an optimal cycle decomposition of $G(A, B)$ has.

A subgraph of $G(A, B)$ is referred to as an *i-cycle candidate*, if it is a cycle which has i black (resp. gray) edges, and in which each vertex is incident with a black and a gray edge. An *i-cycle candidate* in $G(A, B)$ *acts as* an *i-cycle* in a cycle decomposition of $G(A, B)$, if this *i-cycle* uses all the edges of the *i-cycle* candidate. In [4], it has been shown that every 1-cycle candidate in $G(A, B)$ will act as a 1-cycle in an optimal cycle decomposition of $G(A, B)$. Thus a cycle decomposition of $G(A, B)$ is mentioned with all 1-cycle candidates acting as 1-cycles. An edge in $G(A, B)$ is *valid*, if it does not occur in any 1-cycle candidate.

Two subgraphs of $G(A, B)$ are *independent*, if they share no edge of $G(A, B)$. Let $I = [x_i, ..., x_j]$ be a CSP in $G(A, B)$, $V(I) = \{x_i, x_{i+1}, ..., x_j\}$. Then I acts as an SP in a cycle decomposition of $G(A, B)$, if depending on this cycle decomposition, x_i, x_j are assigned positively for $x_i < x_j$, or negatively otherwise. Without loss of generality, let $x_i < x_j$. We denote by $G(A, B, I)$ the subgraph of $G(A, B)$ induced by the vertices in $V(I)$, and $G(A, B, \bar{I})$ the subgraph of $G(A, B)$ induced by the vertices in $V(G) - V(I) + \{x_i, x_j\}$. Since I is a CSP in A, $G(A, B, I)$ is independent with $G(A, B, \bar{I})$. Assigning x_i and x_j positively will not only split x_i and x_j into $l(x_i)$, $r(x_i)$ and $l(x_j)$, $r(x_j)$ respectively, but also transform $G(A, B)$ into two disconnected *components*, which respectively, have the vertex set $\{r(x_i), x_{i+1}, ..., x_{j-1}, l(x_j)\}$ and $V(G) - V(I) + \{l(x_i), r(x_j)\}$. Since that component with the vertex set $\{r(x_i), x_{i+1}, ..., x_{j-1}, l(x_j)\}$ is isomorphic with $G(A, B, I)$ and the other one $G(A, B, \bar{I})$, we also denote by $G(A, B, I)$ the former and $G(A, B, \bar{I})$ the later.

Since I is a CSP, both $G(A, B, I)$ and $G(A, B, \bar{I})$ can be decomposed into alternating cycles alone. Thus combining a cycle decomposition of $G(A, B, I)$ and a cycle decomposition of $G(A, B, \bar{I})$ *forms* a cycle decomposition of $G(A, B)$. A CSP in $G(A, B, I)$ is a CSP within I. A CSP *within* $\bar{I}$ or in $G(A, B, \bar{I})$ is a CSP in A, which is not within I and shares at most one gene with I.

Let respectively, G_1, G_2 be the cycle decompositions of $G(A, B, I)$ and $G(A, B, \overline{I})$, G the cycle decomposition of $G(A, B)$ combined by G_1 and G_2. The TD-snapshot of G is just the TD-snapshot summation of G_1 and G_2.

Let I be a CSP in $G(A, B)$. We turn to bound the TD-snapshot of $G(A, B)$ by the TD-snapshots of $G(A, B, I)$ and $G(A, B, \overline{I})$.

Lemma 2. *If in every optimal cycle decomposition of $G(A, B, I)$, at least one internal minSP occurs, then $G(A, B)$ admits an optimal cycle decomposition combined by a cycle decomposition of $G(A, B, I)$ and a cycle decomposition of $G(A, B, \overline{I})$.*

Moreover, a cycle decomposition of $G(A, B)$ combined by the optimal cycle decompositions of $G(A, B, I)$ and $G(A, B, \overline{I})$ has at least as many cycles as and at most one more $minSP$s than those an optimal cycle decomposition of $G(A, B)$ has.

Lemma 3. *In general, $G(A, B)$ admits a cycle decomposition combined by a cycle decomposition of $G(A, B, I)$ and a cycle decomposition of $G(A, B, \overline{I})$, which has a TD-snapshot at most 1 more than that an optimal cycle decomposition of $G(A, B)$ has.*

Let I be an arbitrary CSP in $G(A, B)$. Then I induces $G(A, B, I)$. A cycle decomposition of $G(A, B, I)$ will be confused as a *cycle decomposition of I*.

Lemma 4. *If an optimal cycle decomposition of a minCSP, say I, has none other than 2-cycles numbered more than 1, then $G(A, B)$ admits an optimal cycle decomposition combined by a cycle decomposition of $G(A, B, I)$ and a cycle decomposition of $G(A, B, \overline{I})$.*

Due to Lemma 4, a $minCSP$ is *safe*, if an optimal cycle decomposition of it has none other than 2-cycles numbered at least 2, and *non-safe* otherwise. In what follows, a cycle decomposition of a safe $minCSP$ always happens optimal to have long cycles as none other than 2-cycles.

Two CSPs in $G(A, B)$ are *independent*, if they share at most one gene. Two subgraphs of $G(A, B)$ must be independent, if they are induced by two independent CSPs respectively. Let P be a set of mutually independent CSPs in $G(A, B)$, $I = [x_i(I), ..., x_j(I)]$ for $I \in P$. Assigning $x_i(I)$, $x_j(I)$ positively if $x_i(I) < x_j(I)$ or negatively otherwise can always make $I \in P$ act as an SP. If all CSPs in P are so made to act as SPs, $G(A, B)$ is decomposed into a breakpoint graph combined by at most $|P| + 1$ mutually disconnected components, where except those $|P|$ ones induced by the CSPs in P, that one with the vertex set $V(G(A, B)) - \{V(I) \mid I \in P\} + \{l(x_i(I)), r(x_j(I)) \mid I \in P\}$, is referred to as the *complement* of P, and denoted as $G(A, B, \overline{P})$. Let $\Im[P] = \{G(A, B, I) \mid I \in P\} \bigcup G(A, B, \overline{P})$.

A CSP in $G(A, B)$ is referred to as a CSP in $G(A, B, \overline{P})$, if for every $I \in P$, it is in $G(A, B, \overline{I})$. A $minCSP$ in $G(A, B, \overline{P})$ is a CSP in $G(A, B, \overline{P})$, for which no internal CSP exists.

Let Q be the set of safe $minCSP$s in $G(A, B)$, G^* an optimal cycle decomposition of $G(A, B)$ in which each $minCSP$ in Q acts as a $minSP$. Let G^{**} be the cycle decomposition of $G(A, B)$ combined by the optimal cycle decompositions of those components in $\Im[P]$. If k components induced by the CSPs in $P - Q$ can provide their optimal cycle decompositions without any internal $minSP$s, then by Lemmas 2 and 3, the TD-snapshot of G^* will be,

$$b - c(G^*) + s(G^*) \geq b - c(G^{**}) + s(G^{**}) - k, \tag{1}$$

where respectively, $c(G^*)$, $c(G^{**})$ and $s(G^*)$, $s(G^{**})$ stand for the cycle and $minSP$ numbers of G^* and G^{**}, b the black edge number of $G(A, B)$.

An optimal cycle decomposition of a component in $\Im[P]$ always has a minimum number of internal $minSP$s over all its optimal cycle decompositions. For $I \in P$, we denote by $c^*(I)$, $c_i^*(I)$, $s^{**}(I)$ the cycle, i-cycle, and minimum internal $minSP$ numbers for an optimal cycle decomposition of $G(A, B, I)$. We denote by $c^*(\overline{P})$, $c_i^*(\overline{P})$, $s^{**}(\overline{P})$ the cycle, i-cycle, and minimum $minSP$ numbers for an optimal cycle decomposition of $G(A, B, \overline{P})$.

Lemma 5. *If $I \in P - Q$ admits a cycle decomposition with $s(I)$ $minSP$s and at least $\frac{5}{8} c_2^*(I) + \frac{2}{8} c_3^*(I) + s(I) - s^{**}(I)$ long cycles; $\overline{P}$ admits a cycle decomposition with $s(\overline{P})$ $minSP$s and at least $\frac{5}{8} c_2^*(\overline{P}) + \frac{2}{8} c_3^*(\overline{P}) + s(\overline{P}) - s^{**}(\overline{P})$ long cycles, then $G(A, B)$ admits a cycle decomposition with a TD-snapshot at most 1.375 times of $d(A, B)$.*

Proof. Assume a cycle in a cycle decomposition of $G \in \Im[P]$ has at most t black edges. Let $c^* = \sum_{I \in P} c^*(I) + c^*(\overline{P})$, $c_i^* = \sum_{I \in P} c_i^*(I) + c_i^*(\overline{P})$, $s^{**} = \sum_{I \in P} s^{**}(I) + s^{**}(\overline{P})$. Then by Lemma 1 and (1), $d(A, B) \geq b - c^* + s^{**} + |Q| + k - k$, where $b - c^* = c_2^* + 2c_3^* + \dots + (t - 1)c_t^*$. Let G_s be the cycle decomposition combined by the cycle decompositions of those components in $\Im[P]$. If G_s has c cycles, c_i i-cycles and s $minSP$s, then $b - c + s = (c_1^* + 2c_2^* + \dots + tc_t^*) - (c_1 + \dots + c_t) + s$. Let $T = \frac{11}{8} d(A, B) - (b - c + s)$. Then,

$$T \geq \frac{11}{8} \sum_{i=2}^{t} (i - 1)c_i^* - \sum_{i=1}^{t} ic_i^* + \sum_{i=1}^{t} c_i + \frac{11}{8} s^{**} + \frac{11}{8} |Q| - s$$

$$= \sum_{i=2}^{t} \frac{3i - 11}{8} c_i^* + \sum_{i=2}^{t} c_i + \frac{11}{8} s^{**} + \frac{11}{8} |Q| - s$$

$$\geq \sum_{i=2}^{t} c_i - \frac{5}{8} c_2^* - \frac{2}{8} c_3^* + \frac{11}{8} s^{**} + \frac{11}{8} |Q| - s, \tag{2}$$

where the equation of the second line follows from $c_1^* = c_1$. Let the cycle decomposition of $G \in \Im[P]$ have $c_i(G)$ i-cycles and $s(G)$ $minSP$s. Then $c_i = \sum_{G \in \Im[P]} c_i(G)$, $s = \sum_{G \in \Im[P]} s(G)$. Since if G is induced by a safe $minCSP$, then $s(G) = 1$, $s = \sum_{I \in P - Q} s(G(A, B, I)) + s(G(A, B, \overline{P})) + |Q|$. It follows the lemma assumption that, if G is induced by a safe $minCSP$, then $\sum_{i=2}^{t} c_i(G) = 2$; if G is induced by $\overline{P}$, then $\sum_{i=2}^{t} c_i(G) \geq \frac{5}{8} c_2^*(\overline{P}) + \frac{2}{8} c_3^*(\overline{P}) + s(G) - s^{**}(\overline{P})$; otherwise, G is induced by a CSP I with $\sum_{i=2}^{t} c_i(G) \geq \frac{5}{8} c_2^*(I) + \frac{2}{8} c_3^*(I) + s(G) - s^{**}(I)$. Adding the inequalities for all $G \in \Im[P]$ leads to,

$$\sum_{i=2}^{t} c_i - \frac{5}{8} c_2^* - \frac{2}{8} c_3^* \geq s - s^{**} - |Q|. \tag{3}$$

Substituting $\sum_{i=2}^{t} c_i - \frac{5}{8} c_2^* - \frac{2}{8} c_3^*$ with $s - s^{**} - |Q|$ in (2) leads to $T \geq 0$.

4   Approximation for Translocation Sorting

We devote to find a set of mutually independent CSPs, say P in $G(A, B)$, then for each component in $\Im[P]$, find a cycle decomposition to meet Lemma 5.

4.1   Decomposing $G(A, B)$ into Disconnected Components

A $minSP$ in a cycle decomposition of $G(A, B)$ is *knotty*, if it has 2 or 3 valid black edges, and *non-knotty*, otherwise. A $minCSP$ in $G(A, B)$ is *knotty*, if it induce a knotty $minSP$, and *non-knotty*, otherwise. A knotty $minCSP$ in $G(A, B)$ cannot provide so many cycles as those Lemma 5 asks. Thus we try to avoid decomposing a knotty $minCSP$ alone into alternating cycles, if it is internal for a CSP.

A CSP in $G(A, B)$ is referred to as a *second* CSP, if it is not minimal, and each of its internal $minCSP$ is knotty. A second CSP is *minimal*, and abbreviated as a 2nd-$minCSP$, if it does not contain any internal second CSP. A $minCSP$ can be found in polynomial time trivially. Starting with a knotty $minCSP$, a 2nd-$minCSP$ can be found in polynomial time, by finding in $G(A, B)$ a shortest length CSP containing an internal knotty $minCSP$ but itself, and checking if its internal $minCSP$s are all knotty. Thus to decompose $G(A, B)$ into disconnected graphs, we try to find a set of mutually independent CSPs, each of which is either a non-knotty $minCSP$ or a 2nd-$minCSP$.

Such a CSP set, say P, can be found by: (1)Find the set of $minCSP$s, say P_1, in $G(A, B)$. (2)For each $I \in P_1$, check if I is non-knotty. If yes, add it into P, otherwise, check if a 2nd-$minCSP$ contains I. If yes, add the 2nd-$minCSP$ containing I into P. We name the subroutine for so as to find a CSP set in $G(A, B)$ as CSP-Decompose($G(A, B)$). Let P be the CSP set returned by CSP-Decompose($G(A, B)$).

Lemma 6. *A member in P is either a non-knotty $minCSP$ or a 2nd-$minCSP$. Any two members in P are independent. No non-knotty $minCSP$ occurs in $G(A, B, \overline{P})$.*

4.2   Finding a Cycle Decomposition of a Component

Let P be a mutually independent CSP set in which a member is either a non-knotty $minCSP$ or a 2nd-$minCSP$. An *independent vertex set* of an undirected simple graph is a set of mutually independent vertices in that graph, where two vertices are *independent* if no edge can take them for two ends. For a graph with each vertex given a weight, an independent vertex set whose vertex weight summation are maximized is a *maximum weight independent set* of that graph. The *maximum weight independent set* problem (abbr. WIS) asks to find a maximum weight independent set in an undirected simple graph with weighted vertices. The problem of finding in a breakpoint graph a 2-cycle and 3-cycle set to achieve as large a cycle weight summation as possible, can be reduced to WIS.

Let $G \in \Im[P]$. Corresponding to each 2-cycle or 3-cycle candidate in G, set a vertex, where we denote by V_2, V_3 the sets of vertices respectively corresponding to 2-cycle and 3-cycle candidates in G. Set an edge between two vertices, if their corresponding cycle candidates in G share an edge, where we denote by E the set of so produced edges. Note that two cycle candidates in G are independent, if they share no edges of G. If setting a positive weight to each vertex in $V_2 \cup V_3$, then there exists an independent vertex set in $\mathcal{G} = (V_2 \cup V_3, E)$ whose total vertex weights are maximized, if and only if there is a set of mutually independent 2-cycle and 3-cycle candidates, whose total cycle weights are maximized.

In [2], by setting a weight $\frac{5}{8}$ to each vertex in V_2 and a weight $\frac{2}{8}$ to each vertex in V_3, Berman et al. proposed a local search algorithm which can, for an arbitrary independent vertex set $J_2 \cup J_3$ with $J_2 \subseteq V_2$, $J_3 \subseteq V_3$, find an independent vertex set of $\mathcal{G}$ which has at least $\frac{5}{8}|J_2| + \frac{2}{8}|J_3|$ independent vertices in $\mathcal{G}$. More formally, we state it as,

Lemma 7. *For any set of mutually independent cycle candidates in $G \in \Im[P]$ with a subset J_2 of 2-cycle candidates and a subset J_3 of 3-cycle candidates, a cycle decomposition of G with at least $\frac{5}{8}|J_2| + \frac{2}{8}|J_3|$ 2-cycles and 3-cycles can be found in polynomial time.*

This implies the algorithm of Berman *et al.* can find a set of mutually independent 2-cycle and 3-cycle candidates, whose cardinality is at least $\frac{5}{8}c_2^*(G) + \frac{2}{8}c_3^*(G)$, where $c_i^*(G)$ stands for the number of i-cycles in an optimal cycle decomposition of G. As long as an optimal cycle decomposition of G has none other than 2-cycles and 1-cycles, it can be shown that there always exist more cycles than those the algorithm of Berman et al. outputs.

Lemma 8. *If no cycle decomposition of $G(A, B, I)$ contains none other than 2-cycles and 1-cycles, $G(A, B, I)$ can be decomposed into at least $\frac{5}{8}c_2^*(I) + \frac{2}{8}c_3^*(I) + 1$ long cycles in polynomial time.*

In [14], a polynomial time algorithm was proposed to decide whether a breakpoint graph with respect to two unsigned genomes can be decomposed into none other than 2-cycles and 1-cycles. If so, it will return a cycle decomposition containing none other than 2-cycles and 1-cycles. Otherwise, it will return an empty set. We denote this subroutine by Two-Cycle($G(A, B, I)$).

Lemma 8 implies that, if Two-Cycle($G(A, B, I)$) returns an empty set, then the cycle decomposition Berman $et\ al.$'s algorithm outputs can meet Lemma 5. We denote this algorithm by Non-safe($G(A, B, I)$). Otherwise, I is safe and Two-Cycle($G(A, B, I)$) provides an optimal cycle decomposition of $G(A, B, I)$. By Lemmas 4 and 5, it suffices to select the cycle decomposition Two-Cycle($G(A, B, I)$) returns.

For finding a cycle decomposition of a 2nd-$minCSP$ in $G(A, B)$, we squeeze enough long cycles by a new reduction different slightly from Berman's. Still, we set an undirected simple graph $\mathcal{G} = (V_2 \cup V_3, E)$ as an instance of WIS. However, if a 2-cycle (resp. 3-cycle) candidate in $G(A, B, I)$ does not occur within a knotty $minCSP$, set a vertex in V_2 (resp. V_3) corresponding to it; for every two vertices in $V_2 \cup V_3$ which correspond to two cycle candidates sharing an edge of $G(A, B, I)$, set an edge in E incident with them. This reduction can help get as many long cycles as those to meet Lemma 5, because if a knotty $minCSP$ happens to act as a knotty $minSP$, an extra cycle can be obtained. Let I be a second 2nd-$minCSP$ in $G(A, B)$.

Lemma 9. *In polynomial time, a cycle decomposition of $G(A, B, I)$ can be found such that, if it has $s^{**}(I)$ minSPs, then it has at least $\frac{5}{8}c_2^*(I) + \frac{2}{8}c_3^*(I) + s(I) - s^{**}(I)$ long cycles.*

We denote by Second(G) the algorithm which can find a cycle decomposition of $G = G(A, B, I)$ for which Lemma 9 holds. We denote by $c_i^*(\overline{P})$, $s^{**}(\overline{P})$ the numbers of i-cycles and $minSPs$ an optimal cycle decomposition of $G(A, B, \overline{P})$ has. No non-knotty $minCSP$ can occur in $G(A, B, \overline{P})$ by Lemma 6. If $G(A, B, \overline{P})$ has no $minCSP$, then $s^{**}(\overline{P}) = 0$. Consequently, $G(A, B, \overline{P})$ can be decomposed into alternating cycles in the same way as for a non-knotty $minCSP$.

Lemma 10. *If $G(A, B, \overline{P})$ has no minCSP, then in polynomial time, $G(A, B, \overline{P})$ can be decomposed into at least $\frac{5}{8}c_2^*(\overline{P}) + \frac{2}{8}c_3^*(\overline{P})$ long cycles.*

If $G(A, B, \overline{P})$ has $minCSPs$, then any cycle decomposition of $G(A, B, \overline{P})$ has, if present, no other $minSPs$ than knotty ones. Thus $G(A, B, \overline{P})$ can be decomposed into alternating cycles in the same way as for a 2nd-$minCSP$.

Lemma 11. *If $G(A, B, \overline{P})$ has minCSPs, then in polynomial time, a cycle decomposition of $G(A, B, \overline{P})$ can be found such that, if it has $s(\overline{P})$ minSPs, then it has at least $\frac{5}{8}c_2^*(\overline{P}) + \frac{2}{8}c_3^*(\overline{P}) + s(\overline{P}) - s^{**}(\overline{P})$ long cycles.*

Our algorithm start with decomposing $G(A, B)$ into disconnected components using CSP-Decompose($G(A, B)$). Then for each component, using Two-cycle(G), non-safe(G) or Second(G) to decompose it into cycles to meet Lemma 5. The algorithm is named as USort(A, B), and depicted in Fig. 1.

Theorem 1. *In polynomial time, USort(A, B) always outputs a translocation distance at most $\frac{11}{8}\ d(A, B) + 2$.*

Algorithm $USort(A, B)$
Input: Unsigned genomes A, B.
Output: A translocation sequence to sort A, B.
1   Set a breakpoint graph $G(A, B)$;
2   Set a 1-cycle for each candidate 1-cycle in $G(A, B)$;
3    $P \leftarrow CSP\text{-Decompose}(G(A, B))$;
4    $\Im[P] \leftarrow \{G(A, B, I) \mid I \in P\} \cup \{G(A, B, \overline{P})\}$;
5   For each $G \in \Im[P]$
6   If G contains no knotty $minCSP$, $CD(G) \leftarrow$ Two-Cycle(G);
7   If $CD(G) = \emptyset$, $CD(G) \leftarrow$ non-safe(G);
8   Else, $CD(G) \leftarrow$ Second(G);
9   End for
10   Combine $CD(G)$ for $G \in \Im[P]$ into one cycle decomposition CD;
11   Sort A, B based on CD.

Fig. 1. Sort unsigned genomes by translocations.

5   Conclusion

To approximate unsigned translocation sorting as well as unsigned reversal sorting to a better extent, it awaits to explore new techniques to decompose an unsigned breakpoint graph into alternating cycles. It should be believed that the promising technique for finding cycle decompositions of unsigned breakpoint graphs lands up at making use of the structural natures of the breakpoint graphs.

Acknowledgement. This paper is supported by National natural science foundation of China No 61472222.

References

1. Bader, D.A., Bernard, M.E.M., Yan, M.: A linear-time algorithm for computing inversion distance between signed permutations with an experimental study. J. Comput. Biol. **8**(5), 483–491 (2004)
2. Berman, P., Hannenhalli, S., Karpinski, M.: 1.375-approximation algorithm for sorting by reversals. In: Möhring, R.H., Raman, R. (eds.) ESA 2002. LNCS, vol. 2461, pp. 200–210. Springer, Heidelberg (2002)
3. Caprara, A.: Sorting by reversals is difficult. In: Proceedings of RECOMB 1997, pp. 75–83 (1997)
4. Cui, Y., Wang, L., Zhu, D.: A 1.75-approximation algorithm for unsigned translocation distance. J. Comput. Syst. Sci. **73**(7), 1045–1059 (2007)
5. Cui, Y., Wang, L., Zhu, D., Liu, X.: A $(1.5 + \epsilon)$-approximation algorithm for unsigned translocation distance. IEEE, ACM Trans. Comput. Bioinform. **5**(1), 56–66 (2008)
6. Pradhan, G.P., Prasad, P.V.: Evaluation of wheat chromosome translocation lines for high temperature stress tolerance at grain filling stage. PLoS ONE **10**(2), 1–20 (2015)

7. Hannenhalli, S.: Polynomial-time algorithm for computing translocation distance between genomes. Discret. Appl. Math. **71**(3), 137–151 (1996)

8. Hannenhalli, S., Pevzner, P.A.: Transforming cabbage into turnip (polynomial algorithm for sorting signed permutations by reversals). In: STOC 1995, pp. 178–189 (1995). (J. ACM **46**(1), 1–27 (1999))

9. Jiang, H., Wang, L., Zhu, B., Zhu, D.: A $(1.408+\epsilon)$-approximation algorithm for sorting unsigned genomes by reciprocal translocations. In: Chen, J., Hopcroft, J.E., Wang, J. (eds.) FAW 2014. LNCS, vol. 8497, pp. 128–140. Springer, Heidelberg (2014)

10. Kececioglu, J., Ravi, R.: Of mice and men: algorithms for evolutionary distance between genomes with translocation. In: SODA 1995, pp. 604–613 (1995)

11. Kaplan, H., Shamir, R., Tarjan, R.E.: A faster and simpler algorithm for sorting signed permutations by reversals. SIAM J. Comput. **29**(3), 880–892 (1999)

12. Bulatov, A.A., Marx, D.: Constraint satisfaction parameterized by solution size. In: Aceto, L., Henzinger, M., Sgall, J. (eds.) ICALP 2011, Part I. LNCS, vol. 6755, pp. 424–436. Springer, Heidelberg (2011)

13. Lou, X., Zhu, D.: A new approximation algorithm for cut-and-paste sorting of unsigned circular genomes. J. Comput. Syst. Sci. **78**(4), 1099–1114 (2012)

14. Pu, L., Jiang, H.: Can a breakpoint graph be decomposed into none other than 2-cycles? In: Zhu, D., Bereg, S. (eds.) FAW 2016. LNCS, vol. 9711, pp. 205–214. Springer, Heidelberg (2016)

15. Sankoff, D., Leduc, G., Antoine, N., Lang, B.F., Cedergren, R.: Gene order comparisons for phylogenetic inference: evolution of the mitochondrial genome. Proc. Natl. Acad. Sci. USA **89**, 6575–6579 (1992)

16. Stuart, J., Lucas, B.A., Simkova, H., Kubalakov, M., Dolezel, J., Budak, H.: Next-generation sequencing of flow-sorted wheat chromosome 5D reveals lineage-specific translocations and widespread gene duplications. BMC Genomics **15**, 1080 (2014)

17. Ozery-Flato, M., Shamir, R.: An $O(n^{\frac{3}{2}}\sqrt{\log n})$ algorithm for sorting by reciprocal translocations. J. Discret. Algorithms **9**(4), 344–357 (2011)

18. Wang, L., Zhu, D., Liu, X., Ma, S.: An $O(n^2)$ algorithm for signed translocation. J. Comput. Syst. Sci. **70**(3), 284–299 (2005)

19. Zhu, D., Wang, L.: On the complexity of unsigned translocation distance. Theor. Comput. Sci **352**(3), 322–328 (2006)

20. Zhu, D., Ma, S.: An improved polynomial time algorithm for signed translocation sorting. J. Comput. (Chin.) **25**(2), 189–196 (2002)

Independent Component Analysis to Remove Batch Effects from Merged Microarray Datasets

Emilie Renard[1]($\boxtimes$), Samuel Branders[2], and P.-A. Absil[1]

[1] ICTEAM Institute, Université catholique de Louvain,
Avenue Georges Lemaître 4, 1348 Louvain-la-Neuve, Belgium
{emilie.renard,pa.absil}@uclouvain.be
[2] Tools4Patient, rue Auguste Piccard 48, 6041 Gosselies, Belgium

Abstract. Merging gene expression datasets is a simple way to increase the number of samples in an analysis. However experimental and data processing conditions, which are proper to each dataset, generally influence the expression values and can hide the biological effect of interest. It is then important to normalize the bigger merged dataset regarding those batch effects, as failing to adjust for them may adversely impact statistical inference. In this context, we propose to use a "spatiotemporal" independent component analysis to model the influence of those unwanted effects and remove them from the data. We show on a real dataset that our method allows to improve this modeling and helps to improve sample classification tasks.

Keywords: Batch effect removal · Expression data · Spatio-temporal independent component analysis

1 Introduction

Genes hold the information to build proteins, which are the structural components of cells and tissues. The translation of gene information into proteins is known as "gene expression". Nowadays, the development of sequencing technologies allows to measure those expression levels at a reasonable cost. The analysis of the resulting data helps to better understand how genes are working, with the goal of developing better cures for genetic diseases such as cancer.

Due to the limited number of samples that can be processed at the same time in an experiment, the size of such datasets is often limited in samples. However, statistical inferences need a high number of samples to be robust enough and generalizable to other data. As more and more of those datasets are available on public repositories such as GEO http://www.ncbi.nlm.nih.gov/geo/, merging and combining different datasets appears as a simple solution to increase the number of samples analyzed and potentially improve the relevance of the biological information extracted.

Expression levels of genes are the result of interactions between different biological processes, which can increase or decrease the expression level measured. However, noise may also be added at each step of data acquisition, due

© Springer International Publishing Switzerland 2016
M. Frith and C.N.S. Pedersen (Eds.): WABI 2016, LNBI 9838, pp. 281–292, 2016.
DOI: 10.1007/978-3-319-43681-4_23

to imprecisions or differences in experiment conditions. Confounding factors, or batch effects, that complicate the analysis of genomic data can be for example differences in dates of experiment, differences in laboratory conditions, or even the fact that two samples subsets were treated by two different technicians. The precise effects of the technical artefacts on gene expression levels is often unknown; however some partial information is usually available, such as the batch number, the date of experiment, ... When merging different datasets, some of the main confounding factors are typically due to the fact that the samples were not processed in exactly the same conditions from one experiment/dataset to another. Those batch effects can be quite large and hide the effects related to the biological process of interest. Not including those effects in the analysis process may adversely affect the validity of biological conclusions drawn from the datasets [7,8,17]. It is then important to be able to combine datasets from different sources while removing the unwanted variations such as batch effects. From here, we will call *aggregated dataset* the bigger dataset resulting of the concatenation of the smaller datasets, and *sub-datasets* or *batches* these smaller datasets.

An additional difficulty in the process of removing batch effects is that the biological process (or phenotype) of interest could partially correlate with the batches. For example, if we want to combine two sub-datasets with respectively 75/25 % and 25/75 % of cases/controls, we should check that what is removed during the normalization step is really only the batch effect and does not contain potential useful information about cases/controls.

Different methods exist to tackle the problem of batch effect removal when merging different sub-datasets, each having its advantages and weaknesses [2,6]. They can be classified in two main approaches: location-scale methods and matrix factorization methods. The location-scale methods assume a model for the distribution of the data within batches, and adjust the data within each batch to fit this model. The goal is to obtain genes with similar mean and/or variance for each batch. A main hypothesis is that by adjusting the gene distributions no biological information is removed. The matrix factorization methods assume that the variations across the sub-datasets (biological or due to confounding factors) can be represented by a small set of rank-one components which can be estimated by means of matrix factorization. The components associated with the batch effects are then removed to obtain the normalized dataset. With this approach, the main hypothesis is that the factorization method is able to pick up the batch effects in some of its resulting components.

In this paper, building on [1,12], we propose to use spatio-temporal Independent Component Analysis (ICA) to remove batch effects when combining microarray datasets. We compare our method to three other normalization methods. We show on a real dataset that spatio-temporal ICA allows to better model the factors influencing gene expression levels, and may improve results in a sample classification task.

The paper is organized as follows. Section 2 presents the method, which is validated in Sect. 3, and conclusions are drawn in Sect. 4.

2   Proposed Method to Reduce Batch Effect

Building on [1,12], we propose to use spatio-temporal Independent Component Analysis (ICA) to remove batch effects when combining microarray datasets. After factorization of the aggregated dataset, components showing some correlation with the sub-datasets are removed in order to obtain a final dataset, hopefully cleaned from the main batch effects. The advantage of a matrix factorization approach is that the removed components are interpretable: it is easy to check that they do not correlate with some biological information of interest. In [1], the authors use singular value decomposition (SVD) to model batch effects. However ICA was shown to better model the different sources of variation [17], so we propose here an ICA based approach. We first describe the spatio-temporal ICA of [12], then we explain how we use it to normalize the dataset.

2.1   Spatio-Temporal Independent Component Analysis

We consider the aggregated dataset as a gene-by-sample matrix X, where $X_{i,j}$ indicates the value of gene i in sample j. Applying an ICA method to matrix X yields a decomposition

$$X \approx AB^T = \sum_{k=1}^{K} A_{:,k} B_{:,k}^T \qquad (1)$$

where component $A_{:,k}$ can be interpreted as the gene activation pattern of component k and component $B_{:,k}$ as the weights of this pattern in the samples.

When computing this decomposition, the question arises whether one should maximize the independence between the columns of A or those of B. Independence across genes means that the activation patterns should be as independent as possible. Independence across samples means that the weights attributed to the activation patterns should be as independent as possible. In earlier times, because of the very vertical shape of matrix X in genetic datasets, independence across genes has been favored in the literature. However aggregating sub-datasets allows to have a more reasonable number of samples. Imposing independence among genes, or samples, or on both was shown to give good results [14]. As both options are justifiable a priori, we use a spatio-temporal ICA; this method introduces a trade-off parameter allowing an easy adaptation to the different options.

We now present the ICA method from [12] that we use to generate matrices A and B from the data matrix $X \in \mathbb{R}^{p \times n}$. The algorithm depends on a *spatiotemporal parameter* $\alpha \in [0, 1]$ that allows it to explore a continuum between imposing independence solely on A ($\alpha = 0$) and solely on B ($\alpha = 1$). The term "spatiotemporal" comes from the pixel-by-time data in medical imaging for which the concept was introduced [16].

The first step consists of centering the gene-by-sample data matrix X by subtracting the row and column means, followed by a dimensionality reduction by means of a K-truncated SVD, yielding a new matrix $\tilde{X} = U_K D_K V_K^T$. All the possible decompositions of $\tilde{X}$ are given by $\tilde{X} = AB^T = AW^{-1}WB^T$ with W a $K \times K$ invertible matrix. The considered decomposition is then:

$$\tilde{X} = \underbrace{U_K D_K^\alpha W^{-1}}_{=:A} \underbrace{W D_K^{1-\alpha} V_K^T}_{=:B^T} \tag{2}$$

where W is restricted to the orthogonal group $O(K) = \{W \in \mathbb{R}^{K \times K} : W^T W = I\}$. Consequently, the columns of A, resp. B, are structurally decorrelated when $\alpha = 0$, resp. $\alpha = 1$.

In the spirit of the JADE ICA algorithm [3], the objective function to minimize is of the form:

$$f_\alpha(W) = \alpha \sum_i \text{Off}(C_i(B^T)) + (1 - \alpha) \sum_i \text{Off}(C_i(A^T)), \quad W \in O(K)$$

where A and B depend on W through (2), $\text{Off}(Y)$ returns the sum of squares of the off-diagonal elements of Y, and the C_i's are fourth-order cumulant matrices, satisfying the property $C_i(WM) = WC_i(M)W^T$. The minimization of f_α is thus a joint approximate diagonalization problem, which is addressed as in JADE using Jacobi rotations. The Jacobi algorithm is initialized with $W = I$, ensuring that both A and B initially have decorrelated columns.

2.2   Dataset Normalization

The normalization process to remove batch effects is detailed in Algorithm 1. Matrices A and B are first computed (line 1), then we can use the components $B_{:,k}$ to remove possible batch effects. For this, we select the components $B_{:,k}$ that correlate with the batch. As batch is a categorical information and the components $B_{:,k}$ are continuous, the usual correlation formula (Pearson or Spearman) can not be used. To estimate which components are related to batch, we compute the R^2 value (line 2) that measures how well a variable x can predict a variable y in a linear model:

$$R^2(x, y) \equiv 1 - \frac{SS_{res}}{SS_{tot}}$$

where

- $SS_{tot} = \sum_i (y_i - \bar{y})^2$ is the sum of squares of the prediction errors if we take the mean $\bar{y} = \frac{1}{n} \sum_{i=1}^n y_i$ as predictor or y,
- $SS_{res} = \sum_i (y_i - \hat{y}_i)^2$ is the sum of squares of the prediction errors if we use a linear model $\hat{y}_i = f(x_i)$ as predictor: if x is continuous the prediction model is a linear regression, if x is categorical we use a class mean.

The R^2 value indicates the proportion of the variance in y that can be predicted from x, and has the advantage to be usable with categorical or continuous variables. So the higher the R^2 value, the better the association between both variables. As the sub-datasets information is categorical, $R^2(\text{sub-datasets}, B_{:k})$

Algorithm 1. ICA based normalization

Require: X $(p \times n)$ the aggregated dataset to be normalized, c (n) a categorical variable indicating the sub-datasets, α the spatio-temporal parameter for the ICA method, $t \in [0, 1]$ the threshold to consider a component associated to c, [optional] c_2 categorical/continuous information that we want to preserve

1: $A, B \leftarrow ICA(X, \alpha)$
2: $R \leftarrow R2(c, B)$
3: $ix \leftarrow which(R \geq t)$
4: $R_2 \leftarrow R2(c_2, B)$ ▷ optional
5: $ix_2 \leftarrow which(R_2 \geq R)$ ▷ optional
6: $ix \leftarrow ix \setminus ix_2$ ▷ optional
7: $X_n \leftarrow X - A[:, ix] * B[:, ix]^T$

---

compares the prediction of B_{ik} by a general mean $\sum_j \frac{B_{jk}}{n}$ or by a sub-dataset mean $\sum_{j \in C_j} \frac{B_{jk}}{n}$ (where C_j represents all samples in the same sub-datasets as sample j).

If a component presents some correlation with the sub-datasets (line 3), then this component is selected. An additional step can be added in the process to check if the selected components do not correlate with some information of interest (lines 4–6, optional). The selected components are then removed from the matrix X to obtain a cleaned dataset (line 7).

3   Results

We tested our normalization method on breast cancer expression. We combined different datasets which can be accessed under GEO numbers GSE2034 [18] and GSE5327 [11], GSE7390 [4], GSE2990 [15], GSE3494 [10], GSE6532 [9] and GSE21653 [13]. All datasets were summarized with MAS5 and represented in log2 scale, except GSE6532 which was already summarized with RMA. With those datasets come different pieces of information: age of the patient, grade and size of the tumor, if a lymphatic node is affected, the estrogen receptor status, the treatment, the subtype, two values estimating the relapse risk (scoreGene76 and scoreODX), and one estimating the proliferation (scoreProlif). The last four are values computed from a model and the expression values, and so are more directly dependent on the dataset.

We took estrogen-receptor status (ER) prediction as the classification task. ER is thus our phenotype of interest, and other pieces of information will be called external information later in this paper. We removed the samples (or genes) with missing information which gives an aggregated dataset of 1361 samples for 22276 genes. The repartition of the ER status is described in Table 1. Proportion of ER positive samples depends on the dataset but is always in majority.

As can be seen on the first subplot in Fig. 1 the expression values in the aggregated dataset are clearly associated to the sub-datasets. Many genes have

Table 1. Repartition of ER status in the sub-datasets

Sub-dataset	1	2	3	4	5	6
ER = 0	135	64	34	34	40	110
ER = 1	209	134	149	213	86	153
% of 1's		0.61	0.68	0.81	0.86	0.68	0.58

a strong association with the sub-datasets, but the maximal R^2 value between a gene and the ER status is about 0.25.

3.1 Comparison with Centering-Scaling, ComBat, and SVD Based Methods

Many methods exist that aim to remove the unwanted variation coming from the batch effects. We compare our method to three different approaches: the very simple standardization method, the well-used ComBat method and an SVD-based method that have a similar approach to our proposition.

The simplest way to normalize a dataset in order to remove batch effect is to standardize each sub-dataset separately. That is, for each gene in each sub-dataset, the expression values are centered and divided by their standard deviation.

Another widely used but more complex method is ComBat [5]. The expression value of gene g for sample j in batch i is modeled as $Y_{ijg} = \alpha_g + X\beta_g + \gamma_{ig} + \delta_{ig}\epsilon_{ijg}$ where α_g is the overall gene expression, and X represents the sample conditions. The error term ϵ_{ijg} is assumed to follow a normal distribution $N(0, \sigma_g^2)$. Additive and multiplicative batch effects are represented by parameters γ_{ig} and δ_{ig}. ComBat uses a bayesian approach to model the different parameters, and then removes the batch effects from the data to obtain the clean data $Y_{ijg}^* = \hat{\epsilon}_{ijg} + \hat{\alpha}_g + X\hat{\beta}_g$.

The third method, which we term SVD, is similar to [1]. The main difference is that a singular value decomposition is computed instead of an independent component analysis. As it is not clear how to systematically infer which components to remove in [1], we use our R^2 criterion.

Effects of the different normalization methods on the association between genes and sub-datasets or ER are visible on Fig. 1. Compared to the initial values, all methods about double the maximum R^2 value associated to the ER factor (from 0.25 to 0.5). Effects on association with sub-datasets are more different. The centering-scaling approach and ComBat remove all association with the sub-datasets. The methods based on matrix factorization are less sharp, the SVD one keeping the higher association with sub-datasets.

In the remaining of this section, we compare all four normalization methods (centering-scaling, ComBat, SVD and ICA based) and the case without normalization regarding possible batch effects. First we look in Sect. 3.2 at how the method works and can be interpreted regarding the external information we have

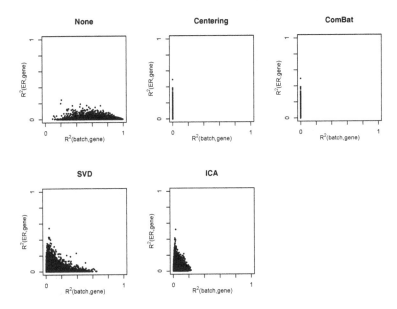

Fig. 1. R^2 values between the gene expression values and the sub-datasets versus the ER factor, for different normalizations of aggregated dataset.

access to. This is only possible for the factorization methods. In a second time, we compare in Sect. 3.3 the results obtained in the context of a classification task.

3.2   Spatio-Temporal ICA to Model Sources of Variations

As a first step to validate our approach, we computed the ICA factorization of the unnormalized aggregated dataset for different values of α, yielding the components $A(\alpha)$ and $B(\alpha)$ as in Eq. 1.

The maximal R^2 values between components of the matrix B and the external information (i.e. $\max_i R^2(\text{info}, B_{:i})$) are represented on Fig. 2. Information related to the sub-datasets appears to be captured quite well in at least one component $B_{:i}$. The quality of the recovering of the external information depends on the α value. If we compare with the SVD components, ICA is at least as good as SVD to recover the external information. For some factors like subType, score-Gene76, scoreProlif, scoreODX, treatment, and even ER in a smaller measure, the ICA factorization improves the modeling.

Influence of sub-datasets appears to be captured in the SVD decomposition, and in all values of α in ICA. However if we examine the relation between components and external information the behaviors differ. On Fig. 3 we represented for different decompositions the R^2 values between all components and the external information, and the correlation between components themselves.

SVD gives two uncorrelated components highly associated to sub-datasets. ICA allows to increase the number of components associated to sub-dataset,

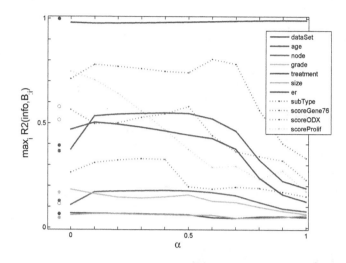

Fig. 2. Maximal R^2 values between components of the ICA factorization and the external information depending on α. The isolated dots on the left hand side give the same information but for the SVD decomposition.

especially for α values close to 0. However, some of those components are redundant: for $\alpha = 0$, components 1,2,4,6,8,9 have high R^2 values. But components 4 and 9 are correlated with component 2, and component 8 with component 1. Increasing the value of α imposes more and more independence on B and so enable to get rid of the redundancy between components. A good trade-off between recovering external information and avoiding redundancy would be an intermediate value of α.

3.3 Validation by Impact on Classification

Classification Process Description. To compare the different methods, we used them in a whole process of classification task. We predicted the ER status using an SVM classifier. The whole process is described in Algorithm 2. The first step is to normalize the aggregated dataset X with the chosen method (here, centering-scaling, ComBat, SVD or ICA based normalization), then center and scale it to be sure to treat all features with the same weight (line 2). Training and testing sets are then separated (line 3). A basic feature selection is performed by selecting the 10 genes with the best association with the ER label based on a Wilcoxon test (line 4). A standard SVM model is trained based on those genes (lines 5 to 8). The labels of testing set are finally predicted using the SVM model (line 9). To keep the SVM model simple, we used a linear kernel, the cost parameter is fixed using the heuristic implemented in the LiblineaR package, and the classes weights are set to $[1 - p_0, 1 - p_1]$ where p_i gives the proportion of samples in class i. In ICA and SVD based normalizations, we computed the $K = 20$ first components and removed the components with an R^2 value higher than $t = 0.5$.

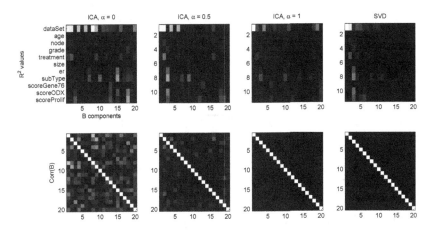

Fig. 3. Top: R^2 values between components resulting from different factorizations and the external information. Bottom: correlation between the different components. Black implies a null correlation (0), and white a perfect one (1).

Algorithm 2. Classification process

Require: X $(p \times n)$ aggregated matrix of genes expression, y (n) the label to predict, c (n) the sub-dataset information

1: $y_{tr}, y_{te} \leftarrow y$
2: $X_n \leftarrow normalize(X, c, y_{tr})$
3: $X_{tr}, X_{te} \leftarrow X_n$;
4: $idx_{bestGenes} \leftarrow Wilcoxon(y_{tr}, X_{tr})$
5: $X_{SVM} \leftarrow X_{tr}[idx_{bestGenes}, :]$
6: $c \leftarrow heuristic(X_{SVM})$
7: $w \leftarrow 1 - [\sharp y_{tr} == 1 \; \sharp y_{tr} == 0]$
8: $model_{SVM} \leftarrow SVM(X_{SVM}, y_{tr}, c, w)$
9: $\hat{y}_{te} \leftarrow prediction(model_{SVM}, X_{te}[idx_{bestGenes}, :])$

Results on Dataset. We kept each time two sub-datasets out of six for testing, and trained the SVM model on the four other sub-datasets, for a total of $C_6^2 = 15$ experiments.

The impact of α in the ICA based normalization on the results are illustrated on Fig. 4. An α closer to 1 tends to predict more positive labels. As discussed in Sect. 3.2, a value of $\alpha = 0.5$ appears to be a good compromise.

The results on the testing set for the five methods (with $\alpha = 0.5$ for the ICA based) are shown on Fig. 5. The case without normalization appears to have a larger variance. ComBat has a smaller variance than the case without normalization, but bigger than the factorization methods. ICA and SVD are closer, ICA being slightly higher.

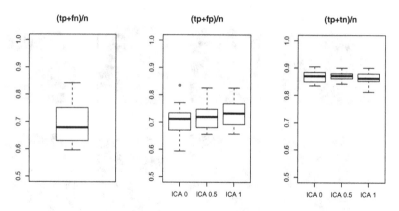

Fig. 4. Impact of α for the validating sets. On the left, proportion of positives in the sub-datasets. In the middle, proportion of positive predictions by the algorithm after applying an ICA based normalization. On the right, proportion of correct predictions.

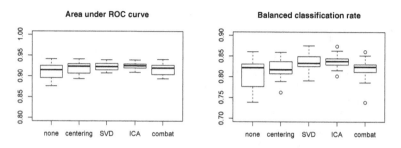

Fig. 5. Area under the ROC curve (obtained by varying the bias b in the separating hyperplane) and balanced classification rate $(0.5(\frac{TP}{TP+FN} + \frac{TN}{TN+FP}))$ for the validating sets.

4   Conclusion

In the context of merging gene expression datasets to increase the number of samples analyzed and so the robustness of extracted information, we have proposed a method to remove batch effects. Inspired from existing methods, we have used a spatio-temporal independent component analysis to model those effects and remove it from the data. We have tested our method on a real breast cancer aggregated dataset in a classification task and compared it to other normalization methods. We have shown that our method can recover external information better than using a simple singular value decomposition. The spatio-temporal parameter α allows to adjust between modeling of external information and redundancy between components. By comparison with ComBat, the factorization approach enables to better understand what is removed in the cleaning process. Results on the classification task on the real dataset shows a slight improvement for the ICA based one. The next step would be to test it on a more difficult dataset where the labels correlate partially with the sub-datasets.

References

1. Alter, O., Brown, P.O., Botstein, D.: Singular value decomposition for genome-wide expression data processing and modeling. Proc. Natl. Acad. Sci. U.S.A. **97**(18), 10101–10106 (2000)
2. Chen, C., Grennan, K., Badner, J., Zhang, D., Gershon, E., Jin, L., Liu, C.: Removing batch effects in analysis of expression microarray data: an evaluation of six batch adjustment methods. PloS ONE **6**(2), e17238 (2011)
3. Cardoso, J.-F.: High-order contrasts for independent component analysis. Neural Comput. **11**(1), 157–192 (1999)
4. Desmedt, C., et al.: Strong time dependence of the 76-gene prognostic signature for node-negative breast cancer patients in the TRANSBIG multicenter independent validation series. Clin. Cancer Res. **13**(11), 3207–3214 (2007)
5. Johnson, W., Li, C., Rabinovic, A.: Adjusting batch effects in microarray expression data using empirical Bayes methods. Biostatistics **8**(1), 118–127 (2007)
6. Lazar, C., Meganck, S., Taminau, J., Steenhoff, D., Coletta, A., Molter, C., Weiss-Solís, D.Y., Duque, R., Bersini, H., Nowé, A.: Batch effect removal methods for microarray gene expression data integration: a survey. Brief. Bioinform. **14**(4), 469–490 (2013)
7. Leek, J.T., et al.: Tackling the widespread and critical impact of batch effects in high-throughput data. Nat. Rev. Genet. **11**(10), 733–739 (2010)
8. Leek, J.T., Storey, J.D.: Capturing heterogeneity in gene expression studies by surrogate variable analysis. PloS Genet. **3**(9), e161 (2007)
9. Loi, S., et al.: Definition of clinically distinct molecular subtypes in estrogen receptor-positive breast carcinomas through genomic grade. J. Clin. Oncol. **25**(10), 1239–1246 (2007)
10. Miller, L.D., et al.: An expression signature for p53 status in human breast cancer predicts mutation status, transcriptional effects, and patient survival. Proc. Natl. Acad. Sci. U.S.A. **102**(38), 13550–13555 (2005)
11. Minn, A.J., et al.: Lung metastasis genes couple breast tumor size and metastatic spread. Proc. Natl. Acad. Sci. **104**(16), 6740–6745 (2007)
12. Renard, E., Teschendorff, A.E., Absil, P.-A.: Capturing confounding sources of variation in DNA methylation data by spatiotemporal independent component analysis. In: 22nd European Symposium on Artificial Neural Networks, Computational Intelligence and Machine Learning (2014)
13. Sabatier, R., Finetti, P., Cervera, N., Lambaudie, E., Esterni, B., Mamessier, E., Tallet, A., Chabannon, C., Extra, J.-M., Jacquemier, J., Viens, P., Birnbaum, D., Bertucci, F.: A gene expression signature identifies two prognostic subgroups of basal breast cancer. Breast Cancer Res. Treat. **126**(2), 407–420 (2011)
14. Sainlez, M., Absil, P.-A., Teschendorff, A.E.: Gene expression data analysis using spatiotemporal blind source separation. In: 17nd European Symposium on Artificial Neural Networks, Computational Intelligence and Machine Learning (2009)
15. Sotiriou, C., et al.: Gene expression profiling in breast cancer: understanding the molecular basis of histologic grade to improve prognosis. J. Nat. Cancer Inst. **98**(4), 262–272 (2006)

16. Stone, J.V., Porrill, J., Porter, N.R., Wilkinson, I.D.: Spatiotemporal independent component analysis of event-related fMRI data using skewed probability density functions. NeuroImage **15**(2), 407–421 (2002)
17. Teschendorff, A.E., Zhuang, J., Widschwendter, M.: Independent surrogate variable analysis to deconvolve confounding factors in large-scale microarray profiling studies. Bioinformatics **27**(11), 1496–1505 (2011)
18. Wang, Y., et al.: Gene-expression profiles to predict distant metastasis of lymph-node-negative primary breast cancer. Lancet **365**(9460), 671–679 (2005)

A Linear Time Approximation Algorithm for the DCJ Distance for Genomes with Bounded Number of Duplicates

Diego P. Rubert[1], Pedro Feijão[2], Marília D.V. Braga[2],
Jens Stoye[2], and Fábio V. Martinez[1(✉)]

[1] Faculdade de Computação, Universidade Federal de Mato Grosso Do Sul,
Campo Grande, MS, Brazil
{diego,fhvm}@facom.ufms.br
[2] Faculty of Technology and Center for Biotechnology (CeBiTec),
Bielefeld University, Bielefeld, Germany
{pfeijao,mbraga,stoye}@cebitec.uni-bielefeld.de

Abstract. Rearrangements are large-scale mutations in genomes, responsible for complex changes and structural variations. Most rearrangements that modify the organization of a genome can be represented by the *double cut and join* (DCJ) operation. Given two genomes with the same content, so that we have exactly the same number of copies of each gene in each genome, we are interested in the problem of computing the *rearrangement distance* between them, i.e., finding the minimum number of DCJ operations that transform one genome into the other. We propose a linear time approximation algorithm with approximation factor $O(k)$ for the DCJ distance problem, where k is the maximum number of duplicates of any gene in the input genomes. Our algorithm uses as an intermediate step an $O(k)$-approximation for the minimum common string partition problem, which is closely related to the DCJ distance problem. Experiments on simulated data sets show that the algorithm is very competitive both in efficiency and quality of the solutions.

1 Introduction

Large-scale mutations or rearrangements can produce complex changes and structural variations in genomes. They include inversions of chromosome segments, translocations of chromosome ends, fusions and fissions of chromosomes. All these rearrangements can be represented by the *double cut and join* (DCJ) operation [15], which basically consists of cutting a genome in two distinct positions (possibly in two distinct chromosomes) and joining the four resultant open ends in a different way.

A basic task in comparative genomics is to find the rearrangement distance between two given genomes, i.e., the minimum number of rearrangements that transform one genome into the other. For genomes without duplicate genes, there are linear time algorithms to compute the distance allowing only DCJ operations [4]. On the other hand, for genomes with duplicate genes, computing

© Springer International Publishing Switzerland 2016
M. Frith and C.N.S. Pedersen (Eds.): WABI 2016, LNBI 9838, pp. 293–306, 2016.
DOI: 10.1007/978-3-319-43681-4_24

the rearrangement distance is NP-hard, even when the genomes have the same content and only DCJ operations are allowed [2,3].

In this paper we study the problem of computing the DCJ distance between two genomes with the same content and possibly duplicate genes, with the restriction that we have exactly the same number of copies of each gene in each genome. We propose a linear time approximation algorithm with approximation factor $O(k)$, where k is the maximum number of duplicates of any gene in the input genomes. The main goal is a construction of a consistent decomposition of the corresponding adjacency graph, which is a disjoint cycle decomposition of this graph. And then, we can easily compute the DCJ distance from this decomposition.

To obtain such a decomposition we use a linear time approximation algorithm for the minimum common string partition problem with approximation factor $O(k)$ [11]. It is an efficient approximation for the breakpoint distance (the number of genes in the genome minus the number of preserved adjacencies), an intermediate step of our proposed algorithm. As we will show, the whole procedure is an approximation algorithm with approximation factor $O(k)$ and linear running time for the DCJ distance problem for genomes with the same content and exactly the same number of copies of each gene in each genome. The proposed algorithm works properly on inputs that are linear unichromosomal genomes.

The next section presents a background for describing the DCJ distance problem and Sect. 3 presents it formally. The subsequent section discusses the algorithm for the minimum common string partition problem and correlates it to the DCJ distance. In Sect. 5 we develop our approach to compute the DCJ distance. Experiments on simulated data sets are presented in Sect. 6. The last section concludes the paper.

2   Preliminaries

A *gene* g in a genome is an oriented DNA fragment that can be represented by the symbol g itself, if it has direct orientation, or by the symbol $-g$, if it has reverse orientation. Genomes can be partitioned into *chromosomes*, that are linear or circular sequences of genes. Each one of the two ends of a linear chromosome is a *telomere*, represented by the symbol $\circ$.

Each chromosome in a genome can be represented by a string of its genes that can be circular, if the chromosome is circular, or linear and flanked by the symbols $\circ$, if the chromosome is linear. Given a gene g, let $m_A(g)$ be the number of occurrences of g in a genome A. To refer to each occurrence of a gene g unambiguously, we number the occurrences of g from 1 to $m_A(g)$. When there exists at least one gene that occurs more than once in genome A, we say that A has *duplicate genes*. Consider for instance the genome $A = \{(\circ\ c_1\ -a_1\ d_1\ \circ), (\circ\ b_1\ -a_2\ c_2\ \circ)\}$, composed of two linear chromosomes (each chromosome is flanked by parentheses). In A we have one occurrence of genes b and d and two occurrences of genes a and c, that is, A has duplicate genes.

We use the notations $\mathcal{G}(A)$ and $\mathcal{G}^N(A)$, respectively, to refer to the set of (non-numbered) genes and to the set of numbered genes of a genome A. Considering again the genome A above, we have $\mathcal{G}(A) = \{a, b, c, d\}$ and $\mathcal{G}^N(A) = \{a_1, a_2, b_1, c_1, c_2, d_1\}$. Observe that the genomes $A' = \{(\circ\ c_1\ -a_2\ d_1\ \circ),\ (\circ\ b_1\ -a_1\ c_2\ \circ)\}$ and $A'' = \{(\circ\ c_2\ -a_2\ d_1\ \circ), (\circ\ b_1\ -a_1 c_1\ \circ)\}$ are equivalent to $A = \{(\circ\ c_1\ -a_1\ d_1\ \circ), (\circ\ b_1\ -a_2\ c_2\ \circ)\}$. Given a genome A, possibly with duplicate genes, we denote by $[A]$ the equivalence class of genomes that can be obtained from A by swapping indices between occurrences of the same gene.

2.1   Balanced Genomes

Given genomes A and B possibly with duplicate genes, if they contain the same number of occurrences of each gene, i.e. $\mathcal{G}^N(A) = \mathcal{G}^N(B)$, we say that A and B are *balanced*. Consequently, $|A| = |B| = n$. Moreover, we define $occ(A) = \max_{g \in A}\{m_A(g)\}$ as the maximum number of duplicates of any gene g in A. Thus, if A and B are balanced genomes then $occ(A) = occ(B)$. For simplicity, in this case we use only occ. For example, genomes $A = \{(\circ\ c_1\ -a_1\ d_1\ \circ), (\circ\ b_1\ c_2\ \circ), (c_3)\}$ and $B = \{(\circ\ a_1\ \circ), (\circ\ c_3\ -c_1\ -b_1\ \circ), (\circ\ d_1\ c_2\ \circ)\}$ are balanced, since $\mathcal{G}^N(A) = \{a_1, b_1, c_1, c_2, c_3, d_1\} = \mathcal{G}^N(B)$, and $occ = 3$.

2.2   DCJ Operations

Rearrangements can change the organization of a genome, i.e., the number of chromosomes in a genome or the order and the orientation of its genes. In general, such a rearrangement cuts a genome in two different positions, creating four open ends, and joins these open ends in a different way. It can be modeled by a *double-cut and join* (DCJ) operation [15]. Consider, for example, a DCJ applied to genome $\{(\circ\ c_1\ -a_1\ d_1\ \circ), (\circ\ b_1\ -a_2\ c_2\ \circ)\}$, that cuts the first chromosome before and after $-a_1\ d_1$, creating the segments $(\circ\ c_1\ \bullet)$, $(\bullet\ -a_1\ d_1\ \bullet)$ and $(\bullet\ \circ)$ (the symbol $\bullet$ represents the open ends). If we then join the first with the third and the second with the fourth open end, we obtain $\{(\circ\ c_1\ -d_1\ a_1\ \circ), (\circ\ b_1\ -a_2\ c_2\ \circ)\}$. This DCJ corresponds to the inversion of contiguous genes $-a_1\ d_1$. DCJ operations can also correspond to other rearrangements, such as translocations, fusions and fissions [15].

2.3   DCJ Distance and Adjacency Graph

Observe that the DCJ operation alone can only sort balanced genomes. We formally define the DCJ distance problem:

Problem DCJ-DISTANCE(A, B): Given two balanced genomes A and B, compute their DCJ distance $d_{DCJ}(A, B)$, i.e., the minimum number of DCJ operations required to transform A into B', such that $B' \in [B]$.

Any sequence of $d_{DCJ}(A, B)$ DCJ operations transforming A into $B' \in [B]$ is called an *optimal* sequence of DCJ operations.

Given two balanced genomes A and B, DCJ-DISTANCE(A, B) can be computed with the help of the following concepts. First note that, since a gene g has an orientation, we can distinguish its two ends, also called its *extremities*, and denote them by g^t (*tail*) and g^h (*head*). An *adjacency* in a genome either is *telomeric* and corresponds to the extremity of a gene that is adjacent to one of its chromosome ends, or it is an unordered pair of consecutive extremities in one of its chromosomes. Thus, a genome A can also be defined as a set of adjacencies adj(A) of its numbered genes. Given genome $A = \{(\circ\ c_1\ -a_1\ d_1\ \circ), (\circ\ b_1\ -a_2\ c_2\ \circ)\}$, for example, we have adj(A) = $\{\ c_1^t\ ,\ c_1^h a_1^h\ ,\ a_1^t d_1^t\ ,\ d_1^h\ ,\ b_1^t\ ,\ b_1^h a_2^h\ ,\ a_2^t c_2^t\ ,\ c_2^h\ \}$.

Given two balanced genomes A and B, the *adjacency graph* $AG(A, B)$ [4] is a bipartite multigraph such that each partition corresponds to the set of adjacencies of one of the two input genomes, and an edge connects the same extremities of adjacencies in both partitions, regardless of their index numbers. We say that the edge *represents* those extremities. The *length* of a path or cycle in $AG(A, B)$ is the number of edges it contains.

Without Duplicate Genes. When the genomes A and B contain no duplicate genes, there is a one-to-one correspondence between the set of edges in $AG(A, B)$ and the set of gene extremities. In this case, vertices have degree one or two and thus the adjacency graph is a collection of disjoint paths and cycles. Here, problem DCJ-DISTANCE can easily be solved in linear time [4] using the formula

$$d_{DCJ}(A, B) = n - c - i/2\ ,$$

where $n = |\mathcal{G}(A)| = |\mathcal{G}(B)|$ is the number of genes in any of the two genomes, c is the number of cycles and i is the number of odd-length paths in $AG(A, B)$.

With Duplicate Genes. When genomes have duplicate genes, problem DCJ-DISTANCE becomes NP-hard [13]. In the same paper, the authors present an exact, exponential-time algorithm for its solution, phrased in form of an Integer Linear Program (ILP).

3   An Approach to Compute the DCJ Distance with Duplicate Genes

Observe that in the presence of duplicate genes, the adjacency graph may contain vertices of degree larger than two. A *decomposition* of $AG(A, B)$ is a collection of vertex-disjoint cycles and paths covering all vertices of $AG(A, B)$. Cycles and paths of a decomposition D are collectively called *components* of D.

There can be multiple ways of selecting a decomposition of the adjacency graph. We need to find one that allows to match each occurrence of a gene in genome A with exactly one occurrence of the same gene in genome B. In order to build such a decomposition, we need the following definitions.

Let g_i and g_j be, respectively, occurrences of the same gene g in genomes A and B. The edge e that represents the connection of the head of g_i to the head of g_j and the edge f that represents the connection of the tail of g_i to the tail of g_j are called *siblings*. Two edges are *compatible* if they are siblings, or they represent the connection of extremities of distinct occurrences of the same gene, or they represent the connection of extremities of distinct genes. Otherwise they are *incompatible*. A set of edges is *compatible* if it has no pair of incompatible edges. A path or cycle C of $AG(A, B)$ is *consistent* if the set $E(C)$ of edges of C is compatible. Note that, when constructing a decomposition, by choosing consistent components one may still select incompatible edges that occur in separate components (see the three dotted cycles of length 2 in Fig. 1). Thus, consistency cannot be taken into account in components separately.

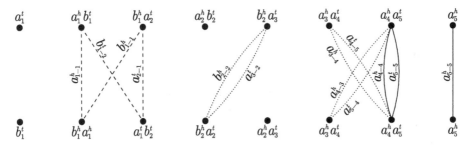

Fig. 1. Examples of an inconsistent cycle (dashed edges) and an inconsistent set of cycles (dotted edges): the adjacency graph for $A = (\circ\ a_1\ b_1\ a_2\ b_2\ a_3\ a_4\ a_5\ \circ)$ and $B = (\circ\ b_1\ -a_1\ b_2\ a_2\ a_3\ a_4\ a_5\ \circ)$, with some edges omitted. For the sake of clarity, edges are labeled with extremities they represent. For example, an edge labeled g^t_{i-j} represents extremities g^t_i from A and g^t_j from B.

A set of paths and cycles $\{C_1, C_2, \ldots, C_k\}$ of $AG(A, B)$ is *consistent* if and only if $E(C_1) \cup E(C_2) \cup \cdots \cup E(C_k)$ is compatible. A *consistent decomposition* D of $AG(A, B)$ is a consistent set of vertex-disjoint cycles and paths that cover all vertices in $AG(A, B)$. Observe that in a consistent decomposition D we have only pairs of siblings, i.e., either an edge e and its sibling f are in D or both e and f are not in D. Thus, a consistent decomposition corresponds to a matching of occurrences of genes in both genomes and allows us to compute the value

$$d_D = n - c_D - i_D/2,$$

where $n = |\mathcal{G}^N(A)| = |\mathcal{G}^N(B)|$ and c_D and i_D are the numbers of cycles and odd-length paths in D, respectively. This provides a way to compute the DCJ distance.

Theorem 1. *Given two genomes A and B, possibly with duplicate genes, the solution for the problem* DCJ-DISTANCE *is given by the following equation:*

$$d_{DCJ}(A, B) = \min_{D \in \mathcal{D}}\{d_D\},$$

where $\mathcal{D}$ is the set of all consistent decompositions of $AG(A, B)$.

Proof. Since a consistent decomposition allows to match duplicates in both genomes, clearly $d_{\text{DCJ}}(A, B) \leq \min_{D \in \mathcal{D}} \{d_D\}$. Now, assume that $d_{\text{DCJ}}(A, B) < \min_{D \in \mathcal{D}} \{d_D\}$. By definition, this distance corresponds to an optimal rearrangement scenario from A to some $B' \in [B]$ and therefore implies a matching between the genes of A and the genes of B'. Furthermore, this matching gives rise to a consistent decomposition D' of $AG(A, B)$ such that $d_{D'} < \min_{D \in \mathcal{D}} \{d_D\}$, which is a contradiction. $\square$

A consistent decomposition D such that $d_D = d_{\text{DCJ}}(A, B)$ is said to be *optimal*.

Once a consistent decomposition D of the adjacency graph $AG(A, B)$ is found, following [4] it is easy to derive in linear time a DCJ rearrangement scenario with d_D DCJ operations transforming A into B. Moreover, an optimal consistent decomposition allows to find all optimal rearrangement scenarios [5].

3.1 Capping Telomeres

A general technique for simplifying algorithms that handle genomes with possibly unequal telomeric adjacencies is called *capping* and consists of transforming each telomeric into a non-telomeric adjacency [9,12,16]. Let *null extremities* be represented by τ and *null adjacencies* be represented by $\tau\tau$. Given two genomes A and B with $2i$ and $2j$ telomeres, respectively, in both genomes each telomeric adjacency x is replaced by the adjacency $x\tau$. Furthermore, in order to add the same number of null extremities to both genomes, $|j - i|$ null adjacencies $\tau\tau$ are added to genome A, if $i < j$, or to genome B, if $j < i$. Let A_τ and B_τ be the new sets of adjacencies obtained by this procedure. Observe that in $AG(A_\tau, B_\tau)$ each null extremity of A_τ must be connected to each null extremity of B_τ.

Observe that any consistent decomposition D of $AG(A_\tau, B_\tau)$ is composed of cycles only, allowing to compute the value

$$d_D = n - c_D \ ,$$

where $n = |\mathcal{G}^N(A)| = |\mathcal{G}^N(B)|$ and c_D is the number of cycles in D.

Theorem 2. *Let A and B be two genomes and let A_τ and B_τ be the genomes obtained from A and B by capping telomeric adjacencies. Then,*

$$d_{\text{DCJ}}(A, B) = \min_{D \in \mathcal{D}_\tau} \{d_D\} \ ,$$

where $\mathcal{D}_\tau$ is the set of all consistent decompositions of $AG(A_\tau, B_\tau)$.

Proof. Each consistent decomposition D of $AG(A, B)$ corresponds to a consistent decomposition D' of $AG(A_\tau, B_\tau)$, such that each path in D becomes a cycle in D'. The null extremities added to both genomes ensure that $d_D = d_{D'}$: in the formula to compute $d_{D'}$ each path adds one to the term c_D but (i) each even path has two new null extremities and adds one to the term n and (ii) each pair of odd paths has two new null extremities and adds one to the term n and decreases two from the term i_D. $\square$

4 Approximating the DCJ Distance by Cycles of Length 2

All definitions and properties for the DCJ distance for balanced genomes presented from the beginning to here work properly for the general case, where genomes are multichromosomal. However, as we will see in this section, to solve the DCJ distance problem we use an intermediate procedure whose inputs are strings. Thus, from now on, we restrict our inputs for linear unichromosomal genomes. The extension to general genomes is left as an open problem.

As mentioned in the previous section, the adjacency graph for balanced and capped genomes is a collection of cycles and thus we have to find a disjoint cycle decomposition of the adjacency graph to compute the DCJ distance according to Theorems 1 and 2. Recall that it is an NP-hard problem [13].

Given a consistent decomposition $D \in \mathcal{D}_\tau$ of an adjacency graph $AG(A_\tau, B_\tau)$, we can see that

$$d_D = n - c_D = n - c_2 - c_> ,$$

where $n = |\mathcal{G}^N(A)| = |\mathcal{G}^N(B)|$, c_2 is the number of cycles of length 2, and $c_>$ is the number of cycles of length longer than 2 in D. A naive approach to solve the DCJ distance problem could be, as a first step, maximizing c_2. However, this strategy is not able to solve properly the DCJ distance problem for two main reasons: (i) finding the maximum number of cycles of length 2 is itself an NP-hard problem, as we will justify below; and (ii) this strategy is not optimal to solve the DCJ distance, as we can see in Fig. 2.

The problem of finding a decomposition maximizing the number of cycles of length 2 is equivalent to the *adjacency similarity problem* [3], the complement of the breakpoint distance problem, where one wants to minimize $n - c_2$. Moreover, from an optimal solution for the adjacency similarity (or the breakpoint distance) problem it is possible to approximate the DCJ distance, as stated in Lemma 1.

Lemma 1. *A consistent decomposition D' of $AG(A_\tau, B_\tau)$ containing the maximum number of cycles of length 2 is a 2-approximation for the* DCJ-DISTANCE *problem.*

Proof. Let c_2^* and $c_>^*$ be the number of cycles of length 2 and longer than 2, respectively, of an optimal decomposition D^* of $AG(A_\tau, B_\tau)$. Let c_2' and $c_>'$ be the numbers analogous to c_2^* and $c_>^*$ with respect to the decomposition D'. It it easy to see that $c_2^* + 2c_>^* \le n$, then

$$0 \le n - c_2^* - 2c_>^*$$
$$n - c_2^* \le n - c_2^* - 2c_>^* + n - c_2^*$$
$$n - c_2^* \le 2(n - c_2^* - c_>^*) . \tag{1}$$

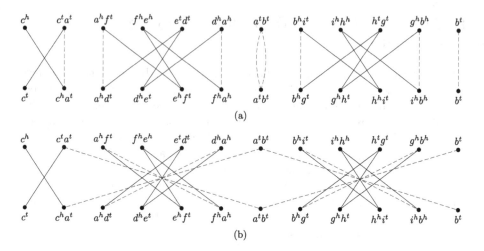

Fig. 2. Consistent decompositions for genomes $A = (\circ\ -c\ a\ f\ -e\ d\ -a\ b\ i\ -h\ g\ -b\ \circ)$ and $B = (\circ\ c\ a\ d\ e\ f\ -a\ b\ g\ h\ i\ -b\ \circ)$, where solid edges are in both decompositions. Gene indexes were omitted. (a) A consistent decomposition D' containing the maximum number of cycles of length 2, composed of 1 cycle of length 2, 1 cycle of length 8, 1 odd path of length 1 and 1 odd path of length 3, resulting in $d_{D'} = 11 - 4 - 2/2 = 7$. (b) An optimal consistent decomposition D^*, composed of 4 cycles of length 4 and 2 odd paths of length 3, resulting in $d_{D^*} = 11 - 4 - 2/2 = 6$.

Therefore, we have

$$\frac{d_{D'}}{d_{D^*}} = \frac{n - c_2' - c_>'}{n - c_2^* - c_>^*}$$

$$\leq \frac{n - c_2^* - c_>'}{n - c_2^* - c_>^*} \qquad (2)$$

$$\leq \frac{n - c_2^*}{n - c_2^* - c_>^*}$$

$$\leq \frac{2(n - c_2^* - c_>^*)}{n - c_2^* - c_>^*} \qquad (3)$$

$$= 2\ , \qquad (4)$$

where (2) holds since $c_2' \geq c_2^*$, and (3) is true from (1). □

Recall that the adjacency similarity and breakpoint distance problems are both NP-hard [3,6]. The former can be approximated by a factor of 4 for balanced genomes [1]. However, an approximation with constant approximation factor for the former problem does not lead to an approximation with constant approximation factor for the latter. The breakpoint distance for balanced genomes has a 1.1037-approximation when $occ = 2$ [8], a 4-approximation when $occ = 3$ [8], and an $O(k)$-approximation when $occ = k$ [11]. Those approximations were developed for the minimum common string partition problem (MCSP) [14], which is equivalent to the breakpoint distance problem [10].

5 Finding Consistent Decompositions

In this section we present a linear time approximation algorithm CONSISTENT-DECOMPOSITION, which receives two linear unichromosomal balanced genomes A and B with $occ = k$ and returns a consistent decomposition for genomes A and B, which is an $O(k)$-approximation for the DCJ distance. The main steps of CONSISTENT-DECOMPOSITION can be briefly described as follows. First, from the input genomes, we obtain capped genomes and then we build the adjacency graph of them. Next, we use an approximation for the (signed) minimum common string partition problem, which gives an approximation for the number of breakpoints in the adjacency graph. After that we clean the chosen cycles of length 2 from the adjacency graph. Following, we iteratively collect arbitrary cycles of length longer than 2, cleaning up the remaining graph after each iteration. Finally, we return the set of collected cycles as a consistent decomposition of the prior adjacency graph.

Algorithm 1. CONSISTENT-DECOMPOSITION(A, B)

Input: balanced genomes A and B such that $occ = k$
Output: a consistent decomposition of $AG(A, B)$
 1: Add null extremities/adjacencies to A and B and obtain A_τ and B_τ, respectively
 2: Build the adjacency graph $AG(A_\tau, B_\tau)$
 3: Obtain an $O(k)$-approximation $\mathcal{S}_2$ for the set of cycles of length 2 in $AG(A_\tau, B_\tau)$ using the $O(k)$-approximation algorithm for the minimum common string partition problem [11]
 4: Remove from the adjacency graph vertices covered by $\mathcal{S}_2$ and all edges incompatible with edges of $\mathcal{S}_2$
 5: Decompose the remaining graph into consistent cycles by iteratively finding a consistent cycle C and then removing from the graph vertices covered by C and edges incompatible with edges of C, collecting them in $\mathcal{S}_>$
 6: Remove null extremities/adjacencies of cycles in $\mathcal{S}_2 \cup \mathcal{S}_>$ and obtain a consistent decomposition D of $AG(A, B)$
 7: Return D

Step 1 of CONSISTENT-DECOMPOSITION consists of capping telomeres from the given balanced genomes A and B as described in Sect. 3.1. In Step 2, CONSISTENT-DECOMPOSITION builds the adjacency graph for capped genomes A_τ and B_τ. After that, Step 3 collects cycles of length 2 using an $O(k)$-approximation algorithm for the minimum common string partition problem [11] as described in Sect. 4. Step 4 removes from $AG(A_\tau, B_\tau)$ vertices covered by cycles in $\mathcal{S}_2$ and edges incompatible with edges of cycles in $\mathcal{S}_2$. Step 5 constructs the set $\mathcal{S}_>$ by decomposing the remaining graph into consistent cycles. Iteratively, it chooses a consistent cycle C and then removes from the remaining graph vertices covered by C and edges incompatible with edges of C. Hence the algorithm does not choose an inconsistent set of components. Further, this guarantees that for every edge in the decomposition, its sibling edge will also be in the decomposition, avoiding for example the selection of the path of length 1 composed of the edge that connects a_5^h of A to a_5^h of B and then the cycle of length 2 composed of the edge that connects a_4^h of A to a_3^h of B and the edge that connects a_5^t of A to a_4^t of B in Fig. 1. In order to obtain the consistent decomposition of $AG(A, B)$, CONSISTENT-DECOMPOSITION removes in Step 6 null extremities/adjacencies of cycles in $\mathcal{S}_2 \cup \mathcal{S}_>$, returning the resulting set D in Step 7.

There is one implicit but important step in the algorithm above, which is to obtain the set $\mathcal{S}_2$ given the output of the k-MCSP approximation algorithm [11]. This algorithm outputs a common string partition $(\mathcal{A}, \mathcal{B})$. Both $\mathcal{A}$ and $\mathcal{B}$ are composed of the same set of substrings, in different orders and possibly reversed. Each substring of length $l > 1$ in $\mathcal{A}$ and $\mathcal{B}$ induces $l - 1$ preserved adjacencies in A and B. First of all, we must normalize strings in $\mathcal{A}$ and $\mathcal{B}$, that is, for each substring s and its reverse r, only s appears in $\mathcal{A}$ and $\mathcal{B}$ (we reverse each occurrence of r, resulting in s). Then we just have to map each substring in $\mathcal{A}$ to the same substring in $\mathcal{B}$ (in case of multiple occurrences, we choose any of them), which can be performed using a prefix tree. Thus this implicit step can be done in linear time on the adjacency graph size.

Lemma 2. *Given balanced genomes A and B such that $|A| = |B|$, the running time of* CONSISTENT-DECOMPOSITION *algorithm is linear on the size of the corresponding adjacency graph.*

Proof. Let m be the size of $AG(A_\tau, B_\tau)$. It is easy to see that Steps 1 and 2 of Algorithm 1 have both linear running time, i.e. $O(m)$. The implementation of the k-MCSP [11] in Step 3 with suffix trees [7] and disjoint sets has running time $O(m)$ (note that $m = O(n^2)$). The running time of Step 4 is $O(m)$ since we have just to traverse vertices and edges of the remaining adjacency graph. Step 5 consists of collecting cycles arbitrarily, and therefore its running time is also linear, we just have to walk in $AG(A_\tau, B_\tau)$ finding cycles. To be sure we walk only in consistent paths, we can use a hash table of size $\Theta(n)$ and store, for each edge of previously chosen cycles in Steps 1 to 5, genes of A_τ (B_τ) associated to genes of B_τ (A_τ). For instance, the selection of an edge representing the connection of extremities a_i^t of A_τ and a_j^t of B_τ is consistent if both a_i and a_j are associated with no gene of B_τ and A_τ, respectively, or both are already associated with each other (this edge is the sibling of a previously chosen edge). This consistency check takes $O(1)$ time. The last step (Step 6) is similar to Step 4 and thus has running time $O(m)$. Therefore, CONSISTENT-DECOMPOSITION has running time $O(m)$.   □

To conclude this section, we present the following result which establishes an approximation factor for DCJ-DISTANCE.

Theorem 3. *Let A and B be balanced genomes such that $occ = k$. Given a common string partition $(\mathcal{A}, \mathcal{B})$ with approximation factor $O(k)$ for the k-MCSP problem, a consistent decomposition D of $AG(A, B)$, containing cycles of length 2 reflecting preserved adjacencies in $(\mathcal{A}, \mathcal{B})$, is an $O(k)$-approximation for the* DCJ-DISTANCE *problem.*

Proof. Let c_2^* and $c_>^*$ be the number of cycles of length 2 and longer than 2, respectively, of an optimal decomposition D^* of $AG(A, B)$. Let $\mathcal{S}_2$ be the set of cycles of length 2 reflecting preserved adjacencies in $(\mathcal{A}, \mathcal{B})$, and let $\mathcal{S}_>$ be an arbitrary decomposition of cycles in $AG(A, B) \setminus \mathcal{S}_2$. Let $D = \mathcal{S}_2 \cup \mathcal{S}_>$, a consistent decomposition, $c_2 = |\mathcal{S}_2|$, and $c_> = |\mathcal{S}_>|$. From [11] we have an $O(k)$-approximation for the k-MCSP problem, and then $n - c_2 \leq \ell(n - c_2')$, where $\ell = O(k)$ and c_2' is the number of cycles of length 2 in a consistent decomposition D' with maximum number of cycles of length 2. Hence,

$$
\begin{aligned}
\frac{d_D}{d_{D^*}} &= \frac{n - c_2 - c_>}{n - c_2^* - c_>^*} \\
&\leq \frac{\ell(n - c_2') - c_>}{n - c_2^* - c_>^*} \\
&\leq \frac{\ell(n - c_2')}{n - c_2^* - c_>^*} \\
&\leq 2\ell \left(\frac{n - c_2' - c_>'}{n - c_2^* - c_>^*} \right) \qquad (5) \\
&\leq 4\ell \, , \qquad (6)
\end{aligned}
$$

where (5) is analogous to (1) and (6) holds from (4), both in the proof of Lemma 1.   □

6   Experimental Results

We have implemented our approximation algorithm in C++, with the addition of a linear time greedy heuristic for the decomposition of cycles not induced by the k-MCSP approximation. The experiments for this approach were performed on an Intel i3 3.3 GHz machine.

We compare our algorithm with Shao *et al.*'s ILP [13] on simulated datasets. Given two genomes, the ILP based experiments first build the adjacency graph, followed by capping of the telomeres, fixing some safe cycles of length two, and finally invoking an ILP solver to obtain an optimal solution with a time limit of 2 h.

Following [13], we simulate artificial genomes with segmental duplications and DCJs. We uniformly select a position to start duplicating a segment of the genome and place the new copy to a new position. From a genome of s distinct genes, we generate an ancestor genome with $1.5s$ genes by randomly performing $s/2l$ segmental duplications of length l, resulting in an average $k = 1.5$.

Then we simulate two extant genomes from the ancestor by randomly performing r DCJs (reversals) independently. Thus, the simulated evolutionary distance between the two extant genomes is $2r$. We set $s = 1000$ and test three different lengths for segmental duplications ($l = 1, 2, 5$). We also vary the r value over a range $200, 220, \ldots, 500$. Figure 3 shows the average difference "*computed number of DCJs − simulated evolutionary distance*", taking as input five pairs of genomes for each combination of l and r, while Fig. 4 shows the average running time. Note that, although the DCJ distance is unknown whenever the ILP solver is stopped after the time limit, it is always less than or equal to the simulated evolutionary distance for these artificial genome pairs.

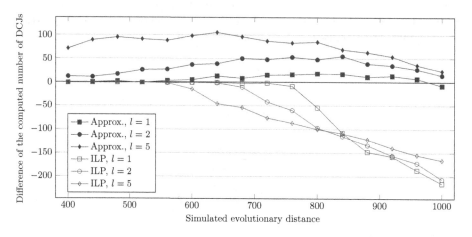

Fig. 3. The computed number of DCJs vs. the simulated evolutionary distance for $s = 1000$. (Color figure online)

The difference of the number of DCJs (blue lines in Fig. 3) calculated by our approximation algorithm remains very close to the simulated evolutionary distance for small values of l. Moreover, it remains roughly the same for the same value of l even for greater values of r. The values obtained by the ILP approach (red lines in Fig. 3) are very close to those obtained by the approximation algorithm and to the simulated evolutionary distance from the simulations for $l \leq 2$ and smaller values of r. However, beyond some point the DCJ distance calculated by the ILP gets even lower than the simulated evolutionary distance in the simulations, showing the limitations of parsimony for larger distance ranges.

Regarding the running time, our implementation time increases slowly from ≈ 0.03 ($2r = 400$) to $\approx 0.08\,\text{s}$ ($2r = 1000$), on average, according to Fig. 4(a), while the ILP approach takes $\approx 0.3\,\text{s}$ to finish for smaller values of r (where the preprocessing step fixes a considerable amount of cycles of length 2 in the adjacency graph), always reaching the time limit of 2 h beyond some point, as displayed in Fig. 4(b).

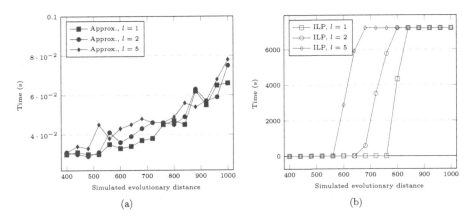

Fig. 4. Execution time for $s = 1000$ of (a) approximation and (b) ILP based programs.

7 Conclusion

In this paper, we have proposed a new approximation algorithm for the DCJ distance for genomes where each gene occurs the same number of times in each input genome and there exists at least one gene that occurs more than once in one of them. This so called DCJ distance with duplicates for balanced genomes problem is NP-hard [13]. Our algorithm works on input genomes where the amount of duplicates is bounded by $O(k)$, where k is the maximum number of duplicates of any gene in the input genomes. The approximation factor of our algorithm is $O(k)$. Furthermore, our algorithm has linear running time on the adjacency graph size. As experiments on simulated genomes have shown, our algorithm is very competitive both in efficiency and quality of the solutions, in comparison to the ILP exact solution.

Due to an intermediate step which approximates the minimum common string partition problem, our algorithm works properly only on linear unichromosomal genomes as input. A natural extension of this work is modifying our algorithm to work with multichromosomal genomes as well. Moreover, we have to extend our experiments, running our algorithm on more simulated data sets and also on biological data sets.

References

1. Angibaud, S., Fertin, G., Rusu, I., Thévenin, A., Vialette, S.: Efficient tools for computing the number of breakpoints and the number of adjacencies between two genomes with duplicate genes. J. Comput. Biol. **15**(8), 1093–1115 (2008)
2. Angibaud, S., Fertin, G., Rusu, I., Thévenin, A., Vialette, S.: On the approximability of comparing genomes with duplicates. J. Graph Algorithms Appl. **13**(1), 19–53 (2009)
3. Angibaud, S., Fertin, G., Rusu, I., Vialette, S.: A pseudo-boolean framework for computing rearrangement distances between genomes with duplicates. J. Comput. Biol. **14**(4), 379–393 (2007)

4. Bergeron, A., Mixtacki, J., Stoye, J.: A unifying view of genome rearrangements. In: Bücher, P., Moret, B.M.E. (eds.) WABI 2006. LNCS (LNBI), vol. 4175, pp. 163–173. Springer, Heidelberg (2006)

5. Braga, M.D.V., Stoye, J.: The solution space of sorting by DCJ. J. Comp. Biol. **17**(9), 1145–1165 (2010)

6. Bryant, D.: The complexity of calculating exemplar distances. In: Sankoff, D., Nadeau, J.H. (eds.) Comparative Genomics, pp. 207–211. Kluwer Academic Publishers, Dortrecht (2000)

7. Farach, M.: Optimal suffix tree construction with large alphabets. In: Proceedings of IEEE/FOCS **1997**, pp. 137–143 (1997)

8. Goldstein, A., Kolman, P., Zheng, J.: Minimum common string partition problem: hardness and approximations. Eletron. J. Comb. **12**, 18 (2005). R50

9. Hannenhalli, S., Pevzner, P.: Transforming men into mice (polynomial algorithm for genomic distance problem). In: Proceedings of FOCS **1995**, pp. 581–592 (1995)

10. Jiang, H., Zheng, C., Sankoff, D., Zhu, B.: Scaffold filling under the breakpoint distance. In: Tannier, E. (ed.) RECOMB-CG 2010. LNCS, vol. 6398, pp. 83–92. Springer, Heidelberg (2010)

11. Kolman, P., Waleń, T.: Reversal distance for strings with duplicates: linear time approximation using hitting set. Electron. J. Comb. **14**(1), R50 (2007)

12. Shao, M., Lin, Y.: Approximating the edit distance for genomes with duplicate genes under DCJ, insertion and deletion. BMC Bioinform. **13**(Suppl 19), S13 (2012)

13. Shao, M., Lin, Y., Moret, B.: An exact algorithm to compute the double-cut-and-join distance for genomes with duplicate genes. J. Comput. Biol. **22**(5), 425–435 (2015)

14. Swenson, K., Marron, M., Earnest-DeYong, K., Moret, B.M.E.: Approximating the true evolutionary distance between two genomes. In: Proceedings of ALENEX/ANALCO **2005**, pp. 121–129 (2005)

15. Yancopoulos, S., Attie, O., Friedberg, R.: Efficient sorting of genomic permutations by translocation, inversion and block interchanges. Bioinformatics **21**(16), 3340–3346 (2005)

16. Yancopoulos, S., Friedberg, R.: DCJ path formulation for genome transformations which include insertions, deletions, and duplications. J. Comput. Biol. **16**(10), 1311–1338 (2009)

A Hybrid Parameter Estimation Algorithm for Beta Mixtures and Applications to Methylation State Classification

Christopher Schröder and Sven Rahmann[(✉)]

Genome Informatics, Institute of Human Genetics, University Hospital Essen,
University of Duisburg-Essen, Essen, Germany
{christopher.schroeder,sven.rahmann}@uni-due.de
http://www.rahmannlab.de

Abstract. Mixtures of beta distributions have previously been shown to be a flexible tool for modeling data with values on the unit interval, such as methylation levels. However, maximum likelihood parameter estimation with beta distributions suffers from problems because of singularities in the log-likelihood function if some observations take the values 0 or 1. While ad-hoc corrections have been proposed to mitigate this problem, we propose a different approach to parameter estimation for beta mixtures where such problems do not arise in the first place. Our algorithm has computational advantages over the maximum-likelihood-based EM algorithm. As an application, we demonstrate that methylation state classification is more accurate when using adaptive thresholds from beta mixtures than non-adaptive thresholds on observed methylation levels.

Keywords: Mixture model · Beta distribution · Maximum likelihood · Method of moments · EM algorithm · Differential methylation · Classification

1 Introduction

The beta distribution is a continuous probability distribution that takes values in the unit interval $[0, 1]$. It has been used in several bioinformatics applications [4] to model data that naturally takes values between 0 and 1, such as relative frequencies, probabilities, absolute correlation coefficients or DNA methylation levels of CpG dinucleotides or longer genomic regions. One of the most prominent applications is for the estimation of false discovery rates (FDRs) from p-value distributions after multiple tests by fitting a beta-uniform mixture (BUM [5]). By linear scaling, beta distributions can be used to model any quantity that takes values in a finite interval $[L, U] \subset \mathbb{R}$.

The beta distribution has two parameters $\alpha > 0$ and $\beta > 0$ and can take a variety of shapes depending on whether $0 < \alpha < 1$ or $\alpha = 1$ or $\alpha > 1$ and $0 < \beta < 1$ or $\beta = 1$ or $\beta > 1$; see Fig. 1. The beta probability density on $[0, 1]$ is

$$b_{\alpha,\beta}(x) = \frac{1}{B(\alpha,\beta)} \cdot x^{\alpha-1} \cdot (1-x)^{\beta-1}, \quad \text{where } B(\alpha,\beta) = \frac{\Gamma(\alpha)\Gamma(\beta)}{\Gamma(\alpha+\beta)}, \quad (1)$$

© Springer International Publishing Switzerland 2016
M. Frith and C.N.S. Pedersen (Eds.): WABI 2016, LNBI 9838, pp. 307–319, 2016.
DOI: 10.1007/978-3-319-43681-4_25

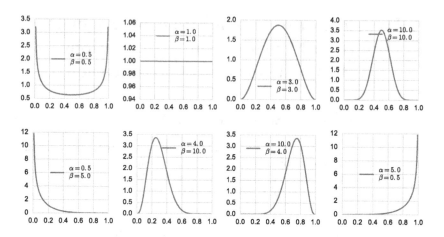

Fig. 1. Different shapes of beta distributions depending on parameters α and β.

and Γ refers to the gamma function $\Gamma(z) = \int_0^\infty x^{z-1} e^{-x} dx$ with $\Gamma(n) = (n-1)!$ for positive integers n. It can be verified that $\int_0^1 b_{\alpha,\beta}(x) dx = 1$. Note that for $\alpha = \beta = 1$, we obtain the uniform distribution. Section 2.1 has more details.

While a single beta distribution can take a variety of shapes, *mixtures* of beta distributions are even more flexible. Such a mixture has the general form

$$f_\theta(x) = \sum_{j=1}^c \pi_j \cdot b_{\alpha_j,\beta_j}(x), \qquad (2)$$

where c is the *number of components*, the π_j are called *mixture coefficients* satisfying $\sum_j \pi_j = 1$ and $\pi_j \geq 0$, and the α_j, β_j are called *component parameters*. Together, we refer to all of these as *model parameters* and abbreviate them as θ. The number of components c is often assumed to be a given constant and not part of the parameters to be estimated.

The *parameter estimation problem* consists of estimating θ from n usually independent observed samples $(x_1, \ldots, x_n)$ such that the observations are well explained by the resulting distribution.

Maximum likelihood (ML) estimation (MLE) is a frequently used paradigm, consisting of the following optimization problem.

$$\text{Given } (x_1, \ldots, x_n), \quad \text{maximize } \mathcal{L}(\theta) := \prod_{i=1}^n f_\theta(x_i),$$

$$\text{or equivalently, } L(\theta) := \sum_{i=1}^n \ln f_\theta(x_i). \qquad (3)$$

As we show in Sect. 2.2, MLE has significant disadvantages for beta distributions. The main problem is that the likelihood function may not be finite (for some parameter values) if some of the observed datapoints are $x_i = 0$ or $x_i = 1$.

For mixture distributions, MLE frequently results in a non-concave problem with many local maxima, and one uses heuristics that return a local optimum from given starting parameters. A popular and successful method for parameter optimization in mixtures is the expectation maximization (EM) algorithm [2] that iteratively solves an (easier) ML problem on each estimated component and then re-estimates which datapoints belong to which component. We review the basic EM algorithm in Sect. 2.3.

Because already MLE for a single beta distribution is problematic, EM does not work for beta mixtures, unless ad-hoc corrections are made. We therefore propose a new algorithm for parameter estimation in beta mixtures that we call *iterated method of moments*. The method is presented in Sect. 3.

Our main motivation for this work stems from the analysis of methylation level data in differentially methylated regions between individuals, not cell types or conditions; see Sect. 4. Our evaluation therefore focuses on the benefits of beta mixture modeling and parameter estimation using our algorithm for methylation state classification from simulated methylation level data.

2 Preliminaries

2.1 Beta Distributions

The beta distribution with parameters $\alpha > 0$ and $\beta > 0$ is a continuous probability distribution on the unit interval $[0, 1]$ whose density is given by Eq. (1).

If X is a random variable with a beta distribution, then its expected value μ and variance σ^2 are

$$\mu := \mathbb{E}[X] = \frac{\alpha}{\alpha + \beta}, \quad \sigma^2 := \mathrm{Var}[X] = \frac{\mu(1 - \mu)}{\alpha + \beta + 1} = \frac{\mu(1 - \mu)}{1 + \phi}, \qquad (4)$$

where $\phi = \alpha + \beta$ is often called a *precision* parameter; large values indicate that the distribution is concentrated.

Conversely, the parameters α and β may be directly expressed in terms of μ and σ^2. First, compute

$$\phi = \frac{\mu(1 - \mu)}{\sigma^2} - 1; \quad \text{then} \quad \alpha = \mu\phi, \quad \beta = (1 - \mu)\phi. \qquad (5)$$

2.2 Maximum Likelihood Estimation for Beta Distributions

The estimation of parameters in a parameterized distribution from n independent samples usually follows the maximum likelihood (ML) paradigm. If θ represents the parameters and $f_\theta(x)$ is the probability density of a single observation, the goal is to find θ^* that maximizes $L(\theta)$ as defined in (3).

Writing $\gamma(y) := \ln \Gamma(y)$, the beta log-likelihood is

$$L(\alpha, \beta) = n(\gamma(\alpha+\beta) - \gamma(\alpha) - \gamma(\beta)) + (\alpha-1) \cdot \sum_i \ln x_i + (\beta-1) \cdot \sum_i \ln(1-x_i). \quad (6)$$

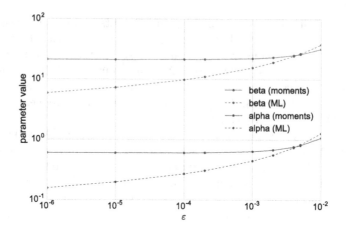

Fig. 2. Estimated parameter values α (blue) and β (red) from a dataset consisting of the ten observations $0.01, \ldots, 0.10$ and ten observations of ε for varying values of ε. Estimation was done using MLE (dotted lines) as implemented in the R package `betareg` and by our (moment-based) method (solid lines).

The optimality conditions $dL/d\alpha = 0$ and $dL/d\beta = 0$ are solved numerically and iteratively. While this is inconvenient (in comparison to a mixture of Gaussians where analytical formulas exist for the ML estimators), the main problem is a different one. The log-likelihood function is not well defined for $\alpha \neq 1$ or $\beta \neq 1$ if any of the observations are $x_i = 0$ or $x_i = 1$. Several implementations of ML estimators for beta distributions (e.g. `betareg`, see below) throw errors then.

Note that, *in theory*, there is no problem, because $x \in \{0, 1\}$ is an event of probability zero if the data are truly generated by a beta distribution. Real data, however, in particular observed methylation levels, may very well take these values. This article's main motivation is the desire to work with observations of $x = 0$ and $x = 1$ in a principled way.

The above problem with MLE for beta distributions has been noted previously, but, to our knowledge, not explicitly attacked. A typical ad-hoc solution is to linearly rescale the unit interval $[0, 1]$ to a smaller sub-interval $[\varepsilon, 1 - \varepsilon]$ for some small $\varepsilon > 0$ or to simply replace values $< \varepsilon$ by ε and values $> 1 - \varepsilon$ by $1 - \varepsilon$, such that, in both cases, the resulting adjusted observations are in $[\varepsilon, 1 - \varepsilon]$.

A simple example, which has to our knowledge not been presented before, will show that the resulting parameter estimates depend strongly on the choice of ε in the ML paradigm. Consider 20 observations, ten of them at $x = 0$, the remaining ten at $x = 0.01, \ldots, 0.10$. For different values of $0 < \varepsilon < 0.01$, replace the ten zeros by ε and compute the ML estimates of α and β. We used the R package `betareg`[1] [3], which performs numerical ML estimation of $\text{logit}(\mu)$ and $\ln(\phi)$, where $\text{logit}(\mu) = \ln(\mu/(1 - \mu))$. We then used Eq. (5) to compute ML estimates of α and β. We additionally used our new approach (Sect. 3) with the

---

[1] https://cran.r-project.org/web/packages/betareg/betareg.pdf.

same varying ε. In contrast to MLE, our approach also works with $\varepsilon = 0$. The resulting estimates for α and β are shown in Fig. 2: Not only is our approach able to directly use $\varepsilon = 0$; it is also insensitive to the choice of ε for small $\varepsilon > 0$.

To summarize, while MLE is known to be statistically efficient for correct data, its results may be sensitive to perturbations of the data. For modeling with beta distributions in particular, the problems of MLE are severe: The likelihood function is not well defined for reasonable datasets that occur in practice, and the solution depends strongly on ad-hoc parameters introduced to rectify the first problem. Before we can introduce our solution to these problems, we first discuss parameter estimation in mixture models.

2.3   The EM Algorithm for Beta Mixture Distributions

For parameters θ of mixture models, including each component's parameters and the mixture coefficients, the log-likelihood function $L(\theta) = \sum_{i=1}^{n} \ln f_{\theta}(x_i)$, with $f_{\theta}(x_i)$ as in Eq. (2), frequently has many local maxima; and a globally optimal solution is difficult to compute.

The EM algorithm [2] is a general iterative method for ML parameter estimation with incomplete data. In mixture models, the "missing" data is the information which sample belongs to which component. However, this information can be estimated (given initial parameter estimates) in the E-step (expectation step) and then used to derive better parameter estimates by ML for each component separately in the M-step (maximization step). Generally, EM converges slowly to a local optimum of the log-likelihood function [6].

E-Step. To estimate the expected responsibility $W_{i,j}$ of each component j for each data point x_i, the component's relative probability at that data point is computed, such that $\sum_j W_{i,j} = 1$ for all i.

$$W_{i,j} = \frac{\pi_j\, b_{\alpha_j,\beta_j}(x_i)}{\sum_k \pi_k\, b_{\alpha_k,\beta_k}(x_i)} \quad \text{and} \quad \pi_j^+ = \frac{1}{n}\sum_{i=1}^{n} W_{i,j}\,. \tag{7}$$

M-Step. Using the responsibility weights $W_{i,j}$, the components are unmixed, and a separate (weighted) sample is obtained for each component, so their parameters can be estimated independently by MLE. The new mixture coefficients' ML estimates π_j^+ in (7) are the averages of the responsibility weights over all samples.

Initialization and Termination. EM requires initial parameters before starting with an E-step. The resulting local optimum depends on these initial parameters. It is therefore common to choose the initial parameters either based on additional information (e.g., one component with small values, one with large values), or to re-start EM with different random initializations. Convergence is detected by monitoring relative changes in the log-likelihood or parameters between iterations and stopping when these changes are below a given tolerance.

Properties and Problems with Beta Mixtures. One of the main reasons why the EM algorithm is predominantly used in practice for mixture estimation is the availability of an objective function (the log-likelihood). By Jensen's inequality, it increases in each EM iteration, and when it stops increasing, a stationary point has been reached [6]. Solutions obtained by two runs with different initializations can be objectively compared by comparing their log-likelihood values.

In beta mixtures, there are several problems with using the EM algorithm. First, the responsibility weights $W_{i,j}$ are not well defined for $x_i = 0$ or $x_i = 1$ because of the singularities in the likelihood function described in Sect. 2.1. Second, the M-step cannot be carried out if the data contains any such point for the same reason. Third, even if all $x_i \in {]}0, 1{[}$, the resulting mixtures are sensitive to perturbations of the data. Fourth, because each M-step already involves a numerical iterative maximization, the computational burden over several EM iterations is significant. We now propose an algorithm for parameter estimation in beta mixtures that does not suffer from these drawbacks.

3   The Iterated Method of Moments

With the necessary preliminaries in place, the main idea behind our algorithm can be stated briefly before we discuss the details.

From initial parameters, we proceed iteratively as in the EM framework and alternate between an E-step, which is a small modification of EM's E-step, and a parameter estimation step, which is not based on the ML paradigm but on Pearson's method of moments until a stationary point is reached.

To estimate Q free parameters, the method of moments' approach is to choose Q moments of the distribution, express them through the parameters and equate them to the corresponding Q sample moments. This usually amounts to solving a system of Q non-linear equations. The method of moments has been applied directly to mixture distributions. For example, a mixture of two one-dimensional Gaussians has five free parameters: two means μ_1, μ_2, two variances σ_1^2, σ_2^2 and the weight π_1 of the first component. Thus one needs to choose five moments, say $m_k := \mathbb{E}[X^k]$ for $k = 1, \ldots, 5$ and solve the corresponding relationships. Solving these equations for many components (or in high dimensions) seems daunting, even numerically. Also it is not clear whether there is always a unique solution.

For a single beta distribution, however, α and β are easily estimated from sample mean and variance by Eq. (5), using sample moments instead of true values. Thus, to avoid the problems of MLE in beta distributions, we replace the likelihood maximization step (M-step) in EM by a method of moments estimation step (MM-step) using expectation and variance.

We thus combine the idea of using latent responsibility weights from EM with moment-based estimation, but avoid the problems of pure moment-based estimation (large non-linear equation systems). It may seem surprising that nobody appears to have done this before, but one reason may be the lack of an objective function, as we discuss further below.

Initialization. A general reasonable strategy for beta mixtures is to let each component focus on a certain sub-interval of the unit interval. With c components, we start with one component responsible for values around $k/(c-1)$ for each $k = 0, \ldots, c-1$. The expectation and variance of the component near $k/(c-1)$ is initially estimated from the corresponding sample moments of all data points in the interval $[(k-1)/(c-1), (k+1)/(c-1)] \cap [0,1]$. (If an interval contains no data, the component is removed from the model.) Initial mixture coefficients are estimated proportionally to the number of data points in that interval.

E-step. The E-step is essentially the same as for EM, except that we assign weights explicitly to data points $x_i = 0$ and $x_i = 1$.

Let j_0 be the component index j with the smallest α_j. If there is more than one, choose the one with the largest β_j. The j_0 component takes full responsibility for all i with $x_i = 0$, i.e., $W_{i,j_0} = 1$ and $W_{i,j} = 0$ for $j \neq j_0$. Similarly, let j_1 be the component index j with the smallest β_j (among several ones, the one with the largest α_j). For all i with $x_i = 1$, set $W_{i,j_1} = 1$ and $W_{i,j} = 0$ for $j \neq j_1$.

MM-step. The MM-step estimates mean and variance of each component j by responsibility-weighted sample moments,

$$\mu_j = \frac{\sum_{i=1}^n W_{ij} \cdot x_i}{\sum_{i=1}^n W_{ij}} = \frac{\sum_{i=1}^n W_{ij} \cdot x_i}{n \cdot \pi_j}, \qquad \sigma_j^2 = \frac{\sum_{i=1}^n W_{ij} \cdot (x_i - \mu_j)^2}{n \cdot \pi_j}. \qquad (8)$$

Then α_j and β_j are computed according to (5) and new mixture coefficients according to (7).

Termination. Let θ_q be any real-valued parameter to be estimated and T_q a given threshold for θ_q. After each MM-step, we compare θ_q (old value) and θ_q^+ (updated value) by the relative change $\kappa_q := |\theta_q^+ - \theta_q| / \max\left(|\theta_q^+|, |\theta_q|\right)$. (If $\theta_q^+ = \theta_q = 0$, we set $\kappa_q := 0$.) We say that θ_q is stationary if $\kappa_q < T_q$. The algorithm terminates when all parameters are stationary.

Properties. The proposed method does not have an objective function that can be maximized. Therefore we cannot make statements about improvement of such a function nor can we directly compare two solutions from different initializations by objective function values. It also makes no sense to talk about "local optima", but, similar to the EM algorithm, there may be several stationary points. We have not yet established whether the method always converges. On the other hand, we have the following desirable property.

Lemma 1. *In each MM-step, before updating the component weights, the expectation of the estimated density equals the sample mean. In particular, this is true at a stationary point.*

Proof. For a density f we write $\mathbb{E}[f]$ for its expectation $\int x \cdot f(x)\,dx$. For the mixture density (2), we have by linearity of expectation that $\mathbb{E}[f_\theta] = \sum_j \pi_j \mathbb{E}[b_{\alpha_j, \beta_j}] = \sum_j \pi_j \mu_j$. Using (8) for μ_j, this is equal to $\frac{1}{n} \sum_j \sum_i W_{ij} x_i = \frac{1}{n} \sum_i x_i$, because $\sum_j W_{ij} = 1$ for each j. Thus $\mathbb{E}[f_\theta]$ equals the sample mean. $\square$

4   Application: Classification of Methylation States

4.1   Motivation

We are interested in explaining differences in methylation levels of genomic regions between individuals by genetic variation and would like to find single nucleotide variants (SNVs) whose state correlates well with methylation state. In a diploid genome, we expect the methylation level of a homgeneously methylated region in a homogeneous collection of cells to be (close to) 0, 0.5 or 1, and the state of the corresponding region may be called unmethylated, semi-methylated or fully methylated, respectively.

When we measure the methylation level of each CpG dinucleotide in the genome, for example by whole genome bisulfite sequencing (WGBS) [1], we observe fractions $M/(M + U)$ from numbers M and U of reads that indicate methylated and unmethylated cytosines, respectively, at each CpG dinucleotide. These observed fractions differ from the true methylation levels for several reasons: incomplete bisulfite conversion, sequencing errors, read mapping errors, sampling variance due to a finite number of reads, an inhomogeneous collection of cells being sequenced, the region being heterogeneously methylated, and others.

Therefore we model the observed methylation level by a probability distribution depending on the methylation state. The overall distribution of the observations is captured by a 3-component beta mixture model with one component representing values close to zero (unmethylated), one component close to 1/2 (semi-methylated), and one component close to 1 (fully methylated).

Thus the problem is as follows. After seeing n observed methylation levels $(x_1, \ldots, x_n)$, find the originating methylation state for each x_i. This is frequently done using reasonable fixed cut-off values (that do not depend on the data), e.g. calling values below 0.25 unmethylated, values between 0.25 and 0.75 semi-methylated and values above 0.75 fully methylated [7]. One may leave x_i unassigned if the value is too close to one of the cut-off values.

An interesting question is whether choosing thresholds adaptively based on the observed sample is advantageous in some sense. Depending on the components' parameters, the value range of the components may overlap, and perfect separation may not be possible based on the value of x_i. Good strategies should be based on the component weights W_{ij}, assigning component $j^*(i) := \text{argmax}_j W_{ij}$ to x_i. We may refuse to make an assignment if there is no clearly dominating component, e.g., if $W_i^* := \max_j W_{ij} < T$, or if $W_i^* - W_i^{(2)} < T$ for a given threshold T, where $W_i^{(2)}$ is the second largest weight among the W_{ij}.

4.2   Simulation and Fitting

We investigate the advantages of beta mixture modeling by simulation. In the following, let U be a uniform random number from $]0, 1[$.

We generate two datasets, each consisting of 1000 three-component mixtures. In the first (second) dataset, we generate 200 (1000) samples per mixture.

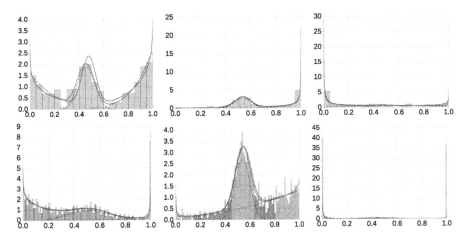

Fig. 3. Examples of generated three-component beta mixtures (green solid lines), data samples (blue histograms) and fitted mixture models (blue solid lines). Dashed lines show estimated weighted component densities (green: unmethylated; red: semi-methylated; magenta: fully methylated). Top row: examples with $n = 200$ samples; bottom row: $n = 1000$.

To generate a mixture model, we first pick mixture coefficients $\pi = (\pi_1, \pi_2, \pi_3)$ by drawing U_1, U_2, U_3, computing $s := \sum_j U_j$ and setting $\pi_j := U_j/s$. This does not generate a uniform element of the probability simplex, but induces a bias towards distributions where all components have similar coefficients, which is reasonable for the intended application. The first component represents the unmethylated state; therefore we choose an $\alpha \leq 1$ and a $\beta > 1$ by drawing U_1, U_2 and setting $\alpha := U_1$ and $\beta := 1/U_2$. The third component represents the fully methylated state and is generated symmetrically to the first one. The second component represents the semi-methylated state (0.5) and should have large enough approximately equal α and β. We draw U_1, U_2 and define $\gamma := 5/\min\{U_1, U_2\}$. We draw V uniformly between 0.9 and 1.1 and set $\alpha := \gamma V$ and $\beta := \gamma/V$.

To draw a single random sample x from a mixture distribution, we first draw the component j according to π and then value x from the beta distribution with parameters α_j, β_j. After drawing $n = 200$ (dataset 1) or $n = 1000$ (dataset 2) samples, we modify the result as follows. For each mixture sample from dataset 1, we set the three smallest values to 0.0 and the three largest values to 1.0. In dataset 2, we proceed similarly with the 10 smallest and largest values.

We use the algorithm described in Sect. 3 to fit a mixture model with a slightly different initialization. The first component is estimated from the samples in $[0, 0.25]$, the second one from the samples in $[0.25, 0.75]$ and the third one from the samples in $[0.75, 1]$. The first (last) component is enforced to be falling (rising) by setting $\alpha_1 = 0.8$ ($\beta_3 = 0.8$) if it is initially estimated larger.

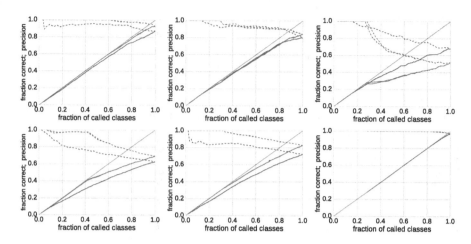

Fig. 4. Performance of several classification rules. Shown is the fraction of called classes N/n (i.e., data points for which a decision was made) on the x-axis against the fraction of correct classes C/n (solid lines) and against the precision C/N (dashed lines) on the y-axis for the three decision rules in Sect. 4.3 (blue: fixed intervals; red: highest weight with weight threshold; magenta: highest weight with gap threshold). The datasets are in the same layout as in Fig. 3.

Figure 3 shows examples of generated mixture models, sampled data and fitted models. The examples have been chosen to convey a representative impression of the variety of generated models, from well separated components to close-to-uniform distributions in which the components are difficult to separate. Overall, fitting works well (better for $n = 1000$ than for $n = 200$), but our formal evaluation concerns whether we can infer the methylation state.

4.3   Evaluation of Class Assignment Rules

Given the samples $(x_1, \ldots, x_n)$ and the information which component J_i generated which observation x_i, we evaluate different procedures:

1. fixed intervals with a slack parameter $0 \leq s \leq 0.25$: point x is assigned to the leftmost component if $x \in [0, 0.25 - s]$, to the middle component if $x \in]0.25 + s, 0.75 - s]$ and to the right component if $x \in]0.75 + s, 1]$. The remaining points are left unassigned. For each value of s, we obtain the number of assigned points $N(s)$ and the number of correctly assigned points $C(s) \leq N(s)$. We plot the fraction of correct points $C(s)/n$ and the precision $C(s)/N(s)$ against the fraction of assigned points $N(s)/n$ for different $s \geq 0$.
2. choosing the component with the largest responsibility weight, ignoring points when the weight is low: point x_i is assigned to component j^* with maximal responsibility $W_i^* = W_{ij^*}$, unless $W_{ij^*} < t$ for a given threshold $0 \leq t \leq 1$, in which case it is left unassigned. We examine the resulting numbers $C(t)$ and $N(t)$ as for the previous procedure.

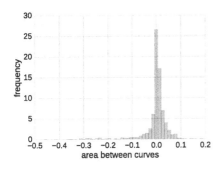 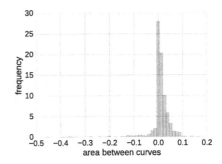

Fig. 5. Signed areas between the red curve and the blue curve as in Fig. 4 for all 1000 simulated mixtures in dataset 1 (left; 200 samples each) and in dataset 2 (right; 1000 samples each).

3. choosing the component with the largest responsibility weight, ignoring points when the distance to the second largest weight is low: as before, but we leave points x_i unassigned if they satisfy $W_i^* - W_i^{(2)} < t$.
4. repeating 2. and 3. with the EM algorithm instead of our algorithm would be interesting, but for all reasonable choices of ε (recall that we have to replace $x_i = 0$ by ε and $x_i = 1$ by $1 - \varepsilon$ for EM to have a well-defined log-likelihood function), we could not get the implementation in betareg to converge; it exited with the message "no convergence to a suitable mixture".

Figure 4 shows examples (the same as in Fig. 3) of the performance of each rule (rule 1: blue; rule 2: red; rule 3: magenta) in terms of N/n against C/n (fraction correct: solid) and C/N (precision: dashed). If a red or magenta curve is predominantly above the corresponding blue curve, using beta mixture modeling is advantageous for this dataset. Mixture modeling *fails* in particular for the example in the upper right panel. Considering the corresponding data in Fig. 3, the distribution is close to uniform except at the extremes, and indeed this is the prototypical case where beta mixtures do more harm than they help.

We are interested in the average performance over the simulated 1000 mixtures in dataset 1 ($n = 200$) and dataset 2 ($n = 1000$). As the magenta and red curve never differed by much, we computed the (signed) area between the solid red and blue curve in Fig. 4 for each of the 1000 mixtures. Positive values indicate that the red curve (classification by mixture modeling) is better. For dataset 1, we obtain a positive sign in 654/1000 cases (+), a negative sign in 337/1000 cases (−) and absolute differences of at most 10^{-6} in 9/1000 cases (0). For dataset 2, the numbers are 810/1000 (+), 186/1000 (−) and 4/1000 (0). Figure 5 shows histograms of the magnitudes of the area between curves. While there are more instances with benefits for mixture modeling, the averages (− 0.0046 for dataset 1; + 0.0073 for dataset 2) do not reflect this because of a small number of strong outliers on the negative side. Without analyzing each instance separately here, we identified the main cause for this behavior as close-to-uniformly distributed data, similar to the example in the upper right panel in Figs. 3 and 4, for which appropriate (but incorrect) parameters are found. In fact, a single beta distribution with $\alpha < 0$ and $\beta < 0$ would fit that data reasonably well, and the three-component model is not well identifiable. Of course, such a situation can be diagnosed by computing the distance between the sample and uniform distribution, and one can fall back to fixed thresholds.

5   Discussion and Conclusion

Maximum likelihood estimation in beta mixture models suffers from two drawbacks: the inability to directly use 0/1 observations, and the sensitivity of estimates to ad-hoc parameters introduced to mitigate the first problem. We then presented an alternative parameter estimation algorithm for mixture models. The algorithm is based on a hybrid approach between maximum likelihood estimation and the method of moments; it follows the iterative framework of the EM algorithm. For mixtures of beta distributions, it does not suffer from the problems introduced by ML-only methods. Our approach is computationally simpler and faster than numerical ML estimation in beta distributions. Although we established a desirable invariant of the stationary points, other theoretical properties of the algorithm remain to be investigated. In particular, how can stationary points be characterized? As we do not have an objective function, we cannot make statements about "local optima", and a key open question is how to rank different stationary points.

With a simulation study based on realistic parameter settings, we showed that beta mixture modeling is often beneficial when attempting to infer an underlying SNV state from observed methylation levels, in comparison to the standard non-adaptive threshold approach. Mixture modeling failed when the samples were close to a uniform distribution without clearly separated components; in practice, we can detect such cases before applying mixture models and fall back to simple thresholding. Our study was restricted to three components, as is appropriate for methylation states. While the algorithm works in principle for any number of components, our results indicate difficulties if the components are not easily separable. A comparison of our algorithm with the EM algorithm (from the `betareg` package) failed because the EM algorithm did not converge and exited with errors (however, we did not attempt to provide our own implementation). Data and Python code can be obtained from the bitbucket repository https://bitbucket.org/genomeinformatics/betamix.

Acknowledgments. C.S. acknowledges funding from the Federal Ministry of Education and Research (BMBF) under the Project Number 01KU1216 (Deutsches Epigenom Programm, DEEP). S.R. acknowledges funding from the Mercator Research Center Ruhr (MERCUR), project Pe-2013-0012 (UA Ruhr professorship) and from the German Research Foundation (DFG), Collaborative Research Center SFB 876, project C1.

References

1. Adusumalli, S., Mohd Omar, M.F., Soong, R., Benoukraf, T.: Methodological aspects of whole-genome bisulfite sequencing analysis. Brief. Bioinform. **16**(3), 369–379 (2015)
2. Dempster, A.P., Laird, N.M., Rubin, D.B.: Maximum likelihood from incomplete data via the EM algorithm. J. Roy. Stat. Soc. Ser. B **39**(1), 1–38 (1977)
3. Grün, B., Kosmidis, I., Zeileis, A.: Extended beta regression in R: Shaken, stirred, mixed, and partitioned. J. Stat. Softw. **48**(11), 1–25 (2012)
4. Ji, Y., Wu, C., Liu, P., Wang, J., Coombes, K.R.: Applications of beta-mixture models in bioinformatics. Bioinformatics **21**(9), 2118–2122 (2005)
5. Pounds, S., Morris, S.W.: Estimating the occurrence of false positives and false negatives in microarray studies by approximating and partitioning the empirical distribution of p-values. Bioinformatics **19**(10), 1236–1242 (2003)
6. Redner, R.A., Walker, H.F.: Mixture densities, maximum likelihood, and the EM algorithm. SIAM Rev. **26**, 195–239 (1984)
7. Zeschnigk, M., et al.: Massive parallel bisulfite sequencing of CG-rich DNA fragments reveals that methylation of many X-chromosomal CpG islands in female blood DNA is incomplete. Hum. Mol. Genet. **18**(8), 1439–1448 (2009)

Author Index

Printed in the United States
By Bookmasters